BERKELEY'S
PHILOSOPHY OF
MATHEMATICS

Science and Its Conceptual Foundations
David L. Hull, Editor

BERKELEY'S PHILOSOPHY OF MATHEMATICS

Douglas M. Jesseph

The University of Chicago Press
Chicago and London

Douglas M. Jesseph is assistant professor in the Department of Philosophy, North Carolina State University.

The University of Chicago Press, Chicago 60637
The University of Chicago Press, Ltd., London
© 1993 by The University of Chicago
All rights reserved. Published 1993
Printed in the United States of America
02 01 00 99 98 97 96 95 94 93   1 2 3 4 5
ISBN: 0-226-39897-8 (cloth)
    0-226-39898-6 (paper)

Library of Congress Cataloging-in-Publication Data

Jesseph, Douglas Michael.
  Berkeley's philosophy of mathematics / Douglas M. Jesseph.
    p.   cm.—(Science and its conceptual foundations)
  Revision of thesis (Ph. D.)—Princeton, 1987.
  Includes bibliographical references and index.
  1. Mathematics—Philosophy. 2. Berkeley, George, 1685–1753.
  I. Title. II. Series.
QA8.4.J47   1993
510'.1—dc20   92-43495
                                                      CIP

⊚ The paper used in this publication meets the minimum requirements of the American National Standard for Information Sciences—Permanence of Paper for Printed Library Materials, ANSI Z39.48-1984.

To R. D. M.

Contents

	Preface	ix
	Works Frequently Cited	xi
	Introduction	1
1	**Abstraction and the Berkeleyan Philosophy of Mathematics**	**9**
	Aristotelian and Scholastic Background	9
	Seventeenth-Century Background	13
	Berkeley's Case against Abstract Ideas	20
	Sources of Berkeley's Antiabstractionism	38
2	**Berkeley's New Foundations for Geometry**	**44**
	The Early View	45
	Abstraction and Geometry in the *Principles*	69
	Geometry in the *New Theory of Vision*	78
	Geometry and Abstraction in the Later Works	83
3	**Berkeley's New Foundations for Arithmetic**	**88**
	Geometry versus Arithmetic	89
	Numbers as Creatures of the Mind	95
	The Nonabstract Nature of Numbers	99
	Berkeley's Arithmetical Formalism	106
	Algebra as an Extension of Arithmetic	114
	The Primacy of Practice over Theory	117
	Berkeley's Formalism Evaluated	118

Contents

4 Berkeley and the Calculus: The Background — 123
Classical Geometry and the Proof by Exhaustion — 124
Infinitesimal Mathematics — 129
The Method of Indivisibles — 132
Leibniz and the Differential Calculus — 138
The Newtonian Method of Fluxions — 143

5 Berkeley and the Calculus: Writings before the *Analyst* — 152
The Calculus in the *Philosophical Commentaries* — 153
The Essay "Of Infinities" — 162
The *Principles* and Other Works — 173

6 Berkeley and the Calculus: The *Analyst* — 178
The Object of the Calculus — 183
The Principles and Demonstrations of the Calculus — 189
The Compensation of Errors Thesis — 199
Ghosts of Departed Quantities and Other Vain Abstractions — 215
The *Analyst* Evaluated — 226

7 The Aftermath of the *Analyst* — 231
Berkeley's Disputes with Jurin and Walton — 233
Other Responses to Berkeley — 259
The Significance of the *Analyst* — 292

Conclusions — 297
Bibliography — 301
Index — 317

Preface

This book is a remote and unintended consequence of a Fulbright grant I received for the academic year 1981–82. I was then fresh out of college and very interested in the philosophy of mathematics, so I spent a year in the Federal Republic of Germany doing research on the work of Gottlob Frege. At the time, I assumed that the philosophy of mathematics had no history before Frege. It seemed evident to me that although Frege's own program was largely a failure he had asked the right kinds of questions and was the first genuine philosopher of mathematics. In between games of chess and late nights in smoke-filled bars, I actually did a fair amount of research. I discovered that the philosophy of mathematics did not begin in the late nineteenth century, and I became more and more interested in pre-Fregean philosophies of mathematics. By the end of my year in Germany, I had more or less abandoned Frege and my interests had shifted to the philosophy of mathematics in the seventeenth and eighteenth centuries.

In the fall of 1982 I enrolled in the graduate program in philosophy at Princeton University. There I had the good fortune to meet Dan Garber, who was a visiting professor for the year and who encouraged my interest in the history and philosophy of mathematics. In those days, first-semester graduate students were required to take an oral examination on the work of a "Great Philosopher." I chose George Berkeley and began a systematic reading of his works. The rest, as they say, is history. Over the next four years I pursued my interest in Berkeley's philosophy of mathematics to the point of writing a dissertation on the subject. In that undertaking, I was very fortunate to have Margaret Wilson as an adviser. She, more than anyone else, is responsible for what knowledge I have of philosophy in the early modern period.

This is a vastly revised, expanded, and improved version of that old Ph.D. dissertation. Parts of chapter 2 appeared as Jesseph (1990), while some of chapter 4 made its way into Jesseph (1989), and parts of chapter 6 are contained in my Editor's Introduction to Berkeley (1992). I am grateful to the *Archiv für Geschichte der Philosophie, Studies in History and Philosophy of Science,* and Kluwer Academic Publishers for their permission to use this material.

Many individuals and institutions have assisted me in this project, and I am happy to record my gratitude. Margaret Wilson played an essential role, both in advising my dissertation and in later conversations and comments on my work. Dan Garber has also helped the project along in many ways for a period of more than a decade. Thanks are due to Roger Ariew, who made me understand the importance of placing philosophical theories in their broader intellectual context. Ken Winkler, whose writings have taught me a great deal, deserves special thanks for helping me sort out my ideas on Berkeley and abstraction. Earlier debts are to Paul Benacerraf and John Burgess of Princeton University, for seminars and discussions on the philosophy of mathematics and related matters. During the summer of 1988 I attended the NEH/Council for Philosophical Studies Summer Institute on the History of Early Modern Philosophy. I learned a lot from many people there but most particularly want to thank Alan Gabbey, Mark Waymack, and Scott Brophy. I must also acknowledge my debts to my old colleagues from the Illinois Institute of Technology—Bob Ladenson, Warren Schmaus, Jack Snapper, Michael Davis, and David Weberman—for commenting on drafts of several parts of the present work and letting no questionable point pass without objection. The National Endowment for the Humanities assisted with Travel to Collections grant #FE-23722-89, which helped send me to the British Library. On a more personal note, my deepest gratitude is to my wife, Doreen. She has had to put up with a lot.

Works Frequently Cited

(See Bibliography for publication details.)

Works by Berkeley

Alciphron.   *Alciphron; or, The Minute Philosopher; in Seven Dialogues. Containing an Apology for the Christian Religion, Against those who are called Free-Thinkers.* Works 3. References are to dialogue number, followed by section number.

Analyst.   *The Analyst; or, A Discourse Addressed to an Infidel Mathematician. Wherein it is examined whether the Object, Principles, and Inferences of the Modern Analysis are more distinctly conceived, or more evidently deduced, than Religious Mysteries and Points of Faith.* Works 4:65–102, or Berkeley (1992, 159–221). References are to section number.

Appendix.   *An Appendix, concerning Mr. Walton's Vindication of Sir Isaac Newton's Principles of Fluxions.* Works 4:139–41. References are to section number.

Arithmetica.   *Arithmetica absque Algebra aut Euclide demonstrata.* Works 4:156–230. References are to chapters.

Commentaries.   *Philosophical Commentaries.* Works 1:9–139 or Berkeley ([1976] 1989). References are to the entry numbers as established by Luce in *Works*.

De Motu.   *De Motu; Sive de motus principio et natura, et de causa communicationis motuum.* Works 4:11–30 or Berkeley (1992, 43–107). References are to section number.

Works Frequently Cited

Defence. — *A Defence of Free-Thinking in Mathematics; In answer to a pamphlet of Philalethes Cantabrigiensis, intituled "Geometry no friend to Infidelity, or a defence of Sir Isaac Newton and the British Mathematicians."* Works 4:103–38. References are to section number.

Introduction. — Introduction to *Principles.* Works 2:25–40. References are to section number.

Principles. — *A Treatise concerning the Principles of Human Knowledge.* Works 2:19–113. References are to section number.

Reasons. — *Reasons for not replying to Mr. Walton's "Full Answer," in a Letter to P. T. P.* Works 4:147–56. References are to section number.

Theory. — *An Essay towards a new Theory of Vision.* Works 1:159–239. References are to section number.

Works. — The standard edition of Berkeley's works (Berkeley 1948–57), followed by volume number and page number.

Works by Other Authors

Elements. — *Elements* (Euclid [1925] 1956). References are to book and proposition or definition number.

Essay. — *An Essay concerning Human Understanding* (Locke 1975). References are to book, chapter, and section number.

Lectiones. — *Lectiones Mathematicæ,* in the *Mathematical Works of Isaac Barrow* (Barrow [1860] 1973, 1:23–414). References are to lecture and page number.

Principia. — *Sir Isaac Newton's Mathematical Principles of Natural Philosophy and his System of the World* (Newton [1729] 1934). References are to book, part, and lemma number.

References to the Revised Oxford Translation of Aristotle (1984) cite the Bekker page numbers, given marginally in that edition.

All translations are by the author unless otherwise indicated in the bibliography.

Introduction

Mathematics has long been the object of philosophical attention, and a moment's thought reveals why this is the case. Our mathematical knowledge is in many ways paradigmatic of knowledge generally, and nearly everyone agrees that mathematical truths are known more clearly and certainly than other items of our knowledge. Small wonder, then, that mathematics has often been taken as a bulwark against the challenges of skepticism or used as a model for philosophical theorizing. For all its exalted status in philosophy, however, mathematics has also been the source of important philosophical problems. It is notoriously difficult to say what (if anything) numbers are, and there is no obvious or easy metaphysical account of the objects of geometry or how they relate to the more familiar contents of the physical world. Even mathematical knowledge, despite its central place in epistemology, is not entirely unproblematic. Accounting satisfactorily for the acquisition of even the most elementary mathematical knowledge is a difficult (not to say daunting) task, and there are notorious differences of opinion over whether and how the finite human mind can acquire knowledge of the infinite. Moreover, the necessity characteristic of mathematics has prompted considerable puzzlement: it seems evident that there could not be a world where 16 is a prime number or triangles have four sides, but it is difficult to explain what it is that makes mathematical truth necessary or how we can come to know this.

Because mathematics is so philosophically intriguing, the history and philosophy of mathematics can provide useful insights for the historian of philosophy. This is particularly true for the philosophy of the seventeenth and eighteenth centuries. The early modern period was an era in which philosophy and mathematics were much more closely connected than they are today, largely because the mathematics of the period was accessible to the philosophical public. It is no accident that Descartes and Leibniz figure prominently in the history of both mathematics and philosophy, nor is it anomalous that mathematicians of the rank of Barrow, Wallis, or Newton should devote many pages to the consideration of issues that we today classify as philosophical. This is not to say that every significant thinker contributed equally to philosophy and mathematics, but it is worth remembering that a well-educated person from the early modern period might read and understand contemporary philosophical and mathematical texts. George Berkeley is a case in point. He was familiar with the mathematical developments of the seventeenth and early eighteenth centuries, and a considerable portion of his work is directed toward issues that are recognizably part of the philosophy of mathematics. It would doubtless be a mistake to characterize Berkeley as a mathematician who dabbled in metaphysics, but it is no less a caricature to depict him as a serious metaphysician and mathematical dilettante.

My purpose in this book is twofold: to study the place of mathematics in Berkeley's philosophy, and to explore his place in the history of mathematics. Historians of philosophy have tended to ignore Berkeley's mathematical writings or to regard them as only worthy of passing notice.[1] This neglect seems to stem from the widespread impression that the "real" Berkeley is to be found in the arguments against matter or the defense of the thesis that *esse* is *percipi*. Similarly, historians of mathematics have, with few exceptions, taken Berkeley to be worthy of little more than a footnote.[2] He is known as a critic of the calculus, and it is often conceded that his criticisms had a point; but his writings are largely disregarded, apparently because he is not taken to have made a real contribution to the development of mathematics. I do not propose that Berkeley scholars abandon their study

---

1. Among recent books in English, at any rate, this is the case. Dancy (1987), Grayling (1986), Moked (1988), Pitcher (1977), and Winkler (1989) say little or nothing about Berkeley's mathematics. Among earlier studies, Johnston (1923) takes Berkeley's mathematical writings seriously. Giorello (1985) and Breidert (1989) are worthwhile recent books dealing with Berkeley's mathematical works, but because the former is in Italian and the latter in German, they are not widely read by English-speaking historians of philosophy.

2. Cajori (1919) is the principal exception here. Standard histories such as Boyer ([1949] 1959) treat Berkeley only as a prelude to the work of others.

of the more famous tenets of his metaphysics, nor do I think that the history of mathematics should be rewritten to place Berkeley at its center. Nevertheless, I think that there is much to be learned by paying attention to Berkeley's mathematical writings and their context. Mathematics was a subject of great interest to Berkeley, and his polemics against the calculus had a significant influence on British mathematics in the eighteenth century.

The starting point of this inquiry is a study of Berkeley's famous critique of abstract ideas. Although this aspect of Berkeley's philosophy has been much written about, it is essential that we look into it because his philosophy of mathematics is largely based on his rejection of the doctrine of abstraction. Chapter 1 explores the background to Berkeley's attack on the doctrine and investigates the strengths and weaknesses of his case against it. I do not offer a great reinterpretation of Berkeley's critique, but I think that chapter 1 casts this part of his philosophy in a new light by showing that he aims not merely at correcting an error in Locke or the Scholastics but at a wholesale repudiation of the received view in the philosophy of mathematics.

Chapters 2 and 3 are concerned with Berkeley's account of geometry and arithmetic, respectively. Among other things, I argue that there are important differences in his interpretation of these two branches of mathematics. Because he could not accept a philosophy of mathematics based on the doctrine of abstract ideas, Berkeley was forced to find some interpretation of mathematics that could avoid abstractions. In this respect there is a unity in his account of the two main branches of mathematics, since both are to be taken nonabstractly. But in working out the details of his antiabstractionist philosophy of mathematics, Berkeley was led to adopt significantly different treatments of geometry and arithmetic: he regards the former as concerned with the properties of (perceived) extension, but takes the latter as a "purely nominal" science concerned with the manipulation of symbols. Furthermore, his attitude toward classical geometry changed in important ways over the course of his career. In his early notebooks or *Philosophical Commentaries*, Berkeley was prepared to adopt a radical program for the revision of geometry. This demanded the rejection of nearly all of classical geometry, and its replacement by a science based on the curious doctrine of sensible minima. In the *Principles* and other works, however, Berkeley took a much more conciliatory attitude toward classical geometry. He continued to reject abstraction, but he was no longer willing to sacrifice classical geometry for the sake of his epistemology.

Chapters 4 through 7 are devoted to Berkeley's critique of the calculus, with special attention to his 1734 work, *The Analyst*. These are organized in roughly chronological order: chapter 4 is a very brief historical introduction to the calculus, chapter 5 investigates Berkeley's pre-1734 writings on the calculus, chapter 6 deals with the *Analyst*, and chapter 7 outlines and evaluates the principal responses to Berkeley's critique of the calculus. This investigation leads into material not usually studied by historians of philosophy, but an adequate treatment of the subject demands a certain convergence between the history of philosophy and the history of mathematics.

Throughout, my emphasis is on matters that bear on the philosophy of mathematics, although I have something to say on other areas of Berkeley's philosophy. In particular, I will try to unravel some difficult issues concerning instrumentalism as a philosophy of science and the extent to which Berkeley was a committed instrumentalist in mathematics. Because my interest is in the philosophy of mathematics, and more particularly in Berkeley's philosophy of mathematics, it will be useful to outline what I take to be the central problems in the philosophy of mathematics, as well as the "received view" that Berkeley sought to overthrow. There are three fundamental issues that can be taken as the "hard core" of any philosophy of mathematics: the ontological question of what mathematical objects are, the epistemological question of how we acquire mathematical knowledge, and the methodological task of distinguishing correct from incorrect mathematical procedures. Although these topics by no means exhaust the subject, they are the starting point for any philosophical study of mathematics, and Berkeley addressed each of them. It is important to understand that answers to these three questions are not generally independent. Thus, a given answer to the ontological question of what mathematical objects are will typically have important consequences for one's theory of how mathematical knowledge is acquired. This, in turn, can be expected to influence one's methodological division between good and bad mathematical procedures.

As is to be expected, the history of the philosophy of mathematics records widely varied answers to these ontological, epistemological, and methodological questions. However, one set of answers was almost universally accepted by philosophers in the seventeenth and early eighteenth centuries, and this set of answers is a doctrine that Berkeley self-consciously opposed. My study of his philosophy of mathematics can thus begin most fruitfully by detailing the theory he rejected before taking up his reasons for repudiating it.

The object of Berkeley's attack was a philosophy of mathematics which characterizes mathematics as a science of abstractions. Although there were differences over points of detail, the broad outlines of this account can be easily sketched. An abstractionist philosophy of mathematics claims that mathematical objects are in some important sense the products of human thought and attempts to link mathematical knowledge with knowledge gained through sense perception. In contract to a "platonistic" ontology, where mathematical objects are supposed to exist independently of human activity, or to a nominalism that denies mathematical vocabulary any extralinguistic reference, abstractionism regards mathematical objects as the result of a mental process which begins with perception but creates a special kind of nonperceptual object.

The key claim here is that the mind can "pare away" irrelevant features of perceived objects and thereby produce an object appropriate for the science of pure mathematics. As an example, Euclid's definition of a line as "length without breadth" would be understood as an abstraction in which the mind mentally separates length from breadth, thereby forming an abstract object appropriate for the science of geometry. Similarly, the number 52 would be taken as an abstraction from the collection of playing cards in a standard deck, in this case an abstraction where the mind removes the individuating features of each card and regards the deck merely as a collection of fifty-two distinct objects.

The abstractionist's mathematical ontology leads to an epistemology where mathematical knowledge begins by framing the relevant abstract ideas. But mere contemplation of such ideas is insufficient for mathematical knowledge, and the abstractionist's epistemology typically requires that knowledge of mathematical theorems is gained through demonstrations. These demonstrations begin with axioms, definitions, and principles which state the fundamental or essential properties of mathematical objects; a process of logical inference then connects these basic properties with more recondite ones. In this way, the security of mathematical knowledge is guaranteed because the theorems are linked to the fundamental truths about the mathematical objects. The axioms and definitions expressing these truths are clearly apprehended by the intellect, which frames the appropriate abstract ideas and infallibly discerns that the axioms are true of them. Moreover, the deductive structure of mathematics assures that the theorems are no less true and certain than the initial axioms or definitions.

An abstractionist criterion of mathematical rigor flows quite naturally from this ontology and epistemology. To count as rigorous, a mathematical procedure must begin with clearly conceived abstract ideas of its appropriate objects—which ideas are typically set forth in definitions. Next, fundamental axioms describing these abstract objects are laid down and seen to be transparently true of their objects. Finally, the theorems are demonstrated by the use of truth-preserving rules of inference which make the most recherché theorem as certain as the axioms from which it is derived.

It should be clear that denial of the doctrine of abstract ideas requires Berkeley to supply some alternative to the ontology, epistemology, and methodology underwritten by abstractionism. In particular, Berkeley must develop a plausible theory of what mathematical objects are if they are not abstractions. Furthermore, he must explain how mathematical proofs can yield demonstrative knowledge if they are not conversant about abstract ideas. In the following chapters, I will examine Berkeley's critique of abstraction and follow the development of his positive doctrines on the nature of mathematical objects and the demonstrative procedures proper to the science of mathematics. Even where the critique of abstraction recedes into the background (as in some aspects of his critique of the calculus), we can find it reemerging in his statement of objections to mathematical procedures or their philosophical interpretation. The rejection of abstract ideas, then, serves as the thread which binds together Berkeley's mathematical writings.

It is interesting to observe that Leibniz clearly discerned the relationship between Berkeley's critique of abstraction and his philosophy of mathematics. In a note written on the last page of his copy of Berkeley's *Principles,* Leibniz evaluates the Irish philosopher's main work.[3] The note is worth quoting in full:

> There is much here that is correct and close to my own view. But it is expressed paradoxically. For it is not necessary to say that matter is nothing, but it is sufficient to say that it is a phenomenon, like the rainbow; and that it is not a substance, but the result of substances, and that space is no more real than time, that is, that space is nothing but the order of coexistents, just as time is the order of things that have existed before [*subexistentia*]. True substances are monads, that is

---

3. This is one of only two occasions where Leibniz is known to have referred to Berkeley. The other is a very brief passage in a letter to Bartholomeus Des Bosses from March of 1715, where Leibniz declares, "The one in Ireland who attacks the reality of bodies does not seem to bring forth suitable reasons, nor does he explain himself sufficiently. I suspect that he is one of that sort of men who wants to be known for his paradoxes" (Leibniz 1989, 306).

perceivers. But the author should have gone further, to the infinity of monads, constituting everything, and to their pre-established harmony. Badly, or at least in vain, he rejects abstract ideas, restricts ideas to imaginations, and condemns the subtleties of arithmetic and geometry. The worst thing is that he rejects the division of extension to infinity, even if he might rightly reject infinitesimal quantities. (Leibniz 1989, 307)

The final two sentences in this passage are the Leibnizian indictment of Berkeley. The denial of matter is not the problem for Leibniz, but he vigorously condemns the rejection of abstract ideas. He also finds fault with the resulting restriction of all ideas to the imagination and Berkeley's attitude toward arithmetic and geometry. But the "worst thing" in the *Principles* is the denial of infinite divisibility, which Leibniz distinguishes from a rejection of infinitesimal magnitudes. All of these topics are closely related to one another, as the following chapters will show. Leibniz is certainly not our best source for the interpretation of Berkeley, but it is remarkable that in reading the *Principles* he found himself most at odds with Berkeley over issues in the philosophy of mathematics.

I should stress that I will not be deeply concerned with defending Berkeley's philosophy of mathematics against what might be called "Leibnizian" objections. Although I think Berkeley ultimately adopted a relatively stable and coherent position in the philosophy of mathematics, I do not claim that his is the best or only option. Nevertheless, it is worthwhile to see how Berkeley managed to find a place for mathematics within the constraints of his more general philosophical tenets, and it is important to understand the context in which Berkeley developed his philosophy of mathematics. Perhaps Leibniz was right to condemn Berkeley's program as mistaken, but we must first understand what the program was and why it was anathema to a thinker like Leibniz.

CHAPTER ONE

Abstraction and the Berkeleyan Philosophy of Mathematics

The brief summary of abstractionism in the Introduction suffices to locate it in the conceptual landscape of philosophies of mathematics, but the task remains of showing that it was indeed the dominant view in Berkeley's day. The best way to do this is to outline the Aristotelian and Scholastic use of the abstractionist philosophy of mathematics, since the doctrine has its roots in Aristotle's writings and was adopted and developed by the Scholastics. I will then turn to some of the seventeenth-century versions of the doctrine. With this material in hand I will investigate and evaluate Berkeley's case against abstract ideas, as well as his alternative to the abstractionist philosophy. Finally, I will consider some possible sources of Berkeley's views on the subject of abstraction.

Aristotelian and Scholastic Background

We can credit Aristotle with the most influential statement of the abstractionist philosophy of mathematics. His views on the subject, as mediated by the writings of the Scholastics, had become firmly entrenched by the seventeenth century. The details of the Aristotelian theory can be left aside here,[1] but its salient features can be set forth fairly simply. One consequence of Aristotle's rejection of the Platonic theory of Forms was his denial of the claim that mathematical objects inhabit a suprasensible world of Forms or serve as an intermediary between the world of sensible objects and the realm of ideal Forms.

---

1. See Apostle (1952), Cleary (1982), Graeser (1987), Jones (1983), Lear (1982), and Mueller (1970) for details on Aristotle and the philosophy of mathematics.

Chapter One

In articulating an alternative to Platonism, Aristotle faced the problem of giving some content to the assertion that mathematical objects exist, without thereby locating mathematical existence in an extramundane realm. But it is equally important to the Aristotelian program that mathematical truth not depend upon the structure or contents of the actual world, since mathematically exact triangles or circles cannot be found in nature and there is little hope of constructing a theory of arithmetic in which numbers are physical objects.

Aristotle's solution to this dilemma was to claim that mathematical objects are constructed out of our ordinary experience but constructed in such a way that mathematics does not depend on specific features of the sensible world. The key to this philosophy of mathematics is the assertion that the mind creates mathematical objects by *abstraction*—the mental removal of extraneous features of perceived objects as a means of isolating selected features for consideration. The result is that

> the mathematician investigates abstractions (for in his investigation he eliminates all the sensible qualities, e.g., weight and lightness, hardness and its contrary, and also heat and cold and the other sensible contrarieties, and leaves only the quantitative and continuous, sometimes in one, sometimes in two, sometimes in three dimensions, and the attributes of things *qua* quantitative and continuous, and does not consider them in any other respect, and examines the relative positions of some and the consequences of these, and the commensurability and incommensurability of others, and the ratios of others; but yet we say there is one and the same science of all these things—geometry). (Aristotle 1984, $1061^a29–1061^b2$)

Aquinas endorsed the Aristotelian conception of mathematics. In the eighty-fifth query of part 1 of the *Summa Theologica*, he accepts an abstractionist treatment of mathematical knowledge, arguing that the intellect constructs mathematical objects by separating out the mathematical features of sensible objects. He holds that mathematicians abstract from the "sensible matter" available to perception, but leave in place the "intelligible matter" contemplated by the intellect. As he puts it:

> Mathematical species, on the other hand, can be abstracted by the intellect from both individual and common *sensible matter*—though not from common (but only individual) *intelligible matter*. For *sensible matter* means corporeal matter as underlying sensible qualities—hot and cold, hard and soft, etc.—whereas *intelligible matter* means substance as underlying quantity. Now

Abstraction and the Berkeleyan Philosophy of Mathematics

it is obvious that quantity inheres in substance before sensible qualities do. Hence quantities—numbers, dimensions, shapes (which are boundaries of quantities)—can be considered apart from sensible qualities, and this is precisely to abstract them from sensible matter. They cannot, however, be considered apart from an understanding of *some* substance as underlying quantity—which would be to abstract them from common intelligible matter—though they can be considered apart from this or that substance—which is to abstract them from individual intelligible matter. (Aquinas 1964, 11:55)

In his commentary on Boethius's *De Trinitate* Aquinas elaborated this account by distinguishing two kinds of mathematical abstraction.[2] The first, called *abstractio totius,* was common to all sciences and involved the abstraction from particulars to form general concepts. Thus, we can abstract from individual men to form the concept "man," abstract further to form the concept "animal," continue abstracting to arrive at "body," and so forth. Of course, the Aristotelian theory of demonstration (especially as propounded in the *Posterior Analytics*) holds that the object of demonstrative science must be such abstract universal notions; in this respect, mathematics does not differ from natural philosophy, since *abstractio totius* is required in both sciences. But mathematics is distinguished from other sciences because it employs a second kind of abstraction, known as *abstractio formæ*. This kind of abstraction involves the separation of a mathematical quantity (i.e., a figure or number) from a perceived object.[3] Thus, *abstractio formæ* from a billiard ball would yield a mathematical sphere when the shape is abstracted from its matter.

Although these abstracting processes begin with ordinary objects of perception, their products are a peculiar kind of purely mental entities which can be grasped by the intellect but not found in nature. By

2. See Aquinas (1948, 37–40) and Smith (1954) for Aquinas's views on mathematical abstraction. For an account of the relation between scholastic theories of mathematical abstraction and Descartes's epistemology, see Olson (1988).

3. Aquinas declares: "there are two kinds of abstraction: the first in which a form is abstracted from matter, the other in which a whole is abstracted from parts. But this form can be abstracted from matter only if its essential nature does not depend on that kind of matter. . . . So quantity can be understood in a substance before the sensible qualities (for which reason it is called sensible matter). Quantity, then, according to its essential nature does not depend on sensible matter, but only on intelligible matter. For when accidents have been removed, substance remains only in the intellect, because the power of sense does not extend so far as to comprehend substance. And mathematics is concerned with abstract things of this kind, for it considers quantities and those things which follow from quantities, such as figure and like things" (Aquinas 1948, 39).

declaring abstractions to be the object of mathematics, Aristotle and his followers could avoid making mathematical truth dependent on the features of the actual world while providing a link between our knowledge of mathematics and our knowledge of more mundane matters. Because these abstractions are not independently existing parts of the natural world, mathematics need not be constrained to deal with things actually existing in nature. Yet by connecting the abstract objects of mathematics to ordinary perception, abstractionism has a tolerably unified epistemology in which our mathematical knowledge is not wholly inexplicable or mysteriously different from our knowledge of other things.

Later, Thomists accepted the basic position that mathematics deals with abstractions, although there were important differences of opinion as to the status of mathematics among the sciences. The Jesuit Benito (Benedictus) Pereira wholeheartedly endorsed an abstractionist reading of mathematics but took this as evidence that mathematics fails to satisfy Aristotle's criteria for perfect scientific demonstration in the *Posterior Analytics*. In Pereira's reading of Aristotle, a true scientific demonstration must appeal to causes, but "mathematical things are abstracted from motion, therefore from all kinds of cause" (Pereira 1576, 70). He thus regarded the abstract nature of mathematics as a hindrance to truly scientific status.[4]

The abstractionist account of mathematics also allows for a simple account of the traditional doctrine that the mathematical sciences occupy a middle ground between metaphysics and physics. This view is of Platonic origin and is prominent in the commentary by Proclus on book I of Euclid's *Elements*, but it was given an Aristotelian reading by taking mathematical objects to be less abstract than the objects of the metaphysical sciences.[5] Christoph Clavius, while defending the scientific status of mathematics against thinkers such as Pereira, accepted this traditional placement of mathematics in the classification of the sciences. He argued that the abstract character of mathematics made it an intermediary between metaphysics and physics. Discussing the nature of mathematical sciences in his commentary on Euclid's *Elements*, he declares:

---

4. Pereira claimed that mathematical disciplines were easily understood, precisely because they abstract from matter, motion, and cause. But this, in his view, was insufficient to make the demonstrations scientific. For more on Pereira and debates surrounding the status of mathematics see Crombie (1977) and Wallace (1984). It is noteworthy that both those who attacked and those who defended the scientific status of mathematics were committed to an abstractionist treatment of the subject.

5. For Proclus's doctrine see Proclus (1970), especially the comments prefatory to the definitions.

Abstraction and the Berkeleyan Philosophy of Mathematics

> Because the mathematical disciplines treat of things which are considered apart from all sensible matter, although they are themselves immersed in matter, this is the principal reason that they occupy a middle position between the metaphysical and natural sciences. If we consider [the sciences each according to its] subject, as Proclus correctly pronounced, the subject of metaphysics is indeed separated from all matter of any kind. But the subject of physics is always conjoined with some kind of matter. Whence, as the subject of the mathematical disciplines is considered apart from all matter, although it is found in matter, it is clear that it constitutes a mean between the other two. (Clavius 1612, 1:5)

This doctrine allows an easy distinction between pure and applied mathematics, while simultaneously explaining how a mathematical theory can be applied to nature. If the objects of mathematics are abstracted from the contents of the physical world, then we can take pure mathematics to be concerned with fully abstract objects and applied mathematics to treat partial abstractions which retain some of the sensible qualities of material objects. The application of mathematics to nature thus involves a reversal of the abstracting process by which the objects of pure mathematics were generated: we begin with a fully abstract result in pure mathematics, but then regard objects in nature as if they were the points, lines, or numbers of pure mathematics.

Another feature of this account of mathematical abstraction is that it underwrites the standard typology of mathematical objects and the traditional division of mathematical sciences. The general object of mathematics is *quantity*, that is, anything that can be measured or counted. But the two main branches of mathematics (namely geometry and arithmetic) can be distinguished by declaring "magnitude" and "multitude" to be two species of quantity, the former being the object of geometry and the latter the object of arithmetic. Magnitudes such as lines, angles, surfaces, and solids are created by abstracting one aspect of a perceived extension from another; a line, for example, is the abstraction of breadth from length while a surface is an abstraction of length and breadth from depth. In contrast, a multitude or "number" is conceived of as an abstract collection of units, where the units are formed by abstraction from the individuating features of particulars.

Seventeenth-Century Background

The philosophical virtues of the abstractionist philosophy of mathematics account for its almost universal acceptance in the seventeenth century. In fact, philosophers and philosophically-minded mathematicians who disagreed on almost every other issue were unanimous in their adherence to an abstractionist philosophy of mathematics. Perhaps the best evidence of this can be taken from the writings of Isaac Barrow and John Wallis. Both were noted British mathematicians from the second half of the seventeenth century, but they disagreed about fundamental questions in the philosophy of mathematics. Wallis was a proponent of the new algebraic methods of analytic geometry, while Barrow revered classical geometry and resisted the departure from classical mathematics.[6] Despite their considerable differences on questions in the foundations of mathematics, both men explicitly endorsed an abstractionist account of mathematical objects, and it is worthwhile to consider their views.

Barrow was appointed the first Lucasian professor of mathematics at Cambridge in 1664—a post he held until 1669 when he resigned to take up a professorship of theology, recommending Newton as his successor. His *Lectiones Mathematicæ* originated as Lucasian lectures and express his general philosophy of mathematics. In his discussion of the object of mathematics in the first lecture, Barrow endorses the abstractionist theory and uses it to distinguish pure from applied mathematics along familiar lines:

> Let us now come to the purpose and consider the object of mathematics, from the various kinds or considerations of which, mathematics itself is divided into different parts. . . . But whatever is the case with regard to the general object, it is agreed that mathematics is principally concerned with two things, namely *quantity* and *quotity* (strictly speaking), or, if you prefer, with *magnitude* and *multitude* (others call these continuous and discrete quantity, we will later consider how appropriate this is). . . . Both of these species (namely magnitude and multitude) can be considered in two respects, either as they are mentally separated or abstracted from all specific matter, material circumstances, and accidents (that is, conceived generally in themselves, having left these out), or as they inhere in some particular subject, and as they are found conjoined with certain physical qualities, actions, and circumstances. Thus emerges the division of mathematics into pure (or abstract) and mixed (or concrete). . . . Consequently, pure

---

6. See Cajori (1929), Mahoney (1990), Pycior (1987), and Sasaki (1985) for a fuller account of the differences between Barrow and Wallis. I will take up some of these issues in chapter 3 when I discuss debates over the relative status of geometry and arithmetic.

mathematics is twofold: geometry, which considers the nature of abstract magnitude, distinguishes its species, and investigates its properties; and arithmetic, which considers and studies multitude or number and its species and affections in the same manner. (*Lectiones* 1:29–30)

Barrow's account of mathematical abstraction downplays the medieval distinction between *abstractio totius* and *abstractio formæ*, since he views mathematical objects as products of a unified abstracting process common to all sciences:

> Moreover, this mental abstraction of which we spoke is by no means peculiar to the mathematical sciences, but is common to all sciences. Indeed, every science considers the nature of its subject abstracted from all special subjects; it forms its own most general precepts and theorems; and it distinguishes its own innate properties from other accidents, examining the former and dismissing the latter. . . . In just the same way geometry assumes as its object of investigation magnitude taken universally (not the peculiar magnitude of this or that body, a magnitude from the heavens, or the earth or the sea) and seeks its general affections . . . declaring these to be inherent to it, and in what manner they are. Next it defines the various species of magnitude (line, surface, solid), eliciting and demonstrating their distinct properties one by one. . . . It seems that true mathematical abstraction is of this sort (nor is it any other, whatever strange things some say of it), which agrees completely with all other sciences and disciplines; the abstraction is evidently from special and singular subjects, or a distinct consideration of certain reasons which are more universal, with others not being considered and, as it were, neglected. (*Lectiones* 1:32–33)

Barrow's commitment to an abstractionist philosophy is equaled by that of John Wallis, Savilian Professor of Geometry at Oxford. Wallis's *Mathesis Universalis* apparently began as Savilian lectures and was an exposition of his philosophy of mathematics. Despite his disagreement with Barrow on many important questions, Wallis adhered to a broadly abstractionist philosophy of mathematics strikingly similar to Barrow's account. Consider, for example, Wallis's highly traditional classification of the mathematical sciences. He declares arithmetic and geometry to be the principal branches of mathematics, defining geometry as "the science of magnitude in so far as it is measurable," and arithmetic as "the science of number in so far as it is numerable" (Wallis 1693–99, 1:19). But the question immediately arises whether there really are magnitudes and numbers in nature, and Wallis considers an argument that because mathematical entities are not found in nature they must be purely imaginary and therefore cannot be the object of a proper science.[7] Wallis grants that the search for a mathematically true triangle or sphere in nature would be in vain. But he insists that mathematical objects are in the world potentially, in the sense that lines, circles, and numbers can be formed by abstraction from material objects. He thus resists the conclusion that mathematical entities are purely fictive or imaginary:

> It is absolutely true that lines and surfaces separated from bodies, or mathematical bodies themselves without physical matter, do not exist (nor indeed can they exist). But when they conclude that the object of mathematics is wholly imaginary, [our opponents] make a serious error. For mathematicians no more suppose or affirm that lines, surfaces or mathematical bodies exist without physical body than a physicist supposes that an animal exists which is neither man nor brute, or that a man in general is ever found who is not Plato or Socrates or some other individual. The mathematician does not deny that his lines, surfaces and figures are in physical body, but these are only attended to and contemplated, while the physical body is not considered. Abstraction is one thing, negation is another. The mathematician abstracts his magnitudes from physical body, but does not deny that they are in it. Mathematicians no more assert that quantity can exist without physical body than the physicist asserts that corporeal substance can exist without quantity, and both contemplate the one abstracted from the other. Therefore, mathematical objects exist, or can exist, and are not wholly imaginary, but rather real. They do not exist *per se*, but in physical body. They exist, although they are considered abstractly. (Wallis 1693–99, 1:21)

This agreement between Barrow and Wallis on an abstractionist approach to the philosophy of mathematics is significant for a study of Berkeley, since it is clear that he read Wallis's *Mathesis Universalis* and

---

7. Wallis's argument is drawn from disputation 14 of Martin Smiglecki's *Logica*, which asks "Whether mathematical demonstrations are most perfect and have the conditions of the most cogent demonstrations?" This question is familiar from the Scholastic discussions of the place of mathematics in the Aristotelian classification of the sciences. Smiglecki considers the objection some have raised that "mathematical objects such as quantities and figures, as they are considered by the mathematician, are not in nature, for indeed there are no lines existing *per se*, or surfaces abstracted from body, or perfect planes, or bodies perfectly round... but it is required for a true and perfect demonstration, that it deal with real beings, not imaginary ones" (Smiglecki 1634, 581).

Abstraction and the Berkeleyan Philosophy of Mathematics

Barrow's *Lectiones Mathematicæ*.[8] Thus, when Berkeley comments disparagingly upon the false metaphysics and absurd doctrines of "the mathematicians" and ridicules those who have been misled by the doctrine of abstract ideas, we can presume that Wallis and Barrow are two of his targets.

Barrow and Wallis are hardly unique in their explicit commitment to an abstractionist philosophy of mathematics. The definition of mathematics as a science of abstractions had become a commonplace in Berkeley's day. Joseph Raphson's *Mathematical Dictionary* defined "Simple or Pure Mathematics" as "those parts of it which contemplate Quantity, simply as such, or abstracted from Matter, or any sensible Object" (Raphson 1702, 2). Locke also characterized mathematics as a science of abstract ideas, arguing that we have real and certain knowledge in geometry because the truths of that science concern the relations between our abstract ideas and are not dependent upon the existence of circles, lines, and triangles in nature (*Essay* IV, iv, 6).[9] In a similar vein, Arnauld argues in the *Port-Royal Logic* that the doctrine of abstraction can guarantee the truth of mathematics by providing an appropriate object for mathematical demonstration, thereby overcoming a kind of skepticism about mathematics:

> A second kind of knowledge by parts [i.e., by abstraction] occurs when we consider a mode without paying attention to its substance or consider two modes which are joined together in

---

8. Berkeley refers to Wallis's *Mathesis Universalis* in a note at the end of the first chapter of his *Arithmetica*. The choice is natural because Berkeley is discussing mathematical notation and Wallis devotes chapters 6–12 of the *Mathesis Universalis* to a study of notation. Berkeley refers to Barrow's *Lectiones Mathematicæ* in several entries of the *Philosophical Commentaries* (notably entries 75, 384, and 470). More will be said about the Berkeley-Barrow connection in chapter 2.

9. Locke declares: "I doubt not but it will be easily granted, that the *Knowledge* we may have of *Mathematical Truths*, is not only certain, but *real Knowledge;* and not the bare empty Vision of some vain insignificant *Chimeras* of the Brain: And yet, if we will consider, we shall find that it is only of our own *Ideas*. The Mathematician considers the Truth and Properties belonging to a Rectangle, or Circle, only as they are in *Idea* in his own Mind. For 'tis possible he never found either of them existing mathematically, i.e. precisely true, in his Life. But yet the knowledge he has of any Truths or Properties belonging to a Circle, or any other mathematical Figure, are nevertheless true and certain, even of real Things existing: because real Things are no farther concerned, or intended to be meant by any such Propositions, than as Things really agree to those *Archetypes* in his Mind" (*Essay* IV, iv, 6). This passage speaks only of "ideas" rather than "abstract ideas" but, read in the context of Locke's declaration that "General and certain Truths, are only founded in the Habitudes and Relations of abstract *Ideas*" (*Essay* IV, xii, 7) and his accompanying praise of mathematical method, it is clear that we should read *Essay* IV, iv, 6 as applying equally to abstract ideas. For more on Locke and mathematics, see Cicovacki (1990).

a single substance, regarding each mode separately. That is what geometers do, who have taken as the object of their science body extended in length, breadth, and depth. In order to understand it better geometers have considered it first along only one dimension, length, and then they have given it the name "line." Next, they have considered it along two dimensions, length and breadth, and called it "surface." And finally considering all three dimensions together, length, breadth, and depth, they have called it "solid" or "body." We see now how ridiculous is the argument of those skeptics who wish to cast doubt on the certainty of geometry because it assumes lines and surfaces which do not exist in nature. For geometers do not suppose that there are lines without breadth or surfaces without depth; they suppose only that we can consider length without paying attention to breadth. We certainly do this when we measure the distance from one city to another; we measure only the length of the road, not its breadth. (Arnauld and Nicole 1981, 55–56)

The medieval distinction between *abstractio formæ* and *abstractio totius* was preserved by some philosophers in Berkeley's day, although the terminology was changed. Thus, John Norris rephrased the old distinction as one between "abstraction by way of modality" and "abstraction by way of habitude," finding both exemplified in mathematics:

Abstraction then is the separate consideration of things intelligibly distinct, really indistinct. And of this there are, as I conceive, two sorts, the one in the way of *Modality,* and another in the way of *Habitude,* according to the different Intelligibility that one and the same thing has, either as 'tis consider'd according to the different manners of Being which it has in it self, or according to the different Respects which it carries to other things. In the way of *Modality,* as when Substance is consider'd without its Mode, or sometimes according to one Mode only, and sometimes according to another. In the way of *Habitude,* as when a thing is consider'd not throughout as it is in it self, but only in so far as it agrees, or according to what it has in common with other Things. As to give an instance of each, which will illustrate the Doctrine of Abstraction in general, as well as these particular kinds of it. A Mathematician considers Body sometimes according to the Dimension of length only, without attending to any breadth, and then he calls it a *Line;* sometimes according to length and breadth without attending to Profundity, and then calls it a *Surface;* and sometimes again according to all three Dimensions, and then he calls it a *Solid*. This is Abstraction in the way of *Modality* or *Manner*. Again, the same Mathematician having before him a Figure terminated with three right Lines, considers it sometimes throughout, according to the full specifick extent of it, as 'tis distinct from all other Figures, and then he calls it a *Triangle*. But sometimes again he considers it no further than according to what it has in common with the rest, and then he calls it a *Figure*. This is Abstraction in the way of *Habitude*. (Norris [1701–4] 1978; 2:174–75)

In addition to providing an appropriate subject matter for the mathematical sciences, abstractionism can also account for the fact that a mathematical demonstration consisting of a finite number of steps can provide knowledge about an infinite number of possible cases. If we consider the Pythagorean Theorem, the doctrine of abstraction has a ready answer to the question of how we can be sure that the theorem holds of all right triangles. The answer proceeds by first claiming that we have an abstract idea of triangularity, and more specifically an abstract idea of right-triangularity. But because the abstract idea of a right triangle is an *abstractio totius* from right triangles (or, as Norris phrases it, an abstraction "by way of Habitude"), all right triangles will conform to this abstract idea. Thus, in showing that the theorem holds for the abstract idea of a right triangle we have in effect shown that it is true of any and all right triangles. In this way, an abstractionist criterion of mathematical rigor can be underwritten: abstract ideas provide adequate objects for mathematical demonstrations, and the doctrine of abstraction thus supports the methodological requirement that mathematical procedures are acceptable only in so far as they are demonstrative.

We should not think that the doctrine of abstraction was confined to the philosophy of mathematics. It was put to many other philosophical uses, most notably in semantic theories which hold that the meaning of a term is an idea. Thus, Locke's thesis that "Words in their primary or immediate Signification, stand for nothing but the *Ideas* in the mind of him that uses them" (*Essay* III, ii, 2) leads him to postulate abstract ideas as the referents of general terms, since the idea signified by a general term must represent equally all particulars to which the term applies. As Locke puts it:

> That then which general Words signify, is a sort of Things; and each of them does that, by being a sign of an abstract *Idea* in the mind, to which *Idea*, as Things existing are found to agree, so they come to be ranked under that name; or, which is all one, be of that sort. (*Essay* III, iii, 12)

Because I am concerned primarily with the uses of abstraction in the philosophy of mathematics, I will have relatively little to say about the semantic employment of abstract ideas beyond the fact that they have an obvious use in providing objects as referents for mathematical terms. It is, however, significant that Berkeley took the doctrine of abstract ideas to be rooted in a false conception of language. He complains that Locke was misled by his analysis of general terms into adopting the doctrine of abstract ideas and insists that language is the "source of this prevailing notion" (Introduction, §18). Berkeley's account of the meaning of general terms is intended to avoid the need for abstract ideas and will concern us when we consider his alternative to abstractionism.

The widespread acceptance of the abstractionist account of mathematics and the various uses to which the doctrine of abstraction was put suffice to show that abstractionism was the received view in the philosophy of mathematics in the early eighteenth century. This is not to say that all philosophers of the seventeenth century agreed in the details of the doctrine. Cartesians, for example, could take abstraction as a way for the intellect to achieve a clear and distinct conception of innate ideas, such as the idea of extension. Those such as Locke who denied the existence of innate ideas would take abstraction to be a process which results in a completely new idea. But despite such differences, the philosophers of Berkeley's era agreed that mathematics is a science which deals with abstract ideas. Berkeley's rejection of abstract ideas therefore demanded that he reject the received view of mathematics, and much of the remainder of this book will be concerned with tracing the consequences which Berkeley's rejection of abstraction had for his philosophy of mathematics. Berkeley must make two different claims in developing his antiabstractionist philosophy of mathematics. First, he must show that the doctrine of abstraction is mistaken; second, he must present an alternative that can answer the fundamental questions in the philosophy of mathematics without invoking abstract ideas. I will first turn my attention to Berkeley's reasons for repudiating the doctrine of abstraction and then consider his alternative to the traditional doctrine of abstract ideas.

Berkeley's Case against Abstract Ideas

Berkeley felt that discrediting the theory of abstract ideas was a matter of the utmost philosophical importance. He frequently claimed that a widespread and unreflective acceptance of the theory of abstract ideas has spread confusion throughout philosophy and the sciences, implying that a philosophy or science free of abstractions would be a significant advance. Thus, in the Introduction to the *Principles* he attributes lack of philosophical progress to the prevalence of false principles "amongst all which there is none, methinks, hath a more wide influence over the thoughts of speculative men, than this of abstract general ideas" (Introduction, §17). In the *New Theory of Vision* he expresses a similar sentiment, implying that the repudiation of the doctrine of abstract ideas would free all branches of learning from a burdensome false doctrine, closing off that "prolific womb which has brought forth innumerable errors in all parts of philosophy and in all the sciences" (*Theory*, §125).

Berkeley's hopes for a reorientation in philosophy (and particularly the philosophy of mathematics) thus hinge on his ability to show that there are no abstract ideas. Thus, it is crucial for a proper understanding of his position that we consider his case against abstraction. Given the prevalence of abstractionist philosophies of mathematics in Berkeley's day, his rejection of the doctrine of abstract ideas would appear to require a wholesale repudiation of the received view of the subject. If it should turn out that Berkeley does not have a convincing case against abstract ideas and can provide no alternative theory to bear the explanatory burden previously assigned to them, his philosophy of mathematics would rest on shaky foundations indeed.

Here I will consider his basic argument against abstract ideas and attempt to evaluate its success. On the whole I think that Berkeley makes a strong case against certain extreme versions of the abstractionist thesis, but it is not clear that he can embarrass every variety of abstractionism. In fact, there are moderate versions of the abstractionist thesis which are close to his own view, so that his claim to have refuted all varieties of the doctrine of abstract ideas is rather overblown. This can best be seen by presenting his central argument and noting how a defender of the doctrine of abstract ideas could avoid much of its force.

The Argument from Impossibility

Berkeley's philosophically engaging argument for the absurdity of abstract ideas has been called the "argument from impossibility."[10] It is intended to show that to frame an abstract idea would be to frame an idea of an impossible object that cannot be consistently described. This claim, in conjunction with the principle that what is (logically)

---

10. Other commentators have discussed this argument at length. See especially Weinberg (1965) and Winkler (1989, chapter 2) for different presentations of it. The anthology edited by Doney (1988) contains essays addressing various aspects of Berkeley's case against abstraction.

impossible cannot be conceived, would suffice to refute the doctrine of abstraction. For the argument to succeed, Berkeley must show that the doctrine of abstraction characterizes abstract ideas in such a way that they must be ideas of things which cannot possibly exist. Later I will be concerned with the question of whether Berkeley's argument applies with full generality to all philosophies of mathematics which use the language of abstraction, but I will begin by looking carefully at the argument from impossibility.

In his attack on abstraction, Berkeley sets forth a typology of abstract ideas which has some important similarities to the medieval distinction between *abstractio totius* and *abstractio formæ*. In the Introduction, he introduces a kind of abstract idea generated by separating out single modes or qualities of perceived objects in the manner of *abstractio formæ*:

> It is agreed on all hands, that the qualities or modes of things do never really exist each of them apart by itself, and separated from all others, but are mixed, as it were, and blended together, several in the same object. But we are told, the mind being able to consider each quality singly, or abstracted from those other qualities with which it is united, does by that means frame to it self abstract ideas. For example, there is perceived by sight an object extended, coloured, and moved: this mixed or compound idea the mind resolving into its simple, constituent parts, and viewing each by itself, exclusive of the rest, does frame the abstract ideas of extension, colour, and motion. (Introduction, §7)

He then introduces another kind of abstract idea corresponding to the *abstractio totius*. His example is the abstract idea of extension, which is gained by abstracting from the particular features of perceived extension:

> Again, the mind having observed that in the particular extensions perceived by sense, there is something common and alike in all, and some other things peculiar, as this or that figure or magnitude, which distinguish them one from another; it considers apart or singles out by it self that which is common, making thereof a most abstract idea of extension, which is neither line, surface, nor solid, nor has any figure or magnitude but is an idea entirely prescinded from all these. (Introduction, §8)

He enlarges this second kind of abstract idea, proposing the abstract idea of man, to exemplify ideas which are not ideas of single qualities or modes:

Abstraction and the Berkeleyan Philosophy of Mathematics

> And as the mind frames to it self abstract ideas of qualities or modes, so does it, by the same precision or mental separation, attain abstract ideas of the more compounded beings, which include several coexistent qualities. For example, the mind having observed that Peter, James, and John, resemble each other, in certain common agreements of shape and other qualities, leaves out of the complex or compounded idea it has of Peter, James, and any other particular man, that which is peculiar to each, retaining only what is common to all; and so makes an abstract idea wherein all the particulars equally partake, abstracting entirely from and cutting off all those circumstances and differences, which might determine it to any particular existence. And after this manner it is said we come by the abstract idea of *man* or, if you please, humanity or human nature. (Introduction, §9)

Clearly, abstract ideas of this second kind are the product of an abstracting process much like the medieval *abstractio totius*. Berkeley frequently refers to such ideas as "abstract general ideas," apparently intending to characterize the generality in this second kind of abstract ideas. Although his précis of the doctrine of abstraction distinguishes these two kinds of ideas, Berkeley gives little weight to the distinction. His case against abstraction is intended to apply equally to abstract ideas of either sort, since he sees them both as ideas of impossible objects and regards both as depending upon the same mistaken principles of human knowledge.

Berkeley employs the argument from impossibility in many of his attacks on abstraction. It is stated explicitly in the manuscript version of the Introduction to the *Principles:*

> It is, I think, a receiv'd axiom that an impossibility cannot be conceiv'd. For what created intelligence will pretend to conceive, that which God cannot cause to be? Now it is on all hands agreed, that nothing abstract or general can be made really to exist, whence it should seem to follow, that it cannot have so much as an ideal existence in the understanding. (Berkeley 1987, 75)

An equally straightforward version of the argument appears in *Alciphron* in an exchange between Berkeley's spokesman Euphranor and the "minute philosopher" Alciphron:

> EUPHRANOR. Not to insist on what you allowed, that everyone might easily know for himself whether he has this or that idea or no, I am tempted to think nobody else can form those ideas any more than I can. Pray, Alciphron, which are those things you would call absolutely impossible?

> ALCIPHRON. Such as include a contradiction.
> EUPHRANOR. Can you frame an idea of what includes a contradiction?
> ALCIPHRON. I cannot.
> EUPHRANOR. Consequently, whatever is absolutely impossible you cannot form an idea of.
> ALCIPHRON. This I grant.
> EUPHRANOR. But can a colour or triangle, such as you describe their abstract general ideas, really exist?
> ALCIPHRON. It is absolutely impossible such things should exist in Nature.
> EUPHRANOR. Should it not follow, then, that they cannot exist in your mind, or, in other words, that you cannot conceive or frame an idea of them?
> (*Alciphron,* dialogue 7, §6)

A variant of the same argument is set out in the *Defence of Free-Thinking in Mathematics:*

> To me it is plain there is no consistent idea the likeness whereof may not really exist: whatsoever therefore is said to be somewhat which cannot exist, the idea thereof must be inconsistent. Mr. Locke acknowledgeth it doth require pains and skill to form his general idea of a triangle. He further expressly saith it must be neither oblique nor rectangular, neither equilateral or scalenum; but all and none of these at once. He also saith it is an idea wherein some parts of several different and inconsistent ideas are put together. All this looks very like a contradiction. But to put the matter past dispute, it must be noted that he affirms it to be somewhat imperfect that cannot exist; consequently the idea thereof is impossible or inconsistent. (*Defence,* §45)

These are all straightforward statements of the argument from impossibility and show the extent to which Berkeley relied upon it in his campaign against abstraction.

One curious fact about Berkeley's use of the argument from impossibility is that it does not appear in an explicit form in the Introduction to the *Principles,* although he comes close to stating its main premise when he writes,

> To be plain, I own myself able to abstract in one sense, as when I consider some particular parts or qualities separated from others, with which though they are united in some object, yet, it is possible they may really exist without them. But I deny that I can abstract one from another, or conceive separately, those qualities which it is impossible should exist so separated. (Introduction, §10)

Abstraction and the Berkeleyan Philosophy of Mathematics

This, however, is not so much an application of the argument from impossibility as a simple first-person report of Berkeley's own failure to frame abstract ideas of single modes or qualities.

The absence of the argument from impossibility from the Introduction to the *Principles* raises the question of whether Berkeley lost confidence in the argument or regarded it as peripheral to his case against abstract ideas, especially because the Introduction is the *locus classicus* of his attack on abstraction.[11] I think that such questions can be answered in the negative. Berkeley's confidence in the argument from impossibility is sufficiently shown by the fact that it appeared explicitly in *Alciphron* and the *Defence of Free-Thinking in Mathematics*, both of which appeared more than twenty years after the publication of the *Principles*. Nevertheless, it is important to explain why the argument is absent from the published Introduction (as well as from the *New Theory of Vision*).

This can be done by considering the tactic which Berkeley employs in the Introduction to show the absurdity of abstract ideas. In place of an explicit statement of the argument from impossibility, Berkeley proposes in §13 to "give the reader a yet clearer view of the nature of abstract ideas and the uses they are thought necessary to" by quoting a passage from Locke's *Essay* in which the abstract general idea of a triangle is discussed. Locke's passage reads in part:

> For example, Does it not require some pains and skill to form the *general Idea* of a *Triangle*, (which is yet none of the most abstract, comprehensive, and difficult,) for it must be neither Oblique, nor Rectangle, neither Equilateral, Equicrural, nor Scalenon; but all and none of these at once. In effect, it is something imperfect, that cannot exist; an *Idea* wherein some parts of several different and inconsistent *Ideas* are put together. (*Essay* IV, vii, 9)

Thus described, the abstract general idea of a triangle is surely the idea of an impossible object, and any theory which contends that the mind can frame such an idea is committed to the claim that we can conceive things whose existence is logically impossible.

In §13 of the Introduction Berkeley is content to let this account of abstraction refute itself, and he does not explicitly invoke the principle that what is impossible is inconceivable. Rather than argue

---

11. Berkeley himself regarded the Introduction as the fullest statement of his objections to abstract ideas. He adds a footnote to the discussion of abstract ideas in dialogue 7 of *Alciphron* recommending that the reader "consult the Introduction to the *Principles of Humane Knowledge* where the absurdity of abstract ideas is fully considered" (*Works* 3:333).

directly against Locke's treatment of abstraction, Berkeley challenges the reader to frame the abstract general idea of a Lockean triangle:

> If any man has the faculty of framing in his mind such an idea of a triangle as is here described, it is in vain to pretend to dispute him out of it, nor would I go about it. All I desire is, that the reader would fully and certainly inform himself whether he has such an idea or no. And this, methinks, can be no hard task for any one to perform. What more easy than for any one to look a little into his own thoughts, and there try whether he has, or can attain to have, an idea that shall correspond with the description that is here given of the general idea of a triangle, which is, *neither oblique, nor rectangle, equilateral, equicrural, nor scalenon, but all and none of these at once?* (Introduction, §13)

Because Locke's description of the abstract idea of a triangle is so obviously incoherent, it provides Berkeley with an ideal way of making his point that abstract ideas are impossible without making a straightforward argument from the principle that what is impossible is inconceivable. The rhetorical advantages of this style of argument are obvious. Berkeley does not need to make his own case against abstract ideas, but can simply let Locke's own characterization of them do the work for him.

Early on, Berkeley saw the rhetorical advantage to be gained by using Locke's own words as a way of discrediting the theory of abstract ideas. In the *Philosophical Commentaries* Berkeley declares, "If men did not use words for Ideas they would never have thought of abstract ideas. These include a contradiction in their nature" (*Commentaries*, 561, dismissing *Essay* IV, vii, 9). He reveals his rhetorical strategy more fully later when he writes, "Mem: to bring the killing blow at the last v.g. in the matter of Abstraction to bring Lockes general triangle at the last" (*Commentaries*, 687). Thus, while the argument in §13 of the Introduction may not be a direct application of the argument from impossibility to the kinds of abstract ideas characterized in §§7–9, this is really irrelevant to Berkeley. He begins by disingenuously suggesting that a passage from Locke's *Essay* will help clarify the doctrine of abstraction, then proceeds to quote a passage that, if taken literally, clearly implies that abstract ideas are impossible and self-contradictory.[12] The effect of this rhetorical strategy is to make the doctrine of abstract ideas look ridiculous.

---

12. The passage Berkeley quotes (*Essay* IV, vii, 9) appears in the context of Locke's attempt to demonstrate that our knowledge, and particularly our geometrical knowl-

Abstraction and the Berkeleyan Philosophy of Mathematics

Indeed, the tactic of quoting Locke's discussion of the abstract idea of a triangle as a means of "clarifying" the doctrine of abstract ideas was Berkeley's favorite weapon in his polemics against abstraction. For example, in the *New Theory of Vision* Berkeley, declaring the abstract idea of a triangle "altogether incomprehensible," facetiously recommends consulting Locke's *Essay* to clarify the matter because "if anyone were able to introduce that idea into my mind, it must be the author of the *Essay concerning Humane Understanding;* he who has so far distinguished himself from the generality of writers by the clearness and significancy of what he says" (*Theory*, §125). Predictably, Berkeley quotes the "killing blow" passage and lets Locke do his work for him by making the doctrine of abstract ideas wholly ludicrous. Berkeley completes the argument by observing that other of Locke's remarks (*Essay* III, x, 33) rule out the possibility of such ideas, and concludes that "had this occurred to his thoughts, it is not improbable he would have owned it above all the pains and skill he was master of to form the above-mentioned idea of a triangle, which is made up of manifest, staring contradictions" (*Theory*, §125).

In the two texts where Berkeley attacks the theory of abstract ideas but does not use the argument from impossibility—namely, the published Introduction and the *New Theory of Vision*—he strategically quotes Locke's account of the abstract idea of a triangle as a means of making his point. But this strategy is, in essence, an application of the argument from impossibility. The passage from Locke which provides the "killing blow" quite explicitly describes abstract ideas as ideas of impossible objects and makes the abstract idea of a triangle seem as conceivable as the idea of a round square. Thus, we can treat the argument from impossibility as the centerpiece of Berkeley's attack on abstraction.

The Argument Evaluated

Stripped of the rhetorical advantages provided by the passage from Locke's *Essay*, Berkeley's argument from impossibility is less compelling than it first seems. To see this, consider the structure of the argument as applied to the abstract idea of man: (*a*) if it is impossible

---

edge, is not founded upon such general maxims as "whatever is, is." Locke is concerned to show that mathematical knowledge depends upon our having first framed and then reasoned about abstract ideas, rather than upon certain maxims supposed to be innate and the foundation for all further knowledge. Taking this into account, it is not surprising that Locke might exaggerate the difficulty of abstraction. Thus, we should be wary of treating this passage as representing Locke's considered view.

that an object exist, then the idea of that object cannot be conceived; (*b*) it is impossible that a purely generic man exist; (*c*) the abstract general idea of man is the idea of a purely generic man; therefore, (*d*) the abstract general idea of man cannot be conceived.

Premises *a* and *b* can be allowed to go unchallenged, since they hardly seem controversial. If we were to treat the impossibility in premise *a* as merely a physical impossibility, the premise would be plainly false because we can grant the physical impossibility of something (say, a perpetual motion machine) while holding the idea of such a thing to be conceivable. But if we take the impossibility in premise *a* broadly, reading it as logical or metaphysical impossibility, it becomes innocuous because rejecting it would commit us to the view that such impossible objects as round squares can be conceived. Premise *b* may not be as plausible as premise *a*, but it has considerable merit and could surely have been accepted by many of Berkeley's opponents. We certainly take premise *b* to be true of ordinary objects in nature, since an "indeterminate object" would be difficult to describe consistently, and premise *b* articulates the intuition that whatever is must be a determinate particular thing.

The problematic step in the argument from impossibility is obviously premise *c*. In essence, premise *c* is a special case of the principle that the abstract idea of a thing is the idea of an abstract thing—a claim which is certainly not obvious and must be defended by showing that it is essential to any theory which uses the language of abstraction. But whether all defenders of abstraction are committed to such a characterization of their doctrine is highly problematic. The passage Berkeley used to deliver his "killing blow" against Locke's account of abstraction does state (among other things) that the abstract idea of a triangle is the idea of a triangle with no determinate characteristics, but Locke's more sober comments on abstraction do not necessarily commit him to this position.[13]

More to the point, we should recall Wallis's remark in the *Mathesis Universalis* that "negation is one thing, abstraction another," reading this as a claim that abstracting does not require us to form an idea of an object which lacks the abstracted qualities. His account of mathematical abstraction treats it as a kind of selective attention in which

---

13. Many commentators have suggested that Berkeley misunderstands Locke, and a careful reading of books II and III of the *Essay* makes it clear that Locke's account of abstraction is considerably more plausible than the passage Berkeley cites (*Essay* IV, vii, 9) would indicate. See Winkler (1989, 49–52) and Aaron (1971, 195–7) for a more sympathetic reading of Locke's doctrine.

· 29 ·
Abstraction and the Berkeleyan Philosophy of Mathematics

"the mathematician does not deny that his lines, surfaces, and figures are in physical body, but *these are only attended to and contemplated, while the physical body is not considered*" (Wallis 1693–99, 1:21, my emphasis). This kind of mental process does not seem to result in the formation of an abstract idea of, say, a figure without determinate characteristics; instead, it is consistent with its being a fully determinate object of thought, some of whose characteristics are not considered.

There is a crucial ambiguity in the doctrine of abstraction which Berkeley overlooks and which permits an abstractionism which does not fall victim to the argument from impossibility. The ambiguity here concerns what kind of mental separation is intended in the claim that abstract ideas are formed when the mind separates features of perceived objects. One might conclude that abstraction involves mentally separating features of a perceived object so as to form the idea of an object which lacks the separated features or, on the other hand, that mental separation implies only a "selective attention" in which the relevant features of the perceived object are attended to while the others are disregarded.[14] This second sense of abstraction would not require that one frame an idea of an object which lacks the irrelevant features and would seem to avoid Berkeley's argument from impossibility.

Consider, for example, two ways of characterizing the *abstractio formæ* which yields the abstract idea of a Euclidean line or "breadthless length." One might take the abstraction to be the framing of an idea of an object which has length without breadth. But, alternatively, the line could be thought of as an abstraction in which a perceived object's length is considered while its breadth is disregarded.[15] In this case, abstraction does not form a new mental object, but features of a pre-

---

14. Kenelm Digby draws a distinction between these two kinds of abstraction in the second of his *Two Treatises*, where he declares that "[abstraction] we do two severall wayes; the one, when our *manner* of apprehension determineth us to one precise notion, which is so summed up within it selfe, as it not only abstracteth from all other notions, but also quite excludeth them, and admitteth no society with them. The other way is, when we consider a thing under a determinate notion, yet we do it in such a manner, that although we abstract from all other notions, yet we do so, rather by neglecting than by excluding them" (Digby [1644] 1978, 362). I thank Kenneth Winkler for drawing my attention to this passage.

15. Hobbes takes a view like this in his essay *De Principiis et Ratiocinatione Geometrarum*, where he defines a point as something divisible, none of whose parts are considered in a demonstration; similarly, he defines a line as the path of a moving body whose quantity is not considered in a demonstration. A straight line will then be something with both length and breadth whose breadth is not considered. See Hobbes ([1845] 1966b, 4:389–397).

existing object are "attended to and contemplated" while others are left out of consideration. As we will see later, this is very much like Berkeley's own theory of general reasoning.

Berkeley's characterization of the doctrine of abstract ideas simply assumes that it requires the first kind of abstraction, but this is not obviously correct. If premise *c*, that an abstract idea is an idea of the purely generic, is not clearly crucial to a theory of abstract ideas, we should expect Berkeley to provide an argument for why his opponents must nevertheless adopt it. Unfortunately, he does not, simply defining abstract ideas as "incomplete" or "general" objects which cannot exist in nature. He begins his summary of the doctrine of abstraction, for example, by saying that abstract ideas of color or motion begin when the mind, "able to consider each quality singly, or abstracted from those others with which it is united, does by that means frame to it self abstract ideas." But the language of "considering separately" quickly gives way to language that implies the consideration of something which is separated: "Not that it is possible for colour or motion to exist without extension: but only that the mind can frame to it self by *abstraction* the idea of colour exclusive of extension, and of motion exclusive of both colour and extension" (Introduction, §7). This shift can only be allowed if something analogous to premise *c* is employed.

Berkeley follows a parallel strategy when he characterizes the abstract general idea of man. He begins with the claim that the mind can observe resemblances between men but then claims that the abstract general idea of man must be the idea of a generic man, or an idea "wherein all the particulars equally partake, abstracting entirely from and cutting off all those circumstances and differences, which might determine it to any particular existence" (Introduction, §9). A similar move appears in the *New Theory of Vision* when Berkeley writes:

> We are therefore to understand by extension in the abstract an idea of extension, for instance, a line or surface intirely stript of all other sensible qualities and circumstances that might determine it to any particular existence; it is neither black nor white, nor red, nor hath it any colour at all, or any tangible quality whatsoever, and consequently it is of no finite determinate magnitude. For that which bounds or distinguishes one extension from another is some quality or circumstance wherein they disagree. (*Theory*, §124)

Here he starts out talking about "extension in abstract," moves quickly to "an idea of extension," and ends up talking about the idea of some-

· 31 ·
Abstraction and the Berkeleyan Philosophy of Mathematics

thing extended (i.e., a line or surface) but lacking any determinate qualities. Although such a shift seemed perfectly natural to Berkeley, it is little more than the assumption that an abstract idea is the idea of an object devoid of determinate characteristics. Thus, despite Berkeley's boast that he has "shewn the impossibility of abstract ideas," it would be fairer to say that he has stacked the deck against defenders of the doctrine of abstraction. I conclude that unless Berkeley can provide some compelling reason for thinking that a defender of abstract ideas must defend premise c and its analogues, the argument from impossibility is insufficient to show that every philosopher who uses the language of abstraction is committed to a nonsensical theory.[16]

It does not follow from the foregoing that Berkeley's critique of abstraction is completely devoid of merit. There were certainly some who held the kind of doctrine he refutes,[17] so his critique of abstraction does not amount to an extended attack on a straw man, and his attempt to develop a philosophy of mathematics which avoids abstractions of this sort is worth our attention. Moreover, in analyzing Berkeley's alternative to the doctrine of abstract ideas we can see how far he maintained his epistemological commitments while developing an account of mathematical reasoning not dependent upon the extreme formulations of abstractionism. The extent to which Berkeley could accept much of the mathematics of his era depended upon his ability to read these theories in terms which did not require the doctrine of abstract ideas. Indeed, there is virtually no aspect of his philosophy of mathematics that is not informed by the desire to avoid an abstractionist treatment of the subject. In some cases Berkeley's attempt to develop a nonabstract alternative to the received view led him into

---

16. Note that any argument for the principle which Berkeley needs would come down to an argument in favor of what Kenneth Winkler has termed the "content assumption"—the assumption that the content of thought is determined by the idea which is the object of thought. Winkler has identified this as a leading principle in Berkeley's attack on abstraction and shows that Berkeley himself is prepared to reject the content assumption. On Berkeley's account of reasoning, we can treat a particular idea as a representative of other ideas, so that the content of our thought is only loosely linked to the ideas which we have as we are thinking. As Winkler observes, Berkeley does not trouble himself with showing that his opponents are committed to the content assumption and he seems to have regarded it as an unquestioned dogma of previous thinkers in the "way of ideas" tradition (Winkler 1989, 39–45).

17. Clearly, Barrow's talk of abstract ideas of objects as being "conceived generally in themselves, having left [specific circumstances] out" looks suspiciously like the kind of theory Berkeley can successfully defeat, and Norris's discussion of "abstraction by way of modality" may well fall victim to Berkeley's argument.

embarrassingly implausible doctrines which he later abandoned, but other aspects of his philosophy of mathematics are strikingly innovative and not wholly unattractive.

An important difference between Berkeley's approach and the "extreme" versions of the abstractionist thesis is that Berkeley denies the existence of a mental faculty of "pure intellection" which contemplates nonsensory ideas such as the abstract idea of extension. On Berkeley's view, sensation and imagination are the only sources from which the object of mathematics can be drawn, but this places him in opposition to those who postulate a pure intellect which frames the abstract ideas that are the true concern of the mathematician. This difference is made clearer in Berkeley's exchange with the French scholar Jean Leclerc, editor of the *Bibliothèque choisie*. Leclerc had printed a summary of the *New Theory of Vision* in the *Bibliothèque*, adding notes in which he raised points of disagreement with Berkeley. In particular, he faulted Berkeley's critique of abstraction in §§122–25 and accused him of failing to distinguish intellection from imagination. Where Berkeley had complained that the abstract idea of extension was "perfectly incomprehensible" because it would be the idea of an extended thing without determinate characteristics, Leclerc insisted that

> It seems to me, nevertheless, that one can form an idea of this sort, and that one can, for example, conceive of a length without breadth or depth, or any other quality, whatever it may be. For myself, it seems that I have a very clear idea of this, and the same is true of breadth and depth considered in general, be they separate or together. (Leclerc 1711, 80–81)

On Leclerc's analysis, the faculty of pure intellection can frame ideas which are literally unimaginable, and he faults Berkeley for making imagination the test of what is conceivable. He grants that "We can imagine nothing which is not clothed in general sensible qualities" (Leclerc 1711, 81) but implies that the pure intellect is not bound by such constraints and can frame ideas of things in isolation from these qualities. The ideas of the pure intellect are fundamentally nonsensory, so that the abstract idea of a triangle is not a quasi-sensational image of an abstract triangle. Instead, the abstract idea of a triangle will be an idea of the kind of "general object" which Berkeley dismisses as impossible:

> In effect, one cannot imagine a triangle without representing to oneself a particular triangle; but it seems to me that one can think in general of a figure which is bounded by three

straight lines, which form three angles where they meet, without thinking in any way of the measure of these angles or of the lengths of their sides. (Leclerc 1711, 81–82)

He concludes that the faculty of pure intellection is the only true source of geometric knowledge, and that Berkeley's attempt to reduce geometric knowledge to sensation and imagination has confused the issue.

In a draft of a letter in reply to Leclerc, Berkeley declares that he simply cannot frame the abstract idea of a triangle and insists that postulating a special faculty of pure intellection does not make the task any easier.[18] Moreover, even if the distinction between intellect and imagination is granted, Berkeley argues that the object of geometry must be an idea of the imagination, since "it nevertheless seems to me that the pure intellect is entirely concerned with spiritual things known by the mind's reflection on itself, but [pure intellect] could in no way deal with the ideas arising from sensation, such as extension" (*Works* 8:50). Significantly, his letter stresses the important difference between conceiving of a sensible quality apart from all other qualities and attending to one quality of a perceived object while neglecting others. He finds the first incoherent and the second unproblematic, for the doctrine of "separate consideration" plays a fundamental role in his alternative to Leclerc's extreme version of the abstractionist thesis.

Berkeley's Alternative to the Doctrine of Abstract Ideas

As an alternative to the theory of abstract ideas, Berkeley proposes to make determinate, particular ideas play the roles assigned to abstract ideas in other theories. Thus, general terms will not denote abstract general ideas, nor will universally true propositions be about the relations between abstract ideas. I call this alternative the theory of representative generalization. The most fundamental aspect of Berkeley's alternative is the claim that we can make one idea go proxy for many others by treating it as a representative of a kind. This is summed up in Berkeley's slogan, "an idea, which considered in itself is particular, becomes general by being made to represent or stand for all other particulars of the same sort" (Introduction, §12).

---

18. The letter is reprinted in *Works* 8:49–50. We do not know whether Berkeley actually sent it, but there is no reason to think that he did not. Berkeley's rejection of the faculty of "pure intellect" is indicated in *Commentaries*, 810: "Pure Intellect I understand not." He also critiques the doctrine in the first of the *Three Dialogues* (*Works* 2:193) and in *Alciphron* (dialogue 7, §6).

At the heart of this account is the intuition that ideas can be partitioned into equivalence classes, each class consisting of ideas which resemble one another in a certain respect. Once such a partitioning has been made, we can reason about a particular idea from one of the classes, and any result obtained will apply equally to any other member of the class.

It is important that in the course of reasoning about the particular idea, we rely only upon properties of it that do not distinguish it from any other member of the equivalence class. The standard example of this procedure (and the one which Berkeley invokes) is a geometric proof. To demonstrate that all triangles have interior angles that sum to two right angles, we begin by drawing a particular triangle, which can be thought of as having been randomly selected from the stock of all possible triangles. Call this triangle T. Now, T has many features that distinguish it from other triangles (its color, the length of its sides, etc.), but the only properties of T used in the proof are those which are explicitly appealed to in the course of the argument. In the case we are imagining, the relevant facts about T are that it has three sides, three angles, and that the sides are all line segments. We then consider the proof of the theorem to be in the first instance only about T. But because the properties of T cited in the proof are common to all triangles, what we have proved of T could just as well have been proved of any other triangle; in other words, what we proved of T can legitimately be generalized so as to hold of all triangles.[19]

Berkeley attached a great deal of importance to his theory of representative generalization, and he regarded it as a fundamental contribution to the theory of signs. He frequently claimed that a proper understanding of representative generalization would solve many

---

19. Not all commentators find this account of geometric proof convincing. Pitcher, for example, objects that the Berkeleyan theory of representative generalization does not lend the appropriate degree of generality to the theorems in a geometric proof, since special features of the figures used in the demonstration might actually be essential to the constructions carried out in the proof. He imagines that "although we can be absolutely sure that the proof would apply to right triangles of any color whatever, we have no right to conclude, from the lone fact that no mention is made of the relative sizes of the three angles and sides, that the proof would apply to triangles of any determinate shape whatever" (Pitcher 1977, 76). This is unconvincing, however. Pitcher overlooks the fact that representative generalization requires that the result be generalized only to cover those figures which share the properties which are used in the course of the demonstration. Should it happen that the relative sizes of angles are used in a demonstration, the result cannot be generalized to cover all cases without engaging in the fallacy of "appeal to the figure." But Berkeley's account of demonstration does not license such fallacious inferences. Indeed, it gives a perfectly acceptable explanation of why such fallacious inferences are mistaken.

problems in the sciences, especially mathematics. In later chapters we will be concerned with the influence of this theory on Berkeley's philosophy of mathematics. In particular, I will argue in the second chapter that when he developed his alternative to the theory of abstract ideas, Berkeley's views on the nature of geometry underwent a significant change. I will also consider whether this theory is as free from abstraction as Berkeley imagined but will postpone this discussion until the end of chapter 3, as it ties in with some difficulties relating to the prospects for a purely nominalistic theory of arithmetic.

This should serve as a brief outline of Berkeley's account of general ideas, but there are a few more points to be clarified before I proceed. In the first place, it is clear that Berkeley regarded representation as largely conventional. In his account of natural languages Berkeley continually stresses that it is matter of human convention what letters we use to represent sounds and what sounds we use to represent our ideas. Similarly, that one idea represents another is the result of it being regarded as so representing by someone, and as such it is purely a matter of convention whether we take one idea to represent another. Moreover, the relation of representation can only be relativized to a property, that is, one idea can represent another only with respect to a property they both share. To see why this is the case consider two Berkeleyan ideas, one of a triangular red patch and another of a triangular blue patch. Either of these could represent the other because they share the property of being triangular. But the triangular red patch could also represent a circular red patch since both ideas share the property of being red. Thus, there is no relation 'x represents y' *simpliciter*, but rather something like: 'x represents y as to shape' or 'x represents y as to color.'

We should note the extent to which this theory of representative generalization avoids the objections which Berkeley directed against the doctrine of abstract ideas. In Berkeley's theory, there is no need to frame the idea of an object which is indeterminate or general, since a perfectly determinate idea is used to represent a class of others. Berkeley frequently speaks of this representation involving a "disregarding" of irrelevant features of the idea which signifies others, so his alternative has a fair amount of similarity with those abstractionist theories which do not require that the abstract idea of a thing is the idea of an abstract thing. Although Berkeley's account is thus not as groundbreaking and revolutionary as he imagined, it is more carefully worked out than any related theory and is therefore worthy of attention.

Once Berkeley's account of representative generalization is understood, it is easy to see why he felt that abstract ideas are unnecessary in explaining how general terms can acquire meaning or how a mathematical demonstration can be made to cover an infinite number of cases. These applications of the theory are worth setting forth in a general way before I attend more specifically to Berkeley's account of geometry and arithmetic. By using the figures or symbols in mathematical demonstration as representatives of other perceivable objects, Berkeley dispenses with abstract ideas in accounting for our ability to grasp demonstrations. Rather than claiming that we come to understand demonstrations by inspecting our abstract ideas and reasoning about them, he claims that we reason only about particular ideas, treating them as signs or representatives of a kind. In this way, mathematical knowledge is made to be general without being abstract. He grants that "all knowledge and demonstration are about universal notions" but insists that the universality of a demonstration is a function of the representative capacity of the particulars used in it. It is thus by representative generalization that "things, names, or notions, being in their own nature *particular,* are rendered *universal*" (Introduction, §15). A geometric demonstration must take a particular figure as its object, but the theorem demonstrated will be proved for all figures which can be represented by that one:

> Thus when I demonstrate any proposition concerning triangles, it is to be supposed that I have in view the universal idea of a triangle; which ought not to be understood as if I could frame an idea of a triangle which was neither equilateral nor scalenon nor equicrural. But only that the particular triangle I consider, whether of this or that sort it matters not, doth equally stand for and represent all rectilinear triangles whatsoever, and is in that sense *universal.* (Introduction, §15)

With regard to the semantics of general terms, Berkeley argues that we need not postulate an abstract idea as the referent of a general term, but can simply think of such a term as having divided reference. In his view, "a word becomes general by being made the sign, not of an abstract general idea, but of several particular ideas, any one of which it indifferently suggests to the mind" (Introduction, §11). Thus, Berkeley can retain the fundamental semantic thesis that words signify ideas, but a general term will signify many particular ideas instead of a single abstract idea.

This means that when a general term appears in a sentence it must be understood as a suppressed universal quantification in which the term refers to any member of a class of ideas, rather than to an abstract general idea which contains only the features common to all members of the class:

> When it is said *the change of motion is proportional to the impressed force,* or that *whatever has extension is divisible;* these propositions are to be understood of motion and extension in general, and nevertheless it will not follow that they suggest to my thoughts an idea of motion without a body moved, or any determinate direction and velocity, or that I must conceive an abstract general idea of extension, which is neither line, surface, nor solid, neither great nor small, black, white, nor red, nor of any other determinate colour. It is only implied that whatever motion I consider, whether it be swift or slow, perpendicular, horizontal or oblique, or in whatever object, the axiom concerning it holds equally true. As does the other of every particular extension, it matters not whether line, surface, or solid, whether of this or that magnitude or figure. (Introduction, §11)

Berkeley justifies this account of the semantics of general terms by invoking his theory of representative generalization. He asserts that we can understand how words become general if we first consider how ideas become general, but the generality of ideas is guaranteed by their representational capacity. If we accept that representative generalization can account for the generality of a geometric proof, it is a short step to conclude that terms such as "line" can gain their generality in a similar fashion:

> And as that particular line becomes general, by being made a sign, so the name *line* which taken absolutely is particular, by being a sign is made general. And as the former owes its generality, not to its being the sign of an abstract or general line, but of all particular right lines that may possibly exist, so the latter must be thought to derive its generality from the same cause, namely, the various particular lines which it indifferently denotes. (Introduction, §12)

We can now sketch the salient features of Berkeley's philosophy of mathematics. Where the traditional account had declared abstractions to be the object of mathematical investigation, Berkeley treats mathematics as a science concerned with objects of sense. These perceived objects are of little interest in themselves, however, because the essential feature of mathematical knowledge is its generality. On Berkeley's view, this generality is to be explained in terms of the capacity of perceived objects to function as signs. Note also that Berkeley can accept the traditional division of mathematics into geometry and arithmetic, since he finds fundamentally different kinds of ideas to be the immediate objects of these two mathematical sciences. Geometry will have extension as its proper object, while arithmetic will be interpreted nominalistically so that its immediate object will be symbols. As we will see, this division leads Berkeley to accept quite different accounts of geometrical and arithmetical truth. The theorems of geometry must answer to the facts of perception (since perceivable extension is its object), while arithmetical truths will have a conventional element (because they will be truths about the symbols themselves and choice of symbolic notation is largely arbitrary). Chapters 2 and 3 will develop this sketch more fully and thereby elucidate Berkeley's philosophy of mathematics, but first I will close this account of Berkeley and abstraction by considering some of the sources of Berkeley's critique of the doctrine of abstract ideas.

Sources of Berkeley's Antiabstractionism

The extent to which Berkeley's attack on abstract ideas derives from the work of earlier philosophers deserves consideration in an account of abstraction and the Berkeleyan philosophy of mathematics. In this section I consider some possible influences and review some of the claims that have been made for locating Berkeley's antiabstractionist polemic within various philosophical traditions.

Some commentators see Berkeley's critique as a direct inheritance of Malebranche. Luce argues for a Malebranchian connection to Berkeley's attack on abstraction, although he admits that the evidence "falls short of proof" (Luce [1934] 1967, 143). Bracken goes much farther when he places Berkeley in the Cartesian tradition of antiabstractionism:

> No step that Berkeley could have taken would so clearly have put him within the Cartesian camp as beginning an examination of the principles of human knowledge with a set of arguments against abstractionism. The historical source of his arguments in Malebranche and his target is the empiricist doctrine of concept formation which, as he clearly spells out in the Introduction, is found in Locke and the Schoolmen. (Bracken 1974, 74)

I think that this overstates the case. It was a commonplace among proponents of the "New Philosophy" in the seventeenth and eighteenth centuries to attack the Scholastics for their use of such abstract terms as "essence," "accident," and "power." For the most part these attacks consisted of the charge that there is nothing in the world to correspond to these terms and that these meaningless words had introduced an intolerable confusion into philosophy. Malebranche spent a good deal of time attacking abuses of abstraction, but antischolastic polemics were hardly unique to Malebranche or the Cartesians.

Both Malebranche and Berkeley reject much of Scholastic philosophy because they see it as a jumble of false abstractions, but this does not go very far toward tracing Berkeley's views on abstraction to Malebranche. In fact, Malebranche and Berkeley attack abstraction on entirely different grounds. Malebranche never denies that the mind can frame abstract ideas; he claims only that certain scholastic terms (such as "matter" and "power") signify nothing more than the abstract idea of "being in general" and are useless in a properly philosophical account of the world (McCracken 1983, 232–34). Berkeley, of course, denies that we can have any abstract ideas at all and goes to some lengths to distance himself from Malebranche's opinions. In the *Three Dialogues* he accuses Malebranche of building "on the most abstract general ideas, which I entirely disclaim" (*Works* 2:214) and, in a letter to Sir John Percival from 27 November 1710, he resisted comparison to Malebranche and Norris with the retort, "I think the notions I advance are not in the least coincident with, or agreeing with, theirs, but indeed plainly inconsistent with them in the main points, insomuch that I know few writers whom I take myself at bottom to differ more from than them" (*Works* 8:41). Moreover, nothing in Malebranche corresponds to the most characteristic features of Berkeley's attack on abstraction—the argument from impossibility and the theory of representative generalization. Thus, it is improbable that Berkeley's views on abstraction are the product of a direct Cartesian (or, specifically, Malebranchian) influence.

There is, however, some reason for treating Hobbes's account of the meaning of general terms as a precursor of Berkeley's theory of representative generalization.[20] Hobbes's theory of language treats the word as the fundamental unit of meaning, and he claims that all significant words are names. A name, for Hobbes, is

> a word taken at pleasure to serve for a mark, which may raise in our mind a thought like to some thought we had before, and which being pronounced to others, may be to them a sign of what thought the speaker had, or had not before his mind. (Hobbes [1845] 1966a, 1:16)

---

20. Robert Baum (1969, chap. 2) claims a Hobbesian background for Berkeley's philosophy of mathematics, particularly in its treatment of signs and representation.

Chapter One

Names can denote either bodies, accidents (i.e., properties of bodies), phantasms (Berkeleyan "ideas"), or other names. Important for our purposes is Hobbes's discussion of common or universal names:

> And a common name, being the name of many things severally taken, but not collectively of all together (as man is not the name of all mankind, but of every one, as of Peter, John, and the rest severally) is therefore called an *universal name;* and therefore this word *universal* is never the name of any thing existent in nature, nor of any idea or phantasm formed in the mind, but always the name of some word or name; so that when *a living creature, a stone, a spirit,* or any other thing, is said to be *universal,* it is not to be understood, that any man, stone, &c. ever was or can be universal, but only that these words *living creature, stone, &c.* are *universal names,* that is, names common to many things; and the conceptions answering them in our mind, are the images and phantasms of several living creatures, or other things. And therefore, for the understanding of the extent of an universal name, we need no other faculty but that of our imagination, by which we remember that such names bring sometimes one thing, sometimes another, into our mind. (Hobbes [1845] 1966a, 1:19–20)

This account of the meaning of general terms is obviously quite close to Berkeley's theory of representative generalization. Notably, Hobbes's insistence that "we need no other faculty but that of imagination" to reason about general terms is reminiscent of Berkeley's denial of the faculty of pure intellection whose business it is to frame abstractions. A universal name, to use Hobbes's terminology, does not denote an abstract idea or a universal object. Instead, the term has divided reference and denotes each individual member of a certain class. The similarity of this account with Berkeley's theory of representative generalization is strengthened by Hobbes's claim in the *Leviathan* that to understand a universal name one need only call to mind the idea of a member of a class of objects which resemble one another in some important respect, so that "One Universall name is imposed on many things, for their similitude in some quality . . . and whereas a Proper Name bringeth to mind one thing onely; Universals recall any one of those many" (Hobbes [1845] 1966a, 3:21).

It is not clear whether Berkeley's views on abstraction were significantly influenced by Hobbes's writings. There is nothing similar to the Berkeleyan argument from impossibility in Hobbes, so it is apparent that this aspect of Berkeley's critique of the doctrine of abstract ideas was not taken from him. Hobbes and Berkeley agree, as does Malebranche, in their dismissal of Scholasticism and its dependence on abstractions, but this is hardly enough to establish a direct influence. Berkeley was obviously familiar with Hobbes's works, since he mentions him several times in the *Philosophical Commentaries*.[21] But none of these entries deal with abstraction or universals, so it may well be that Berkeley did not take Hobbes's theory as a model for his own account of representative generalization. Moreover, Berkeley speaks disparagingly of Hobbes (the archmaterialist) whenever he has occasion to refer to his works, and seems to regard him as an atheist and proponent of dangerous views.

Berkeley certainly worked out the details of his theory of representative generalization more carefully than Hobbes developed his account of the meaning of universal names. I know of nothing in Hobbes's works to parallel Berkeley's account of how geometric proofs can be construed as dealing with particulars and the way in which the generality of the theorem proved can be explained in terms of the representative capacity of the relevant particulars. Thus, even if Berkeley's views on abstraction betray a Hobbesian influence, it is clear that Berkeley went well beyond Hobbes in developing his account of general ideas.

Peter Browne, an Irish cleric and contemporary of Berkeley's, may well have had the most profound influence on Berkeley's critique of abstraction. Browne was some twenty years Berkeley's senior (his date of birth is unknown), and taught philosophy at Trinity College, Dublin, from 1682 to 1710, also serving as provost from 1699 to 1710 (Winnett 1974). Berkeley entered Trinity College in 1700, and there is no reason to think that Browne did not teach Berkeley. Luce, at any rate, thinks that Browne "probably taught Berkeley in college" (Luce [1934] 1967, 74).

That Browne influenced Berkeley is suggested by the very strong similarity in their views on abstraction. Consider the following long passage from Browne's *Procedure, Extent and Limits of Human Understanding*, in which Browne attacks the thesis that pure intellect can frame general ideas by abstracting from particulars:

> *Logical* Abstraction in order to form *General* Ideas . . . is thought to be performed by withdrawing the Mind intirely from all the *Individuals,* and then forming one single Idea which shall represent the *Whole* Kind or Species at once; as when we remove our Thoughts intirely from all the Indi-

---

21. See *Commentaries*, 795–799, 806, 822, 824–825, 827, 834, and 837.

viduals of *Men*, and frame to our selves one general Idea distinct from them all to represent the whole Race. And these *General abstract* Ideas shall, in the modern refined Method of forming them, prove the strangest and most inconsistent *Monsters* in the World. Thus the *General abstract Idea* of *Man*, shall not be of a black or white, short or tall, thick or slender Man; but shall be *All* these and *None* of them at the *Same* time: The general abstract Idea of a Triangle shall be neither of an *Equilateral*, nor *Equicrural*, nor *Scalenum;* neither *Oblique* nor *Rectangle*, but all and none of these at once. (Browne [1728] 1976, 186–87)

Browne is convinced that such an account of abstraction is complete nonsense, and he continues with a polemic that could have come straight out of Berkeley's Introduction to the *Principles:*

> Now this is utterly impracticable, and therefore Abstraction in this Sense is a hard Word and without any determinate Meaning; for if the Intellect abstracts from all the Ideas of every Individual, it can have no Idea at all left to operate upon. When we say *Mankind*, it expresseth no one *Distinct, Abstract, General* Idea which stands in the Mind for all the Individuals at once; but it signifies the Idea of *One Individual*, which is no otherwise made general, than by our conceiving all the rest of the same Kind *By* that one; so that in truth it is the single Idea of any one Individual which is made to stand for and represent the whole Species. There is no such thing in Nature as any *Universal* really *Existing*, either to strike upon our Senses, or to be an Object of our Reason; and consequently there can be no such general abstract Idea in the Mind. (Browne [1728] 1976, 187)

These passages show a nearly complete agreement between Browne and Berkeley on the question of abstract ideas, with striking similarities in language and doctrine. Both conclude that abstract general ideas must be absurd and impossible, and both rely upon Locke's discussion of the abstract general idea of a triangle in order to make a case against abstract ideas. Similarly, both rely upon a version of representative generalization in order to account for our ability to use and understand general terms. Perhaps most interesting here is the fact that both authors base their attack on the suspect premise that the abstract idea of, say, man is the idea of an abstract man.

If we agree that both Browne and Berkeley share a certain conception of abstraction, the question arises whether Browne influenced Berkeley's views on abstraction or *vice versa*. Because the first edition

Abstraction and the Berkeleyan Philosophy of Mathematics

of Browne's *Procedure* appeared in 1728, nearly twenty years after Berkeley's first published attack on abstract ideas, it seems quite plausible that it was Berkeley who influenced Browne. The only problem with this theory is that the *Procedure* was Browne's first publication in philosophy, and it is not unreasonable to think that he may have developed his views on abstraction while he was teaching at Trinity, long before the actual publication of his *Procedure*. Whatever we take to be the true story of the sources behind Berkeley's critique of abstraction, it is clear that it grows out of concerns which are by no means unique to him. The application of this antiabstractionist sentiment to problems in the philosophy of mathematics is interesting and original, and we are now in a position to undertake a study of Berkeley's account of mathematics.

CHAPTER TWO
Berkeley's New Foundations for Geometry

The central thesis in Berkeley's philosophy of geometry is a negative one, namely that the object of geometry is not "extension in the abstract" or "the abstract idea of extension." This in itself is not terribly noteworthy, since the rejection of abstract ideas which I examined in chapter 1 mandates such a rejection of the abstractionist conception of geometry. More significant, however, is the fact that Berkeley's views on geometry changed in important ways even as he maintained his fundamental thesis on the nonabstract nature of geometry. The changes I will trace here follow both from his denial of abstract ideas and from his attempts to provide a nonabstract object for geometrical investigation.

We saw earlier that philosophers were inclined to explain the possibility of geometrical propositions being true of objects by construing geometric objects as abstractions from experience. This approach makes mathematical truth independent of the structure or contents of the actual world and construes mathematical demonstrations as concerned only with the properties of abstract ideas. As I have indicated, this orthodox account of mathematics is not without philosophical virtues, in that it gives at least some kind of answer to fundamental problems in the philosophy of mathematics. Because Berkeley insists that the human mind is incapable of framing abstract ideas, we can expect him to maintain that the received view of geometry is fundamentally mistaken. This expectation is fulfilled, and in Berkeley's attempts to provide an alternative to the abstractionist conception of geometry we can trace an interesting development in his philosophical position.

Berkeley's New Foundations for Geometry

In the *Philosophical Commentaries* (ca. 1707–8) Berkeley rejects the standard view and advocates a wholesale revision of classical geometry designed to bring geometrical practice into conformity with his austere epistemological principles. By the time of the publication of the *Principles of Human Knowledge* (1710) the program for revision of geometry has been abandoned. Instead, Berkeley goes beyond the mere assertion that there are no abstract ideas and articulates his theory of representative generalization as an alternative account of geometric knowledge. This permits him to offer a reinterpretation of classical geometry in which the object of geometry is not "extension in the abstract" but does not demand the rejection of all of traditional geometry. By the time of the publication of the *Analyst* (1734), Berkeley's views have evolved even further and he wholeheartedly endorses classical geometry, upholding the classical standards of rigor against the infinitesimal methods introduced in the seventeenth century. This chapter is concerned with tracing the development of Berkeley's account of abstraction and geometry.

The Early View

The *Philosophical Commentaries* provide a natural starting place for a discussion of Berkeley's views on geometry. Even a cursory reading of the work shows that Berkeley was profoundly concerned with philosophical questions in mathematics when he wrote the *Commentaries*, and a closer look shows that he was working on a plan for the complete revision of classical geometry as he wrote them. Nearly one-third of the entries in the *Commentaries* deal with mathematics, and almost all are highly critical of the established mathematical practices of the late seventeenth and early eighteenth centuries. Although Berkeley's commitment to a philosophical program cannot be inferred from an isolated note in the *Commentaries*, the number of entries dealing with geometry and their uniform hostility toward the accepted account of the subject clearly reveal Berkeley's attitude toward geometry as that of a critic bent on the complete revision of the subject. Furthermore, entries attacking classical geometry appear throughout both of the notebooks which constitute the *Commentaries* and present a consistent picture of a program for the overthrow of standard geometry.

These notebooks are not, however, the first Berkeleyan works to deal with mathematics. In the years before he wrote the *Commentaries* Berkeley's interest in the subject had led to his first publication, a small collection of mathematical works entitled *Arithmetica et Miscella-*

nea Mathematica.[1] This unusual book contains an expository piece in the science of arithmetic, an "algebraic game" designed to help students master techniques for solving equations, papers on notation for irrational roots and the "atmospheric tide," and a demonstration of a geometric theorem concerning the proportion between the surfaces, altitudes, volumes, and bases of an equilateral cone and cylinder circumscribed about a sphere. The theorem states that the volume, surface, altitude, and area of the equilateral cone stand in the ratio of $3:2$ to those of the cylinder. The proof is the trivial application of a lemma which asserts that the ratio of the side of an equilateral triangle to the diameter of an inscribed circle is $\sqrt{3}:1$ while a perpendicular dropped from any angle to the opposite side stands in a ratio of $3:2$ to the diameter. No proof is offered for the lemma, and Berkeley assumes that it will be easily admitted by all who understand anything of algebra or geometry. In discussing the theorem being demonstrated, Berkeley refers to the work of Archimedes, Wallis, André Tacquet and Claude Dechales while immodestly proclaiming his own prowess in geometry.[2] His geometrical work here testifies to Berkeley's understanding of the standard methods of proof from the seventeenth century and betrays no reservation about the reliability of traditional geometry. He does, however, suggest that infinitesimal methods may be objectionable, and in this respect his essay foreshadows some of his objections to the calculus. After noting that the theorem demonstrated can be obtained by the infinitesimal technique known as the "method of indivisibles," he remarks "But both this method of indivisibles, and the arithmetic of infinities based on it, have been deemed not to be geometrical by some" (*Works* 4:213). Hardly a critique of traditional geometry, this sentiment represents an opinion widely shared among British mathematicians of the early eighteenth century. In the *Commentaries*, however, Berkeley's attitude changes drastically to that of an uncompromising opponent of the accepted approach to geometry.

When he wrote the *Commentaries* Berkeley was convinced that the doctrine of abstract ideas was a philosophical blunder that had led to

---

1. Published in 1707, these papers were composed between 1704 and 1707 and apparently published as part of Berkeley's campaign for a fellowship at Trinity College, Dublin. See Luce's "Editor's Introduction" (*Works* 4:159) and Luce (1963, 40–44). I will be concerned with the *Arithmetica* and with Berkeley's *De Ludo Algebraico* in chapter 3 when I discuss his approach to arithmetic and algebra.

2. He specifically refers to Wallis's *Treatise of Algebra* and *Arithmetica Infinitorum*, Tacquet's *Selecta ex Archimede theoremata*, and Dechales's *Cursus seu mundus mathematicus*. See Breidert (1989, 85–89) for more on this Berkeleyan proof and its background.

a serious misunderstanding of the nature of geometry. "No Idea of Circle, etc. in abstract," he declares (*Commentaries*, 238) and later elaborates, "Extension without breadth i.e. invisible, intangible length is not conceivable tis a mistake we are led into by the Doctrine of Abstraction" (*Commentaries*, 365a). Here he is obviously challenging the claim that a Euclidean line can be conceived by abstracting breadth from length. As he expresses it, the doctrine of abstraction has led to the mistake of thinking that there could be such a thing as breadthless length, and the result of this mistake is a wholly misguided theory of the object of geometry. This entry also connects Berkeley's critique of abstract ideas and his *esse* is *percipi* thesis in an interesting way, because the "mistake" to which the doctrine of abstraction leads is that of supposing there could be an invisible and intangible length.

Implicit in this entry is a rejection of the entire orthodox view of geometry as a science of abstractions, because the abstraction of breadth from length is a paradigm of the way in which the standard view of geometry addresses questions in the epistemology and ontology of geometry. Berkeley voices similar complaints throughout the *Commentaries:* "A meer line or distance is not made up of points, does not exist, cannot be imagin'd or have an idea fram'd thereof no more than meer color without extension" (*Commentaries*, 253), and "We can no more have an idea of length without breadth or visibility than of a General figure" (*Commentaries*, 483). The conception of a line as "breadthless length" is repeatedly attacked as requiring us to frame an impossibly abstract idea analogous to "meer color" or a "general figure." Berkeley extends this rejection of the received view when he insists that the doctrine of abstract general ideas has made a hash of the philosophy of mathematics generally: "No general Ideas, the contrary a cause of mistakes or confusion in Mathematiques, etc. this to be intimated in ye Introduction" (*Commentaries*, 401).[3]

It is evident that Berkeley opposes the standard account of geometry and that his opposition is based upon his rejection of abstraction, but it is not yet clear just what (if any) consequences follow from this denial of abstractionism. The relevant entries reveal that, whereas the traditional conception of geometry declared geometric magni-

---

3. It might be thought that the use of the term 'general ideas' rather than 'abstract ideas' undermines my claim that Berkeley's attack on traditional geometry is tied to his critique of abstraction. However, as we saw in chapter 1, Berkeley employed the terms 'abstract ideas,' 'general ideas,' and 'abstract general ideas' with roughly equal frequency with reference to the same doctrine. The doctrine he opposes is set forth in §§7–9 of the Introduction to the *Principles,* where a taxonomy of the species of abstract ideas is given.

tudes such as lines, angles, surfaces, and arcs to be infinitely divisible, Berkeley repeatedly and emphatically asserts the contrary. It is therefore worthwhile to investigate this issue in some detail.

Euclidean Geometry and its Commitment to Infinite Divisibility

To begin with, we should be clear about what is supposed to be infinitely divisible. Berkeley frequently writes as if the issue is the infinite divisibility of *extension,* while among geometers it is generally the infinite divisibility of *geometric magnitude* that is discussed. As we saw in chapter 1, the mathematicians and philosophers of Berkeley's era contrasted magnitude and multitude, declaring the former to be the object of geometry while designating the latter as the object of arithmetic. For the most part, the pronouncements of mathematicians and philosophers of the period on the nature of magnitude are uninformative, and they apparently relied upon an intuitive notion according to which a magnitude was something continuous and capable of measure. What is clear is that geometric magnitudes were conceived of as divided into species, including line segments, areas, volumes, and angles. Divisibility was considered an essential property of such magnitudes. In his *Introductio ad Veram Physicam* John Keill claims that divisibility "is a property of extension which necessarily pertains to all species of magnitude, as well to lines as surfaces, as much to space as to body" (Keill 1739, 25).[4] Berkeley's remarks about infinite divisibility are clearly intended to apply to just such geometric magnitudes. On the Berkeleyan view, all such magnitudes are ultimately "ideas of extension," so that we can read his critique of the doctrine of infinite divisibility of extension as applying to geometric magnitudes generally.

The thesis of infinite divisibility can be stated as the assertion that any magnitude of a given kind can be divided into two magnitudes of the same kind. If we confine our attention for the moment to geometric line segments, then the thesis of infinite divisibility asserts that any finite line segment can be divided into two line segments, while a similar claim would hold for angles, surfaces, and solids. In denying

---

4. Like Barrow's *Lectiones Mathematicæ,* Keill's *Introductio* was originally a series of university lectures dealing with philosophical problems in mathematics and natural science. It was first published in 1702, and Berkeley refers to it at entries 308, 322, and 364 of the *Philosophical Commentaries.* The third and fourth lectures in Keill's *Introductio* are of special significance for Berkeley's views on infinite divisibility. Lecture 3 is a statement of the thesis of infinite divisibility and lecture 4 is concerned with answering various objections that were commonly raised against it. See Neri (1980) for a study of Berkeley's reaction to Keill and the doctrine of infinite divisibility.

Berkeley's New Foundations for Geometry

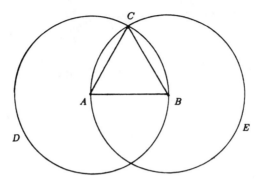

Fig. 2.1

the doctrine of infinite divisibility, Berkeley requires that a finite line segment admit only a finite number of divisions before indivisible minimal parts are reached. From the standpoint of traditional geometry the consequences of this doctrine are radical indeed, since essentially all of traditional geometry is based upon the assumption that geometric magnitudes can be divided without end.

It should be clear from the start that Euclidean geometry requires some version of the thesis of infinite divisibility. Barrow gives a succinct statement of the thesis when he declares:

> [Divisibility] prevails of all magnitude, and likewise of all its parts. As all magnitudes are composed of parts, so all can be resolved into one kind of parts or another. There is no absolute minimum in any species of magnitude. Whatever is divided into parts is divided into parts that are further divisible. (*Lectiones* 9:140)

It is well known that Euclid's *Elements* presuppose a strong principle of continuity sufficient to guarantee the infinite divisibility of finite geometric magnitudes.[5] The first proposition in book I of the *Elements* invokes this principle in the proof that an equilateral triangle can be constructed on any given finite line segment (see fig. 2.1). The proof is quite simple:

---

5. Essentially, the principle of continuity is the Dedekind continuity axiom which states that every set of real numbers with an upper bound has a least upper bound. Applied to the case of points on a line ab, the principle states that if we have an ordering of the points on the line such that $a < b$ and there is a condition ϕ such that ϕa, $\sim \phi b$, and $\forall x \forall y (\phi x \cdot (-\phi y)) \to x < y)$ then there is a point c such that ϕc and $\forall x (c < x \to \sim \phi x)$. A discussion of the principle of continuity and its role in Euclidean geometry can be found in Heath's commentary to the *Elements* (Euclid [1925] 1956, 1:234–240) and Mueller (1981).

Chapter Two

On any given finite straight line to construct an equilateral triangle.
Let *AB* be the given finite straight line. Thus it is required to construct an equilateral triangle on the straight line *AB*.

With centre *A* and distance *AB* let the circle *BCD* be described. Again, with centre *B* and distance *BA* let the circle *ACE* be described.

From the point *C*, in which the circles cut one another,[6] to the points *A*, *B* let the straight lines *CA*, *CB* be joined.

Now, since the point *A* is the centre of the circle *CDB*, *AC* is equal to *AB*.

Again, since the point *B* is the centre of the circle *CAE*, *BC* is equal to *BA*.

But *CA* was also proved equal to *AB*.

Therefore each of the straight lines *CA*, *CB* is equal to *AB*. And things which are equal to the same thing are also equal to one another; therefore *CA* is also equal to *CB*. Therefore the three straight lines *CA*, *AB*, *BC* are equal to one another.

Therefore the triangle *ABC* is equilateral and it has been constructed on the given finite straight line *AB*. (*Elements* I, 1).

This construction is used throughout the *Elements* to prove a number of theorems which are versions of the thesis of infinite divisibility—to prove, for example, that any angle can be bisected, with that construction used to prove the following proposition, that any finite line segment can be bisected (*Elements* I, 9–10).

Some who argued for the thesis of infinite divisibility took these basic Euclidean theorems to settle the issue and asserted that infinite divisibility had been proved geometrically, just as geometers had proved that the sum of the interior angles of any triangle is equal to two right angles. Denial of infinite divisibility clearly requires abandoning the continuity principle which generates the various versions of the thesis of infinite divisibility. This, in turn, amounts to a rejection of the most fundamental principle in Euclidean geometry and abandonment of all the theorems from books I and II of the *Elements*. Berkeley is unimpressed by any appeal to results from Euclidean geometry in support of infinite divisibility, and I will now consider just how radically he departs from the Euclidean model.

The importance of the thesis of infinite divisibility for traditional geometry can be seen in the theory of incommensurable ("surd") magnitudes. A standard problem in classical geometry is to establish the proportion between two magnitudes, but such proportions can-

---

6. The principle of continuity is tacitly invoked here to assure the existence of the point *C*, and without the assumption of continuity the proof would not go through.

Berkeley's New Foundations for Geometry

not generally be expressed as ratios of integers. The most familiar example, the proportion between the diagonal δ and side σ of a square—$\delta:\sigma::\sqrt{2}:1$—is notoriously incapable of expression as a ratio of integers. But if the thesis of infinite divisibility were false, then the side and diagonal of the square could be resolved into finite collections of minimal "linelets" and the ratio between the two magnitudes could be expressed as the ratio between the number of minimal parts in the side and diagonal of the square, i.e., as a ratio of integers.[7]

Acceptance of the thesis of infinite divisibility in Berkeley's day was so widespread as to be all but universal,[8] and the extent to which traditional geometry is committed to the thesis should make this very unsurprising. Indeed, infinite divisibility seems to be a deliverance of common sense, at least when it is stated as the claim that any magnitude is capable of division into two lesser magnitudes. Although there is surely a limit to the number of divisions we can make in any line drawn in the construction of a geometric figure, the thesis of infinite divisibility appears unproblematic precisely because we regard lines drawn in chalk or ink as mere physical representations of geometry's true objects and these true geometric objects as abstractions from ordinary experience which do not suffer such physical limitations.

Moreover, in upholding the thesis of infinite divisibility, it is not necessary to assume that any procedure of infinite division is ever actualized, but merely that there is no end to the possibility of further division. It is for this reason that we state the thesis of infinite divisibility, not as the claim that every magnitude contains an infinite number of parts, but as the assertion that any magnitude can be bisected. Characterization of the thesis of infinite divisibility by appeal to the possibility of continued division has its roots in the Aristotelian theory of the potential infinite and was widespread in Berkeley's day. Thus

---

7. To make this resolution of magnitudes work it is necessary to assume that all of these minimal parts are of the same size, but this added condition raises no difficulty as regards Berkeley's account of geometry. As we will see, Berkeley rejected the entire theory of incommensurables in the *Commentaries,* and his doctrine of the minimum sensible requires that all minima be of the same size. In §80 of the *New Theory of Vision* he declares that "the *minimum visibile* is exactly equal in all beings whatsoever that are endowed with the visive faculty."

8. Bayle was something of an exception. See notes F and G to the article "Zeno of Elea" in Bayle (1697), where he discusses the thesis of infinite divisibility and dismisses it as paradoxical. There is reason to believe that Berkeley read Bayle, so that the attack on infinite divisibility may well derive in part from Bayle. See Popkin (1951–52) for an account of the influence of Bayle on Berkeley. The connection between Bayle and Berkeley in regard to the question of infinite divisibility is also examined in Bracken (1974, chapter 2), Bracken (1977–78), and Luce (1963, chapter 4).

Locke, rather than claiming that we can imagine a line as actually containing an infinite number of subdivisions, relies upon a concept of infinity as an "endless growing idea" that is somehow incomplete. In discussing our idea of space he casually endorses the thesis of infinite divisibility when he speaks of the mind "being able to also to shorten any Line it imagines, by taking from it $\frac{1}{2}$, or $\frac{1}{4}$, or what part it pleases, without being able to come to an end of any such Division" (*Essay* II, xiii, 6). He similarly endorses the infinite divisibility of matter:

> And since in any bulk of Matter, our Thoughts can never arrive at the utmost *Divisibility,* therefore there is an apparent infinity to us also in that, which has the Infinity also of Number, but with this difference, That in the former Considerations of the Infinity of Space and Duration, we only use Addition of Numbers; Whereas this is like the division of an Unite into its Fractions, wherein the Mind also can proceed *in infinitum,* as well as in the former Additions, it being indeed but the Addition still of new Numbers: though in the Addition of the one, we can have no more the positive *Idea* of a space infinitely great, than in the Division of the other, we can have the *Idea* of a Body infinitely little; our *Idea* of Infinity being, as I may so say, a growing and fugitive *Idea,* still in a boundless Progression, that can stop no where. (*Essay* II, xvii, 12)

Another statement of this thesis, which also does not require the actualization of an infinite process of division, can be found in John Keill's *Introductio ad Veram Physicam:*

> But here we do not understand by divisibility the actual separation of parts from each other, which supposes motion, and which the nature of space does not admit, nor indeed is it proved in the demonstrations taken from geometry; for the divisibility which we want to show here is simply the resolution of magnitude into its parts, or their distinction and assignability. (Keill 1739, 25)

The thesis of infinite divisibility can also be maintained without licensing talk of infinitesimal magnitudes. If we characterize an infinitesimal magnitude as one which is greater than zero but less than any positive real number, it is clear that the classical conception of geometry will have the thesis of infinite divisibility as a theorem without admitting infinitesimal magnitudes. In a classical proof only finite differences between finite magnitudes are considered and the use of infinitesimals is explicitly barred, but the infinite divisibility of lines, angles, and other magnitudes are proved as elementary theorems. Thus, Berkeley's attack on infinite divisibility should not be conflated with this critique of the infinitesimal calculus. Indeed, in his early essay "Of Infinities" Berkeley himself echoes the received view when he appears to endorse the thesis of infinite divisibility while rejecting the Leibnizian doctrine of infinitesimals:

> 'Tis plain to me we ought to use no sign without an idea answering to it; & 'tis plain that we have no idea of a line infinitely small, nay 'tis evidently impossible there should be any such thing, for every line, how minute soever, is still divisible into parts less than itself; therefore there can be no such thing as a line *quavis data minor* or infinitely small. (*Works* 4:235–36)

"Of Infinities" was read before the Dublin Philosophical Society on 19 November 1707.[9] This means that it would have been prepared at about the same time Berkeley was writing the *Commentaries*, although he seems to endorse the thesis of infinite divisibility in "Of Infinities" while definitely rejecting it in the *Commentaries*.[10] I will have more to say about "Of Infinities" in chapter 5, when I consider Berkeley's critique of the calculus. But it should be clear that in denying the thesis of infinite divisibility in the *Commentaries*, Berkeley launched a frontal assault on well-established philosophical theory and entrenched mathematical practice.

The Case against Infinite Divisibility in the *Commentaries*

Berkeley's discussion of infinite divisibility in the *Commentaries* is both extensive and illuminating, especially for the light it throws on the connection between his epistemological principles and his attack on the thesis of infinite divisibility.[11] Entries 21 and 21a are important, not only because they are the first mention of infinite divisibility in the *Commentaries*, but also because they a show a curious relation

---

9. British Museum Additional MS. 4812 (Sloane Papers) contains the Dublin Philosophical Society's Register. A copy of "Of Infinities" is entered on fol. 30, dated 19 November 1707.

10. I say "seems to endorse" because it is not entirely clear that the passage from "Of Infinities" is a statement of Berkeley's view. He could be arguing *ad hominem*, or he could hold that a line consisting of two minimal parts is divisible into "parts less than itself" without being divisible into two *lines*. This reading (although somewhat strained) would make the passage consistent with the denial of infinite divisibility.

11. Some thirty entries scattered throughout the *Commentaries* deal with infinite divisibility. I will discuss only the most important of them here. A study of some of Berkeley's arguments against infinite divisibility can be found in Fogelin (1988).

between Berkeley's views on geometry and his famous thesis *esse* is *percipi*. They read:

> Demonstrations of the infinite divisibility of extension suppose length without breadth $\wedge$ wch is absurd
> or invisible length (*Commentaries*, 21, 21a)

Here he begins with the insistence that the thesis of infinite divisibility can only be demonstrated by assuming there is length without breadth—an assumption he finds unintelligible. As we have seen, the Euclidean line or "breadthless length" is a paradigmatic abstract idea and it is clear that Berkeley is faulting the standard demonstrations of infinite divisibility (and presumably the proofs in book I of Euclid) precisely because they rely upon such absurd abstractions. The addendum—"or invisible length"—connects the denial of infinite divisibility with the *esse* is *percipi* thesis by rejecting invisible or unperceived length as a similar absurdity. This connection is presaged by an earlier note which reads "Extension a sensation, therefore not without the mind" (*Commentaries*, 18). Criticisms of attempted proofs of infinite divisibility are repeated elsewhere in the *Commentaries*, the best statement reading:

> 3 faults occur in the arguments of the Mathematicians, for divisibility ad infinitum. 1. they suppose extension to exist without the mind or not perceiv'd. 2. they suppose that we have an idea of length without breadth. * or that length without breadth does exist. 3. that unite is divisible ad infinitum.
> *or rather that invisible length does exist. (*Commentaries*, 342, 342a)

As presented in the *Commentaries*, then, Berkeley's case against infinite divisibility runs along several independent lines. The first argument asserts that there can be no abstraction of length from breadth of the kind required to prove the thesis of infinite divisibility, then concludes that the thesis cannot be demonstrated. The second contends that because there is no extension except perceived extension, and perceived extension is not infinitely divisible, the thesis of infinite divisibility is trivially false.[12]

The third objection set forth in entry 342 will not be of great concern to us. It is clearly a claim that demonstrations of the thesis of infinite divisibility are question-begging in an important way, espe-

---

12. This line of reasoning is illustrated in other entries, for example, "Infinite divisibility of extension does suppose ye external existence of extension but the latter is false, ergo ye former also" (*Commentaries*, 26).

Berkeley's New Foundations for Geometry

cially in that they assume that unity is divisible. Berkeley's idea must be that whatever is a unit is one thing and cannot be composed of parts, but the thesis of infinite divisibility is forced to claim that unity is divisible. The argument first appears earlier where Berkeley argues that

> Unite in abstracto not at all divisible it being as it were a point or wth Barrow nothing at all in concreto not divisible ad infinitum there being no one idea diminishable ad infinitum. (*Commentaries*, 75)

The reference to Barrow is to the end of the third lecture in the *Lectiones Mathematicæ*, where Barrow dismisses the opinion that the unit can be characterized as a geometric point. Berkeley pursues a similar line of thought when he writes:

> Mem: To Enquire most diligently concerning the Incommensurability of Diagonal & side. whether it Does not go on the supposition of unit being divisible ad infinitum, i.e. of the extended thing spoken of being divisible ad infinitum (unit being nothing also V. Barrow Lect. Geom:). & so the infinite divisibility deduc'd therefrom is a petitio principii. (*Commentaries*, 263)

This entry cites Barrow's *Lectiones Geometricæ*, although there is nothing in those lectures to parallel the opinion set out in lecture 3 of the *Lectiones Mathematicæ*.

Berkeley hints at another line of argument when he suggests in several entries that the thesis of infinite divisibility itself is inherently paradoxical and inconsistent:[13]

> I do not think that surfaces consist of lines *i.e.* meer distances. Hence may perhaps be solv'd that sophism wch would prove the oblique line equal to the perpendicular between 2 parallels. (*Commentaries*, 259)
>
> Mem. to prove against Keil y^t the infinite divisibility of matter makes the half have an equal number of equal parts with the whole. (*Commentaries*, 322)

---

13. Neri presents an analysis of this argument, concluding that "the hypothesis that extension is divisible to infinity leads to the absurd consequence that every extension must be infinite: this is Berkeley's paradox. Once such a consequence has been deduced, clearly in conflict with the evidence of the senses, Berkeley can easily reject infinite divisibility (or the Archimedean property) and impose a threshold (characterized first logically and then psychologically) beyond which division cannot be actualized" (Neri, 1980, 79–80).

Keils filling the world with a mite thus follows from the Divisibility of extension ad infinitum. (*Commentaries*, 364)

The references to "Keil" are obviously to Keill's *Introductio ad Veram Physicam*, especially lecture 4 where various allegedly paradoxical consequences of the doctrine of infinite divisibility are stated and then resolved. One objection that Keill considers is that if extension is infinitely divisible, then every extended thing will have an infinite number of parts and will be infinitely large, so the smallest thing (such as a mite) will be equal to the largest. This is clearly the sort of argument Berkeley has in mind at entry 364, and he is there claiming that Keill's paradoxes are in fact unresolvable and the thesis of infinite divisibility must therefore be abandoned.

As I interpret these claims of paradox, they all deal with the concept of cardinality and its application to geometric magnitudes. Restricting ourselves to the statement of the problem in terms of line segments, we have three jointly inconsistent claims which constitute the "paradox": (*a*) if two line segments are of different lengths, then the number of points on the longer line is greater than the number of points on the shorter line; (*b*) if there is a 1:1 correspondence between the points on two lines, then both lines contain the same number of points; and (*c*) given line segments A and B where the length of A is twice the length of B, there is a 1:1 correspondence between the points on A and the points on B. Berkeley clearly thinks that (*c*) is the questionable principle here, and his solution to the paradox was to reject (*c*) because he saw it as depending upon the belief that geometric lines are infinitely divisible. He was unwilling to call (*a*) into question, although this principle is denied in all of traditional geometry. The denial of (*a*) amounts to the claim that the number of points on a line is independent of its length—an assertion which Berkeley apparently found unintelligible. As I will show, it is central to Berkeley's early views on geometry that the lengths of line segments be measured by the number of minima they contain.

Berkeley's attempt to challenge the thesis of infinite divisibility by citing such "paradoxes" cannot be considered very effective and was apparently not central to his program in the *Commentaries*. The paradox in question can be resolved by denying either (*a*) or (*c*),[14] and Berkeley gives no independent reason to deny (*c*) rather than (*a*). It may be that Berkeley included versions of Zeno's paradoxes among the "amusing geometrical paradoxes" which he took to show the in-

---

14. I assume that principle *b* is so obviously true that it is not a candidate for rejection.

coherence of infinite divisibility, but I will not consider them here because Berkeley makes no explicit mention of them and it is clear that they can be resolved in much the same way as the above paradox.

Berkeley's Geometry of Minima

Berkeley in the *Commentaries* urges a thorough revision of the traditional conception of geometry. His new geometry requires neither the doctrine of abstract ideas nor the infinite divisibility of geometric magnitudes and is based upon the concept of the minimum sensible—the smallest perceivable part of extension.[15] Of course, for Berkeley, "perceivable extension" is redundant and we could call these minima the smallest parts of extension *simpliciter*. Berkeley actually distinguishes between the minimum visible and the minimum tangible, so that there is a minimal extension for each sensory modality. This distinction is rooted in Berkeley's insistence on the heterogeneity of visible and tangible space, a claim most clearly presented in §§88–120 of the *New Theory of Vision*. There has been some controversy as to whether Berkeley conceived of these minima as extended or unextended. Raynor argues that Berkeley must hold that the minima are unextended because he "would never have described his minima as extended indivisible points, for this would have been a contradiction in terms, given assumptions then current" (Raynor 1980, 197). Similarly, Flage holds that "the geometrical notion of extension holds that anything that is extended is divisible. Thus, if minimum visibles are indivisible, it follows that they are unextended" (Flage 1987, 50). I find this line of reasoning unpersuasive. Berkeley challenged many of the "assumptions then current" about the nature of extension or the nature of geometry, so it seems odd to presume that he would describe his minima as extensionless simply to avoid contradicting contemporary assumptions. As I will show, Berkeley is more than willing to reject the whole of traditional geometry and there is no reason to think that he found "the geometrical notion of extension" more intelligible than its other assumptions. Bracken (1984) notes that Berkeley's denial of abstraction makes it impossible for him to separate color from extension, so the fact that minima have color implies that they must be extended. Moreover, it is difficult to imagine Berkeley accepting the doctrine that a collection of unextended points could create extension, any more than he could accept that something visible could be composed of invisible parts. I therefore assume that Berkeley understood minima as extended.

---

15. See Moked (1988, appendix A) for a study of this doctrine.

Essentially, Berkeley's minima play the role the point plays in Euclidean geometry, but his results conflict completely with the Euclidean system. The root of this conflict is the fact that in Berkeley's system lines and figures are composed of finite collections of minima, whereas Euclidean geometry requires that they contain an infinite number of points. Berkeley does not work out his alternative to classical geometry systematically in the *Commentaries,* but we can sketch its salient features by drawing together the relevant entries and noting how little of traditional geometry would survive if the Berkeleyan account of the subject were correct. The starting point for this radical new geometry is Berkeley's reflection on the fourth "common notion" of book I in Euclid's *Elements:* "Things which coincide with one another are equal to one another" (in the standard Latin editions, "quæ sibi mutuo congruunt æqualia sunt"). Berkeley begins by asking how congruence or coincidence of geometric lines is to be determined, and decides that congruence exists only when two lines contain the same number of minima or "points." This leads him to declare, in the following sequence of entries, that the traditional problem of determining arc lengths can be easily solved:

> Mem: to Endeavour most accurately to understand w^t is meant by this axiom: Quae sibi mutuo congruunt aequalia sunt.
>
> Qu: w^t the Geometers mean by equality of lines & whether according to their definition of equality a curve line can possibly be equal to a right line.
>
> If wth me you call those lines equal wch contain an equal number of points, then there will be no difficulty. that curve is equal to a right line wch contains as [many] points as the right one doth. (*Commentaries,* 514–16)

By parity of reasoning, the definition of congruence for figures would hold that two figures are congruent when the minima of which they are composed can be made to overlap. Similarly, the definition of area for a geometric figure could then be given in terms of the number of minima which compose it. Berkeley arrives at this definition by the same kind of reflection on the definition of geometric congruence which we saw above:

> The Mathematicians should look to their axiom Quae congruunt sunt aequalia. I know not what they mean by bidding me put one Triangle on another. the under Triangle is no Triangle, nothing at all, it not being perceiv'd. I ask must sight be judge of this Congruentia or not. if it must then all Lines seen under the same Angle are equal wch they will not ac-

Berkeley's New Foundations for Geometry

knowledge. Must the Touch be judge? But we cannot touch or feel Lines & Surfaces, such as Triangles etc according to the Mathematicians themselves. Much less can we feel a line or Triangle that's cover'd by another Line or Triangle.[16]

Do you mean by saying one triangle is equall to another that they both take up equal spaces. But then the Question recurs w^t mean you by equal spaces, if you mean spatia congruentia answer the above difficulties.

I can mean (for my part) nothing else by equal Triangles than Triangles containing an equal number of points. (*Commentaries*, 528–30)

Here again, familiar themes are echoed. Berkeley faults the standard approach to geometry because it depends upon a criterion of congruence which cannot be judged by the senses, appears to deny the *esse* is *percipi* thesis, and is apparently applicable only to abstract lines and surfaces which "according to the mathematicians themselves" we cannot see or touch.

Granting that geometric distances and areas are to be defined in this way, a circle can be defined in the standard fashion as a collection of points equidistant from a given center point. But this understanding of circles leads to some wildly nonstandard results. First, if the radius and circumference of a circle are both measured in terms of the number of minima which compose them, then the ratio of diameter to circumference would always be expressible as a rational number, but the number would not be the same for all circles. Thus, while classical geometry has the irrational π as the unique ratio of circumference to diameter in all circles, Berkeley requires different rational numbers for circles of different sizes. He was aware of this fact and seems not to have been bothered by it:

It seems all Circles are not similar figures, there not being the same proportion betwixt all circumferences and their diameters. (*Commentaries*, 340)

D & P are not proportional in all Circles. dd is to $(1/4)$ dp as d to $P/4$ but d & $P/4$ are not in the same proportion in all Circles. Hence 'tis nonsense to seek the terms of one general proportion whereby to rectify all peripheries or of another whereby to square all Circles. (*Commentaries*, 457)[17]

---

16. These entries may be a commentary on the eleventh lecture in Barrow's *Lectiones Mathematicæ*, where Barrow discusses "superposition" of one magnitude upon another as a means of determining congruence.

17. For d and p in this entry read diameter and periphery. The point of the entry is that the proportion $d^2 : dp/4 :: d : p/4$ holds for any given circle, but the right term in the proportion will not be constant for every circle.

Circles on several radius's are not similar figures they having neither all nor any an infinite number of sides. Hence in Vain to enquire after 2 terms of one & y^e same proportion that should constantly express the reason of the d to the p in all Circles. (*Commentaries*, 481)[18]

Mem: to remark on Wallis's harangue that the aforesaid proportion can neither be express'd by rational numbers nor surds. (*Commentaries*, 482)[19]

These entries show that the traditional theory of incommensurable magnitudes cannot be accommodated within the approach to geometry developed in the *Commentaries*. Indeed, the very concept of incommensurable magnitudes seems to be incoherent if we adopt Berkeley's approach. He acknowledges that there will be a problem here when he declares, "Diagonal incommensurable wth ye side Quaere how this can be in my doctrine?" (*Commentaries*, 29). He later decides that the theory of incommensurables can simply be rejected: "Diagonal of particular square commensurable wth its side they both containing a certain number of M[inima] V[isibilia]" (*Commentaries*, 258).

Berkeley provides a good overview of his approach to those problems in traditional geometry where incommensurable magnitudes had been introduced:

> I say there are no incommensurables, no surds, I say the side of any square may be assign'd in numbers. Say you assign unto me the side of the square 10. I ask w^t 10, 10 feet, inches, etc. or 10 points. if the later; I deny there is any such square, tis impossible 10 points should compose a square. if the former, resolve y^r 10 square inches, feet, etc into points & the number of points must necessarily be a square number whose side is easily assignable. (*Commentaries*, 469)

Another result of this account of geometry is the falsity of the Pythagorean theorem: "One square cannot be double of another. Hence the Pythagoric Theorem is false" (*Commentaries*, 500). Berkeley also declares the falsity of other Euclidean theorems in the *Commentaries*,

---

18. The term "reason" in this entry is an antiquated term for ratio or proportion; the reference to an "infinite number of sides" is Berkeley's rejection of an understanding of circles which treats them as regular polygons with an infinite number of infinitesimal sides, as in L'Hôpital (1696).

19. The reference to "Wallis's harangue" is presumably to chapter 24 of the *Mathesis Universalis*, where Wallis remarks on Hobbes's failed attempts to square the circle and expresses the opinion that π cannot be expressed as the root of an algebraic equation but only as an infinite product.

Berkeley's New Foundations for Geometry

but we need not investigate these in detail. Among other things, he declares that the classical problem of squaring the circle can be solved (at least for any given circle) by constructing a square whose area will contain as many minima as the circle (*Commentaries*, 249–51, 395),[20] denies that there is always a mean proportional between any two lines (*Commentaries*, 470), and declares that only a line with an even number of points can be bisected (*Commentaries*, 267, 276). From what has been said so far it should be clear that nothing of use or interest in Euclidean geometry would survive if Berkeley's program were realized.

Not only does Berkeley's geometry of minima deny basic Euclidean theorems, but it is also anti-Euclidean in its methodology. In several entries in the *Commentaries*, Berkeley claims that rather than treating geometry as a deductively organized theory in which transparently true first principles lead to necessarily true theorems, we should instead view geometric problems as something to be solved by appeal to the senses rather than reason. Thus, he writes, "Sense rather than Reason & demonstration ought to be employ'd about lines & figures, these being things sensible, for as for those you call insensible we have prov'd them to be nonsense, nothing" (*Commentaries*, 466). A similar position is voiced earlier: "The folly of the Mathematicians is not judging of sensations by their senses. Reason was given us for nobler uses" (*Commentaries*, 373). It is unclear what "nobler uses" Berkeley has in mind here, but these entries indicate a definite hostility toward the notion of geometric demonstration. Reason, it seems, is out of place in geometry and priority must be given to the senses.

Such an approach to geometry is, of course, a complete departure from the traditional view in which geometry is seen as a deductive science based upon self-evident axioms, and Berkeley's rejection of this view is obviously the consequence of his commitment to the thesis that the object of geometry is perceived extension. Where the traditional account of geometric reasoning had declared geometric demonstrations to have abstractions as their objects, Berkeley insists that the objects of geometrical investigation can only be figures which are immediately perceivable by sense. Moreover, he rejects the traditional criterion of geometric rigor that sanctions proofs which begin with axioms that describe abstractions and proceed by logical inference to

---

20. It is worth observing that Berkeley can only claim to provide a quadrature for circles whose area is measured by a number of minima equal to a square integer. Any circle with a nonsquare number of minima can have its area measured by a rectangle or approximated by a square but cannot be "squared" exactly.

reveal additional properties of abstract objects. In the *Commentaries* Berkeley seems to accept only as much of geometry as can be made consistent with actual sense experience, and he makes perception the standard of rigor against which geometric practices are to be judged. Such a criterion of geometric intelligibility leaves nothing of Euclidean geometry. Berkeley eventually abandoned this project and endorsed the traditional standards of rigorous proof, but this evolution was completed only with the publication of the *Analyst*.

In the *Commentaries* Berkeley anticipates great benefits once his program overthrows Euclidean geometry, claiming that the new geometry will be both easier to learn and more useful. He insists that "All might be demonstrated by a new method of indivisibles, easier perhaps & juster than that of Cavallerius" (*Commentaries*, 346),[21] and later asks, "If at the same time we shall make the Mathematiques much more easie & much more accurate, w^t can be objected to us?" (*Commentaries*, 414). What we can gather from the remarks in the *Commentaries* suggests that Berkeley was quite happy to dismiss Euclidean geometry as a confused jumble of false dogma. Euclid's *Elements* were, in the eighteenth century, generally regarded as a paradigm of knowledge, but Berkeley asks us to believe that the whole subject rests upon a mistaken, illegitimate abstraction from sense perception. The conflict between Berkeley's epistemological commitments and traditional geometry raises serious questions about the status of the entire Berkeleyan enterprise. If our epistemology undermines classical geometry as a secure body of knowledge, then we are faced with the choice of revising our analysis of knowledge or revising our geometry. I think that the only sane reaction to this predicament is to revise the epistemology in order to save the geometry. In fact, I am convinced that this is the choice Berkeley made, although the evidence does not emerge until the publication of the *Principles*.

21. The reference to "Cavallerius" is to Bonaventura Cavalieri, an Italian mathematician of the seventeenth century whose "method of indivisibles" was a precursor to the integral calculus. I will discuss some aspects of Cavalieri's work in chapter 4. Moked holds that Cavalieri "was regarded by Berkeley as a well meaning but rather clumsy ally against the majority of 'the Mathematicians'. Most of the 'Mathematicians' relied on the concept of mathematical point, and, accordingly, assumed that lines 'are divisible *ad infinitum*', i.e. do not consist of indivisible units. Cavalieri, on the other hand, endorsed a partly finitist view" (Moked 1988, 210). Elsewhere he claims that Berkeley "of course supports Cavalieri against the then prevailing theories of infinite divisibility, according to which lines are composed of mathematical, i.e. non-physical points" (Moked 1988, 138). Moked adduces no support for this strange reading of Cavalieri and his relationship to Berkeley, and we will see in chapter 4 that he is mistaken in taking Cavalieri to be opposed to infinite divisibility.

Berkeley's New Foundations for Geometry

Berkeley and Barrow

It is interesting to note that many of Berkeley's claims about the falsity of Euclidean geometry derive directly from his reading of Barrow's *Lectiones Mathematicæ*, especially lecture 9 where Barrow discusses infinite divisibility. Barrow defended the thesis of infinite divisibility against the arguments of atomists,[22] citing numerous difficulties with the atomistic conception of geometric magnitudes and attempting to show that the denial of the thesis of infinite divisibility was simply unintelligible. Berkeley responds to several of these arguments in the *Commentaries*, and it is instructive to observe his responses.

To make his case for the thesis of infinite divisibility, Barrow relies heavily upon the fact that most of Euclidean geometry is inconsistent with the "doctrine of indivisibles," at one point claiming:

> Additionally, there is the necessary agreement of all mathematicians [to the thesis of infinite divisibility]; for although they hardly ever suppose it openly, they often assume it covertly, and if it were not true, many of their demonstrations would fall to the ground. (*Lectiones*, 9:141)

In the *Principles* Berkeley echoes this declaration when he writes:

> The *infinite* divisibility of *finite* extension, though it is not expressly laid down, either as an axiom or theorem in the elements of that science, yet is throughout the same every where supposed, and thought to have so inseparable and essential a connexion with the principles and demonstrations in geometry, that mathematicians never admit it into doubt or make the least question of it. (*Principles*, §123)

Barrow's case for infinite divisibility includes a geometric argument against indivisibles:

> And if, following our adversaries, we suppose a circle whose radius consists of only three points, the radius will be equal to the side of a hexagon inscribed within that circle. For the sixth

---

22. Barrow specifically mentions Epicurus, Lucretius, and Zeno as philosophers who reject the doctrine of infinite divisibility. He seems, however, to regard the issue as one of current interest, declaring "I am not unaware, for indeed nobody is unaware, that this doctrine of the perpetual divisibility of quantity is admitted by some only reluctantly, and simply rejected by others, and that the controversy over composition of magnitude (whether it be from indivisibles, or from homogeneous parts) is everywhere conducted with great obstinacy" (*Lectiones*, 9:140–41). Perhaps Barrow intends the work of Jacobus Fontialis, particularly *De idea mirabilis matheseos de Ente* (1660?), in Fontialis (1740, 437–512).

part of the circle's circumference will not be as great as four points (otherwise the whole circumference would consist of 24 points, and thus be four times the diameter, contrary to the most manifest demonstrations of Archimedes, and to common sense, by which the circumference of a circle is known to be less than the perimeter of its circumscribed square). But if the sixth part of the circle's circumference is less than four points, it cannot (according to the hypotheses of the proponents of indivisibles) be more than three. And therefore, the arc will not be greater than the chord, nor therefore will a right line be the shortest of all lines which can be drawn between the same points. And a similar difficulty will follow if the radius of the circle is supposed to be 5 points. . . . Which consideration alone should suffice to overthrow the contrary opinion, or at least manifestly declare its repugnance to mathematical principles. (*Lectiones*, 1:142)

Berkeley, in response, accuses Barrow of begging the question by applying Euclidean theorems in a context where they simply do not hold:

> Barrows arguing against indivisibles, lect. i. p. 16 is a petitio principii, for the Demonstration of Archimedes supposeth the circumference to consist of more than 24 points. Moreover it may perhaps be necessary to suppose the divisibility ad infinitum, in order to Demonstrate that the radius is equal to the side of the Hexagon. (*Commentaries*, 462)

A later entry suggests much the same line of thought:

> Archimedes's proposition about squaring the Circle has nothing to do with circumferences containing less than 96 points. & if the circumference contain 96 points it may be apply'd but nothing will follow against indivisibles. v. Barrow. (*Commentaries*, 510)

Another Barrovian argument against the hypothesis of indivisibles is stated as follows:

> If the circumference of a circle be supposed to consist of any number of points, to every one of which radii are drawn from the Center, it is very evident that the circumferences of more concentric circles will either consist of the same number of points with the former, and consequently be equal to it, which is most absurd: or otherwise these radii must be supposed to touch, meet, or intersect one another at some place other than the Center. (*Lectiones*, 9:142)

Berkeley's New Foundations for Geometry

This may well have prompted Berkeley to write the following sequence of notes:

> Most certainly no finite Extension divisible ad Infinitum.
> Mem: Difficulties about Concentric Circles. (*Commentaries*, 314–15)

Barrow's question "And concerning geometrical conclusions, how can a right line consisting of an unequal number of points be bisected?" (*Lectiones*, 9:142) is echoed by Berkeley;

> Q: How can a line consisting of an unequal number of points be divisible [ad infinitum] into two equals? (*Commentaries*, 267)
> It seems all lines cannot be bisected in 2 equall parts, Mem: to examine how Geometers prove the contrary. (*Commentaries*, 276)

And to Barrow's demand:

> How, according to our adversaries, can a mean proportional be found between two right lines, one consisting of seven points and the other of nine? Or how can a third proportional be exhibited to the same seven and nine points? For if in finding the above mean the lines are joined together so as to form the diameter of a semicircle, and from the common term of the segments a perpendicular is erected, this perpendicular cannot be in mean proportion between the said right lines, as according to the contrary position no such line can be given. (*Lectiones*, 9:143)

Berkeley presents the following rejoinder:

> A mean proportional cannot be found betwixt any two given lines. It can onely be found betwixt those the numbers of whose points multiply'd together produces a square number. Thus betwixt a line of 2 inches & a line of 5 inches, a mean geometrical cannot be found except ye number of points contain'd in 2 inches multiply'd by ye number of points contain'd in 5 inches make a square number. (*Commentaries*, 470)

Barrow concludes that the denial of infinite divisibility is a disaster for Euclidean geometry, and claims:

> I could devise and produce an infinity of such instances, with which it may be shown that the whole of geometry is altogether subverted and destroyed by this assertion of composition out of indivisibles, having nothing left in it sound and solid. And thereby a great and deplorable ruin, confusion,

Chapter Two

and inconsistency is brought upon this most divine science. (*Lectiones*, 9:144)

Berkeley's response to Barrow's protest is direct. "Barrow owns the Downfall of Geometry However I'll Endeavour to Rescue it. so far as it is usefull or real or imaginable or intelligible, but for the nothings I'll leave them to their admirers" (*Commentaries*, 384).

I will repeat two other entries I mentioned when I considered Berkeley's arguments against infinite divisibility which connect Barrow with Berkeley's concerns about the doctrine of indivisibility:

> Unite in abstracto not at all divisible it being as it were a point or wth Barrow nothing at all in concreto not divisible ad infinitum there being no one idea diminishable ad infinitum. (*Commentaries*, 75)
>
> Mem: to Enquire most diligently concerning the Incommensurability of Diagonal & side. whether it Does not go on the supposition of unit being divisible ad infinitum, i.e. of the extended thing spoken of being divisible ad infinitum (unit being nothing also V. Barrow Lect. Geom:). & so the infinite divisibility deduc'd therefrom is a petitio principii. (*Commentaries*, 263)

Presumably, Berkeley's comments refer to lecture 3 of Barrow's *Lectiones Mathematicæ*, where Barrow contests the comparison of a geometric point to an arithmetical unit:

> But this comparison of a point in geometry with unity in arithmetic is the very worst of all, and brings the worst consequences upon mathematics. For unity answers really to some part of every magnitude, but not to a point. Thus if a line be divided into six equal parts, as the whole line answers to the number six, so every sixth part answers to unity but not to a point which is no part of this line. A point is rightly termed indivisible, but not unity (for how, to take an example, can $\frac{2}{6} + \frac{4}{6}$ equal unity, if unity is indivisible and incomposite and represents a point?). Rather, only unity is properly divisible, and numbers arise from the division of unity. (*Lectiones*, 3:62)

Of course, the coincidence between Barrow's writings and Berkeley's outlines of his geometry of the minimum sensible does not mean that Barrow was the only writer who influenced Berkeley's views on geometry. We know that Berkeley also read Keill's *Introductio ad Veram Physicam* and was aware of his objections to the doctrine of indivisibles. In fact, Keill presents a variety of geometric arguments for the thesis of infinite divisibility in the third lecture of his *Introductio,* one of which is the "difficulty about concentric circles" which Barrow presents. Nevertheless, it should be clear that a great deal of what Berkeley had to say about geometry in the *Commentaries* was a direct and self-conscious response to Barrow's *Lectiones Mathematicæ,* and it is probable that Berkeley was studying Barrow's work carefully as he filled in his early notebooks.

Rejection of infinite divisibility and resistance to traditional geometry did not originate with Berkeley.[23] In fact, there was a long tradition of "mathematical atomism" in ancient and medieval philosophy, and Barrow's reference to "our opponents" must surely be to such mathematical atomists. But Berkeley seems to have been largely unaware of this tradition and never mentions thinkers who might have shared his doctrine.

The Coherence of the Doctrine of Minima

Barrow's arguments raise the important question of whether there can be a coherent geometry of minima. Although he sketches responses to arguments such as Barrow's, Berkeley seems not to worry much about the coherence of his doctrine. He was prepared to abandon Euclidean geometry and to dismiss any argument against indivisibles as question-begging. Thus, after accusing Barrow of a *petitio principii,* he declares, "Shew me an argument against indivisibles that does not go on some false supposition" (*Commentaries,* 463).

It is true that arguments which invoke results from Euclidean geometry are unconvincing to anyone who rejects traditional geometry. Nevertheless, I think Berkeley takes such objections far too lightly in the *Commentaries.* Indeed, the doctrine of the minimum sensible raises problems which cannot be easily solved even if we accept Berkeley's fundamental assumptions. It is worthwhile to mention a few of the more salient problems as a way of ending our study of Berkeley's early geometry of minima. The first problem can be brought out by attempting to answer the question What shape is a minimum sensible? It seems that the minima must have some shape or figure: they have color and extension, and these are necessarily connected with figure. But it is not clear that there is any figure that can satisfy the conditions placed on them. Because the length of a line is defined by the number of minima contained in it, a minimum must be the same size in all directions, i.e., it must be circular. Consider, for example, what would

---

23. For a discussion of the medieval debates on infinite divisibility and the reliability of classical geometry, see Duhem (1985, chapter 1) and Lasswitz (1890).

happen if the minima were square and the visual plane were a grid like a chessboard. We would have a square whose side is equal to its diagonal, since both contain the same number of minima. A similar problem would arise if we assumed that the minimum is a polygon of any finite number of sides. Thus, we are forced to conclude that the minimum is circular.[24]

But it is equally obvious that the minima must cover the plane, because the plane can be resolved into a finite collection of nonoverlapping minimal points. And all of the minima must be the same size. What figure could satisfy all of these requirements? It is a theorem of Euclidean geometry that circles of the same radius cannot cover the plane without overlapping. While Berkeley would presumably treat this theorem as one of the many false doctrines in standard geometry, it is not at all clear what shape the minima can possibly have. Indeed, the task of conceiving a shape that could satisfy the requirements placed on the minima rivals that of conceiving an abstract triangle that is neither equilateral, scalene, nor right-angled, but all and none of these at once.

Another problem arises when we ask whether the points on a line are in contact with one another. If they are not, then we apparently have space in between the points of an unbroken line (a difficult supposition to comprehend). If they do, then any point between two neighboring points would seem to have distinguishable sides or parts—the part in contact with the point on the left and the part in contact with the point on the right. But minima are supposed not to have parts, so we seem compelled to say that the three points must coincide completely if they touch at all. This is at least as implausible as assuming that there is space between the points on a line, so the concept of a line composed of minima appears to be unintelligible.

A third difficulty for Berkeley's conception of the *minimum sensibile* is that we can provide a Berkeleyan argument to prove that sensible extension is not composed of minima.[25] Many of the lines and surfaces we perceive appear continuous and not composed of minima. Pointilist art notwithstanding, a clear sky does not have the phenomenal character of a composition of small patches of blue. But if we follow Berkeley and acknowledge that nothing in our perceptions is hidden from us, we conclude that what looks continuous is continuous, so minima do not exist because they are not perceived. That we do not perceive minima directly presumably forces Berkeley to claim

---

24. See Gray (1978) for a similar analysis of this problem.
25. An argument much like this appears in Armstrong (1961, 43–44).

that we err when we judge perceived lines and surfaces to be continuous. The truth, as Berkeley would have it, is that we overlook the minima and rashly conclude that what we see is continuous. This forces Berkeley to claim that we are mistaken in the overwhelming majority of our ordinary judgments about what we perceive, an embarrassing position for one who spends much of his time deriding his opponents for "despising their senses" and not accepting that things are what they appear to be.

Although it may be possible to rescue Berkeley's doctrine of the minimum sensible from complete incoherence, it should be clear that the doctrine is highly problematic. For Berkeley to advocate a doctrine as questionable as this makes his program for the revision of Euclidean geometry especially unappealing. If the *Commentaries* were Berkeley's only contribution to the philosophy of geometry, his writings on the subject would be justly ignored. However, Berkeley's views on the nature of geometry did eventually change, and this change in view will be my concern in the remainder of this chapter.

Abstraction and Geometry in the *Principles*

The philosophy of geometry expressed in the *Commentaries* is the work of a man who had no serious interest in accommodating traditional geometry within the framework of his metaphysics and epistemology. The views set forward in the *Principles* show a substantial difference in doctrine and tenor. Where Berkeley had earlier shown a fanatical desire to jettison classical geometry, the *Principles* demonstrate an interest in finding a place for geometry within his philosophy. The main reason for the difference is that, by the time of the *Principles*, Berkeley has developed his theory of representative generalization, which gives him the conceptual resources to deny that geometry is a "science of abstractions" without thereby being forced to conclude that the object of geometry is immediately perceived extension.[26] In the *Commentaries* Berkeley insists that geometric truths can only refer to actually perceived lines and figures, to the point of denying that "reason and demonstration" are proper means of solving geometric problems. He therefore advocates a complete revision

---

26. Berkeley's earliest statement of the theory of representative generalization appears as paragraphs 25 and 26 of the manuscript version of the Introduction to the *Principles* (Berkeley 1987, 89–93). This text is substantially the same as §16 of the published Introduction, but Belfrage (1985) notes some important differences between manuscript and printed versions, which shows that Berkeley's views on abstraction were in a state of flux as he wrote the *Principles*.

of classical geometry in which geometric problems are solved by inspecting the sensible properties of perceived lines and figures.

The most obvious difference in the *Principles* is that Berkeley presents a theory of demonstration which admits that geometric proofs are capable of generalization, still without relying upon the traditional account of abstract ideas. The other major change concerns the object of geometrical investigation. The Berkeley of the *Principles* admits not only actually perceived lines and figures but also any possibly perceived lines and figures as objects of geometry because his theory of representative generalization allows "perceivables" to be represented by the diagrams employed in a geometric proof.

If we take seriously the claim that one line can represent many others, then we can regard a given particular line as representative of all possible lines and need not be concerned with the exact number of minima in any particular line. This conception of geometry frees Berkeley from his earlier view that the truths of geometry must be about what we immediately perceive and his thesis that visual inspection of a perceived figure suffices to establish a geometric result. As might be expected, a good deal more of traditional geometry can be accommodated within Berkeley's revised philosophy of geometry. The most striking example occurs in his attitude toward the classical theorem that any line segment can be bisected (*Elements* I, 10). Where he had once insisted that not every line is bisectable, he now appears to accept this theorem, even using it to illustrate his theory of representative generalization:

> To make this plain by an example, suppose a geometrician is demonstrating the method, of cutting a line in two equal parts. He draws, for instance, a black line of an inch in length, this which in it itself is a particular line is nevertheless with regard to its signification general, since as it is there used, it represents all particular lines whatsoever; for that what is demonstrated of it, is demonstrated of all lines or, in other words, of a line in general. (Introduction, §12)

This attitude is also reflected in Berkeley's claim that the proposition "whatever has extension is divisible" should be understood to apply to extension in general, that is, to any particular extension (Introduction, §11). As I have formulated it, the thesis of infinite divisibility is equivalent to the claim that any geometric magnitude can be bisected, and these passages might lead one to expect that in the *Principles* Berkeley would also accept infinite divisibility. This is not the case, and to see why I will now turn to an investigation of the critique of the doctrine of infinite divisibility in the *Principles*.

Berkeley's New Foundations for Geometry

The Case against Infinite Divisibility in the *Principles*

Berkeley's attack on the thesis of infinite divisibility in the *Principles* contains much that we have already seen in the *Commentaries* but with some interesting additions. As before, Berkeley connects his denial of infinite divisibility with the *esse* is *percipi* thesis by arguing that because there is no extension except perceived extension and perceived extension is not infinitely divisible, then the thesis of infinite divisibility must be rejected. His strategy in the *Principles* is to appeal to the manifest properties of perceived extension as a way of discrediting the thesis of infinite divisibility. He insists that:

> Every particular finite extension, which may possibly be the object of our thought, is an *idea* existing only in the mind, and consequently each part thereof must be perceived. If therefore I cannot perceive innumerable parts in any finite extension that I consider, it is certain they are not contained in it: but it is evident, that I cannot distinguish innumerable parts in any particular line, surface, or solid, which I either perceive by sense, or figure to my self in my mind: wherefore I conclude they are not contained in it. (*Principles*, §124)

The key step here is Berkeley's identification of "extension" with "perceived extension," which he takes to have been abundantly demonstrated in his earlier arguments for the *esse* is *percipi* thesis. He also reads the thesis of infinite divisibility as the claim that geometric magnitudes actually contain an infinite number of distinguishable parts, rather than the more modest thesis that any perceived extension has two distinguishable parts. This means that Berkeley's interpretation of the thesis of infinite divisibility departs from the formulation that I have adopted, and he is in effect rejecting a much stronger thesis than the claim that every magnitude can be bisected. We can grant that Berkeley is right to claim that no perceived extension is infinitely divisible in the sense that it contains an infinite number of perceived parts. But to turn this into an argument against infinite divisibility, we must follow Berkeley both in holding that the only concept we can have of extension is that of perceived extension and in his reading of the thesis of infinite divisibility. He does not deem these crucial assumptions to stand in need of a detailed defense and is content to observe that "If by *finite extension* be meant something distinct from a finite idea, I declare I do not know what it is, and so cannot affirm or deny any thing of it" (*Principles*, §124).

Of course, the standard view of geometry simply denies this fundamental Berkeleyan assumption and declares the object of geometry to be "abstract magnitude" or the abstract idea of extension. Berkeley naturally rejects such an attempt to vindicate the thesis of infinite divisibility, but he spends little time discussing the matter. In fact, he dismisses the thesis with the remark that "He whose understanding is prepossessed with the doctrine of abstract general ideas, may be persuaded, that (whatever be thought of ideas of sense), extension in *abstract* is infinitely divisible" (*Principles*, §125). He declines, however, to examine and refute the various arguments for the thesis of infinite divisibility. Instead, he simply asserts that "it were no difficult thing" to show that geometric arguments for the thesis depend upon the mistaken doctrine of abstract ideas or the claim that the objects of sense exist "without the mind."

In the *Principles* as in the *Commentaries* Berkeley hints at the supposedly paradoxical nature of the thesis of infinite divisibility but does not develop this theme into a full-scale attack on infinite divisibility. Berkeley mentions such paradoxes in passing while suggesting that a geometry without infinite divisibility would be easier to learn:

> And as this notion is the source from whence do spring all those amusing geometrical paradoxes, which have such a direct repugnancy to the plain common sense of mankind, and are admitted with so much reluctance into a mind not yet debauched by learning: so is it the principal occasion of all that nice and extreme subtlety which renders the study of *mathematics* so difficult and tedious. (*Principles*, §123)[27]

Similarly, he complains that "The several absurdities and contradictions which flowed from this principle might, one would think, have been esteemed so many demonstrations against it" (*Principles*, §127) but he does not elaborate his objections.

In the *Principles* Berkeley attempts to account for the prevalence of the principle of infinite divisibility and to show how its denial need not lead to the abandonment of traditional geometry. The theory of representative generalization is crucial in this enterprise. Berkeley argues that acceptance of infinite divisibility results from misunderstandings of the object of geometry and the nature of geometric proof—specifically from failures to distinguish the line actually employed in a geometric demonstration from the lines which it represents:

---

27. Interestingly, the manuscript version of the second part of this passage differs from the printed version: "So is it the Principal occasion of all y^t Subtilty, and affected ἀκρίβεια which renders the Study of Mathematics so very difficult, and boring" (British Library Additional MS 39304, fol. 85^r). Berkeley uses the term ἀκρίβεια ("precision" or "exactness") in entry 313 of the *Commentaries*—"What shall I say? dare I pronounce the admir'd ἀκρίβεια Mathematica, that Darling of the Age a trifle?"—and in §19 of the *Analyst*.

Berkeley's New Foundations for Geometry

> It hath been observed in another place, that the theorems and demonstrations in geometry are conversant about universal ideas. *Sect.* 15. *Introd.* Where it is explained in what sense this ought to be understood, to wit, that the particular lines and figures included in the diagram, are supposed to stand for innumerable others of different sizes: or in other words, the geometer considers them abstracting from their magnitude: which doth not imply that he forms an abstract idea, but only that he cares not what the particular magnitude is, whether great or small, but looks on that as a thing indifferent to the demonstration: hence it follows, that a line in the scheme, but an inch long, must be spoken of, as though it contained ten thousand parts, since it is regarded not in it self, but as it is universal; and it is universal only in its signification, whereby it represents innumerable lines greater than it self, in which there may be distinguished ten thousand parts or more, though there may not be above an inch in it. After this manner the properties of the lines signified are (by a very usual figure) transferred to the sign, and thence through mistake thought to appertain to it considered in its own nature. (*Principles*, §126)

On Berkeley's analysis, then, the thesis of infinite divisibility must be read as the claim that every geometric magnitude actually contains an infinite number of parts. But this claim, he thinks, can be rejected without requiring a full-scale overhaul of traditional geometry, because lines in geometric proofs serve as representatives of other, larger lines. Hence, we can assume no limit on the number of parts contained in any magnitude we consider in geometry. Berkeley ultimately concludes that we are forced to make false assumptions about the lines and figures used in any geometric demonstration, because the truth of the theorem depends upon the properties of the figures represented and not those actually employed in the proof: "Because there is no number of parts so great, but it is possible there may be a line containing more, the inch-line is said to contain parts more than any assignable number; which is true, not of the line taken absolutely, but only for the things signified by it" (*Principles*, §127). Strictly speaking, Berkeley concludes, the thesis of infinite divisibility is false, yet we are tempted to say that a finite line contains innumerable parts.[28]

---

28. This recalls a famous passage from the *Three Dialogues* in which Berkeley's spokesman Philonous insists that, strictly speaking, it is incorrect to say that I heard a coach in the street: "in truth and strictness, nothing can be *heard* but *sound:* and that the coach is not then properly perceived by sense, but suggested from experience." (*Works* 2:204). Analogously, no geometric figure contains innumerable parts; but it can suggest or represent figures of any size whatever.

Such false suppositions do not, however, mandate a revision of geometry, because the results obtained by their means are not only reliable but are unobtainable without them. The account of geometry presented in the *Principles* thus contains an element of instrumentalism, because the generality of geometric theorems can only be attained at the cost of treating the lines and figures in a proof as if they had properties which they actually lack:

> From what hath been said the reason is plain why, to the end any theorem may become universal in its use, it is necessary we speak of the lines described on paper as though they contained parts which really they do not. In doing of which, if we examine the matter thoroughly, we shall perhaps discover that we cannot conceive an inch it self as consisting of, or being divisible into a thousand parts, but only some other line which is far greater than an inch, and represented by it. (*Principles*, §128)

Berkeley's theory of representative generalization can be illustrated by considering an application of the Pythagorean theorem and the theory of incommensurable magnitudes. Because the square we construct in a proof of the theorem is supposed to represent all possible squares, we need not be concerned with the number of minima contained in the side and diagonal of any particular one. For any given square, there will be a rational proportion between the number of minima along the diagonal and the number along the side, but this proportion is not the same for all squares. However, as we consider larger and larger squares, the proportion will approach a limit, namely $\sqrt{2}$.

But because there is an upper bound to the number of minima a person can perceive, the claim that the ratio between diagonal and side will approach $\sqrt{2}$ is true only if we admit squares that are too big to be objects of our experience, and this might be thought to pose a problem within the Berkeleyan framework. I am not sure how Berkeley would respond to this problem, for there is no evidence that he ever considered it. In allowing an actually perceived line to represent all possible lines, Berkeley does not confine geometry to figures which have actually been measured. There are plenty of familiar lines whose length is too great for us to express in terms of minima, for example, a line traced by a light ray from the sun to the earth. Presumably a square formed on this line would have a side and diagonal which approximate $\sqrt{2}$ to several million decimal places, though no person could ever count the number of minima in its side and diagonal. I take this to indicate that we can form the idea of a square whose exact measure in terms of minima is inaccessible to us.

Although there cannot be a Berkeleyan square whose diagonal δ and side σ actually stand in the irrational proportion $\sigma:\delta::\sqrt{2}:1$, we can posit squares large enough to approximate this proportion to any desired degree of accuracy. Of course, we do not measure things in terms of minima but rely upon such conventional aggregations of minima as meters or inches. Thus, when we wish to build a bridge or survey a field, we make use of approximations accurate to within some small fraction of an inch or meter. The Pythagorean theorem is quite useful for this purpose, as a method for finding a suitable approximation for the diagonal of a square given the measure of a side. In essence we could read the theorem as saying, If you want a value for the diagonal to within one one-thousandth of your unit of measure (where the unit of measure is understood to contain more than one thousand minimal points), then multiply the given value of the side by 1.414.

The Status of Geometry in the *Principles*

The philosophy of geometry put forth in the *Principles* is clearly a departure from the program for revision of classical geometry which dominates the *Commentaries*. In the *Principles* Berkeley reconsiders the extent to which his denial of infinite divisibility would require a revision of standard geometrical practice:

> But you will say, that if this doctrine obtains, it will follow the very foundations of geometry are destroyed: and those great men who have raised that science to so astonishing an height have all the time been building a castle in the air. To this it may be replied, that whatever is useful in geometry and promotes the benefit of human life, doth still remain firm and unshaken on our principles. That science considered as practical, will rather receive advantage than any prejudice from what hath been said. But to set this in a due light, may be the subject of a distinct inquiry. For the rest, though it should follow that some of the more intricate and subtle parts of *speculative mathematics* may be pared off without prejudice to the truth; yet I do not see what damage will be thence derived to mankind. (*Principles*, §131)

This is a shift from zealous revisionism to a brand of instrumentalism which treats geometric theorems as false when taken as statements about perceived lines and figures in a proof, but true and useful when applied to the solution of practical problems. Thus, where Berkeley was prepared to sacrifice all of classical geometry in 1707–8, by 1710 he had convinced himself that only "some of the more intricate and subtle" parts of geometry need be tossed aside in favor of philosophical principle.

I have characterized this view as a kind of instrumentalism, but more should be said as to what kind of instrumentalism Berkeley intends.[29] As I will show in chapter 6, Berkeley is certainly not a thoroughgoing instrumentalist in the *Analyst*. Instrumentalism, broadly speaking, is the doctrine that a theory can be accepted and applied for reasons of utility, even if the claims made in the theory or its application are not accepted as literally true. An instrumentalist will typically regard a given body of theory as a useful device for making predictions, but deny that the theory itself makes true claims about the world. The strongest version of instrumentalism regards a theory as a "black box" which delivers reliable results, although the theory itself is taken to be false, meaningless or unintelligible. To adopt this strong version of instrumentalism (as some have done with quantum mechanics) is to hold that theories should be accepted or rejected only on the basis of their predictive success and that the search for a true account of the world should be abandoned. Berkeley's attitude toward physical theories, particularly in *De Motu* and dialogue 7 of *Alciphron*, tends strongly toward this reading (Brook 1973; Buchdahl 1969). In these texts Berkeley takes physical theories as useful devices which can be applied to make reliable predictions, despite the fact that they falsely attribute force, attraction, or causal power to bodies.

A weaker version of instrumentalism arises when a certain body of theory is regarded as false but is nevertheless used for purposes of simplicity and economy. Newtonian physics, for example, is employed in building construction although it is not strictly correct to take steel girders as rigid bodies invariant under transport from the ground to the seventy-fifth floor. The errors arising from neglecting relativistic effects in building construction are sufficiently small as to fall well within engineering tolerances. Moreover, the inconvenience of using relativistic considerations is sufficiently great that nobody would seriously propose including them in calculating load and stress on girders. This second kind of instrumentalism accepts that there is a true theory that in principle might be applied, but a false theory is used for reasons of simplicity.

An even weaker version of instrumentalism concerns, not a theory,

---

29. Failure to do this is a weakness in my article (Jesseph 1990), where I attribute instrumentalism to Berkeley without specifying what this really entails.

Berkeley's New Foundations for Geometry

but certain background assumptions used in the application of a theory. For example, in working out the reaction equation for a chemical process, one assumes that the reagents are pure and ignores the potential effects of trace compounds when the reaction is set up. In this case, the chemical theory is accepted as true, but convenient assumptions are made regarding its application to a specific case. This last kind of instrumentalism thus treats certain facts as insignificant for reasons of simplicity, but insists that the theory applied is literally true. A chemist presumes to understand the structure and bonding properties of different molecules, but in the laboratory must neglect certain unpleasant complicating facts. I take Berkeley's instrumentalism about geometry in the *Principles* to be of this weakest kind. That is to say, I read Berkeley as claiming that geometry should be regarded as true, at least for the most part, but holding that it is not fully accurate as a description of what we actually perceive.

There is, however, a genuine puzzle here, for it is far from obvious how Berkeley's account of geometric truth can allow much of geometry to come out true. If we accept the doctrine of the minimum sensible, most geometric theorems will be literally false, not only of particular lines and figures employed in a proof but, because no irrational proportion between any two finite collections of minima can be established, of any perceivable or imaginable lines or figures which they might represent. Arbitrarily close approximations to an irrational proportion are possible with ever larger collections of minima, but geometric theorems still reveal no literal truth. If this is the case, then Berkeley's apparent acceptance of classical geometry might be motivated by its utility rather than its truth and would seem to entail an instrumentalism more thoroughgoing that I have indicated.

One way for Berkeley to avoid this problem and give content to the notion of geometric truth would be to treat the truths of geometry as claims about the limiting cases of any possible approximations. Thus, whereas traditional accounts of geometry treat theorems as descriptive of abstract objects, Berkeley could read them as assertions which identify the limits on the values of any possible measurement. This would allot the theorems of classical geometry a measure of truth short of taking them as truths describing objects of our actual (or even possible) experience: what we can measure will approximate the irrational proportions of geometry, but geometric truth depends upon the limits toward which these approximations tend. In the *Principles* Berkeley does not address the problem of geometric truth, although in the manuscript version he hints that a solution along these lines will be set forward elsewhere: "But to set this in a due light and shew how

Lines and Figures may be measured and their properties investigated without supposing Finite Extension to be infinitely divisible will be the proper business of another place" (*Principles,* §131).[30] Presumably, Berkeley is claiming that his theory of representative generalization will allow him to show how the measurement and investigation of lines and figures can proceed. His program would involve the familiar process of taking a perceived line or figure as representative of others, and the figures represented would provide a sequence of approximations which could be carried out to any desired degree of precision. As we will see in chapter 6, Berkeley's procedures in the *Analyst* reflect this conception of geometry as a science of approximations.

Geometry in the *New Theory of Vision*

Given the dramatic shift in Berkeley's philosophy of geometry between the *Commentaries* and the *Principles,* it is clear that his thinking on the nature of geometry underwent a substantial change between 1708 and 1710. His *Essay towards a New Theory of Vision* was first published in 1709, not long after the *Commentaries* were written, and much of the material in the notebooks makes its way into this published work. Given what we know about the change in Berkeley's account of geometry and given that he devotes several sections of the *New Theory of Vision* to a discourse on the object of geometry, it is natural to ask how much, if any, of his early plan for the overthrow of classical geometry remained intact in this work. There is evidence that Berkeley's account of geometry had already undergone some change, but he still seems ready to jettison classical geometry in favor of the doctrine of the minimum sensible.

The three most significant features of the discussion of geometry in the *New Theory of Vision* are Berkeley's insistence that tangible (as opposed to visible) extension is the object of geometry, his rejection of the thesis of infinite divisibility, and his dismissal of the doctrine of abstract ideas. In this section I want briefly to outline these main points and to assess, on the basis of textual evidence, how much of classical geometry Berkeley was willing to accept at the time he wrote the *New Theory of Vision.*

The case against the thesis of infinite divisibility is not argued directly in the *New Theory of Vision.* Early in the work he raises the issue of the infinite divisibility of extension when he remarks that "whatever may be said of extension in the abstract, it is certain sensible extension is not infinitely divisible. There is a *Minimum Tangibile* and

---

30. British Library Additional MS 39304, fol. 95ʳ.

a *Minimum Visibile,* beyond which sense cannot perceive. This everyone's experience will inform him" (*Theory,* §54). Taken together with the rejection of the abstract idea of extension in §§122–125, this passage implies the same kind of objection to the thesis of infinite divisibility which we found in the *Commentaries* and *Principles.*

Berkeley does not employ the familiar argument against infinite divisibility that declares the thesis false because it depends upon the claim that extension exists "outside the mind or unperceived." The reason for this is fairly obvious. As Luce has argued, Berkeley saw the *New Theory of Vision* as an opening move in a campaign for the philosophy of immaterialism and was prepared to accept tacitly the "vulgar error" of assuming that objects of touch exist "without the mind" even though he was a convinced immaterialist when he wrote it (*Works,* 1:149–50). But it would naturally serve no purpose to argue against infinite divisibility by employing the premise that *esse* is *percipi* in a work which seems to accept the negation of this very principle, so the absence of the argument need not surprise us. Although he makes a case for the denial of infinite divisibility in the *New Theory of Vision,* Berkeley spends no time there working out the consequences of that denial. Instead, he devotes his energies to arguing for the unusual claim that tangible extension is the object of geometry.

Tangible Magnitude as the Object of Geometry

In claiming that tangible extension is the object of geometry, Berkeley first argues that visible and tangible extension are numerically distinct. The arguments for this conclusion, presented in the *New Theory of Vision* (§§121–148), form the background for his thesis that only tangible extension can be the object of geometrical investigation. The details of the Berkeleyan case for the heterogeneity of visible and tangible magnitudes are analyzed in Atherton (1990, chapters 10 and 11) and will be left aside here, except to note that Berkeley regards visible objects as *signs* of tangible objects. Thus, the visible diagrams used in a geometric proof would, on Berkeley's account, be regarded as signs of tangible objects. This conception of geometric diagrams as signs is of obvious importance for the treatment of geometry in the *Principles,* although it is put to very little use in the *New Theory of Vision.*[31]

---

31. One difference worth noting is that in the *New Theory of Vision* Berkeley is concerned only with visible objects as signs of tangible objects, while in the *Principles* he treats visible geometric objects as signs of other visible as well as tangible objects. Thus, where his concern in the *New Theory of Vision* was to use the concept of signification to link one sensory modality to another, in the *Principles* he is concerned with using it to link a perceived object to other perceivable objects, without regard to sensory modality.

Chapter Two

In arguing that tangible extension is the proper object of geometry, Berkeley places most of his emphasis upon the claim that visible extension is not appropriate for measuring distance and magnitude:

> To come to a resolution of this point we need only observe what hath been said in sect. 59, 60, 61 where it is shewn that visible extensions in themselves are little regarded, and have no settled determinate greatness, and that men measure altogether, by the application of tangible extension to tangible extension. All which makes it evident that visible extension and figures are not the object of geometry. (*Theory*, §151)

Berkeley extends this basic argument with a thought experiment in which he considers the geometric knowledge attainable by a disembodied spirit lacking the sense of touch. He concludes that such a "pure intelligence" could know very little of our geometry, largely because his two-dimensional visual field would provide no basis for the geometry of solids and his inability to handle geometric instruments such as the compass would deprive him of any conception of geometric construction.[32]

Abstraction and Extension in the *New Theory of Vision*

In the *New Theory of Vision*, Berkeley's critique of the doctrine of abstract ideas is presented as rebuttal to the claim that the mind is capable of framing an abstract idea of extension. The details of the argument, covered in the first chapter, do not bear repeating here, but it is important to recall that his attack on abstract ideas is clearly intended as a rejection of the received view of geometry. In asserting the impossibility of abstraction, Berkeley argues that the abstract idea of extension would correspond to an impossible, completely indeterminate object, incapable of existing either in nature or in the imagination. He concludes, regarding the nature of geometry and the viability of the received view:

> It is commonly said that the object of geometry is abstract extension: but geometry contemplates figures: Now, figure is the termination of magnitude: but we have shewn that extension in abstract hath no finite determinate magnitude. Whence it clearly follows that it can have no figure, and consequently is not the object of geometry. It is indeed a tenet as well of the modern as of the ancient philosophers that all gen-

---

32. See *Theory*, §§153–158, and Atherton (1990, chapter 11) for the details of this argument.

eral truths are concerning universal abstract ideas; without which, we are told, there could be no science, no demonstration of any general proposition in geometry. But it were no hard matter, did I think it necessary to my present purpose, to shew that propositions and demonstrations in geometry might be universal, though they who make them never think of abstract general ideas of triangles or circles. (*Theory*, §124)

The admission here that "propositions and demonstrations in geometry might be universal" suggests that Berkeley's earlier hostility toward the use of demonstrations in geometry has been tempered and that he has found the means to define an epistemology of demonstrations which will not rely upon the traditional theory of abstract ideas. It further suggests that Berkeley has already developed his theory of representative generalization and, given the strong connection between this theory and his abandonment of the doctrine of the *Commentaries*, that his plan for the revision of geometry might already have been abandoned when he wrote the *New Theory of Vision*. However, as the sections dealing with abstraction were apparently a late insertion in the manuscript,[33] it seems clear that the theory of representative generalization was not part of the *New Theory of Vision* as it was originally planned. If we conclude that Berkeley developed his theory of representative generalization while writing the *New Theory of Vision*, it is not at all clear that he saw how it could be used to accommodate the main results of Euclidean geometry within his epistemological framework.

The Status of Geometry in the *New Theory of Vision*

Thus far, we have found nothing in the *New Theory of Vision* which directly addresses the question of how much of classical geometry Berkeley was willing to accept. There is, however, a revealing passage at the end of the first two editions (both from 1709) which suggests that Berkeley remained very much committed to the revision of classical geometry when he published the *New Theory of Vision:*

> By this time, I suppose, it is clear that neither abstract nor visible extension makes the object of geometry; the not discerning of which may perhaps have created some difficulty and useless labour in mathematics. Sure I am, that somewhat

---

33. The reason for characterizing the attack on abstraction in the *New Theory of Vision* as a late insertion is that the relevant sections (§§122–125) lack any obvious connection with the surrounding material. As Luce notes, the insertion is "clumsily made" and can hardly have been part of the original plan (Luce 1963, 104–6).

relating thereto has occurred to my thoughts, which, tho' after the most anxious and repeated examination I am forced to think it true, doth, nevertheless, seem so far out of the common road of geometry, that I know not, whether it may not be thought presumption, if I should make it publick in an age, wherein that science hath received such mighty improvements by new methods; great part whereof, as well as of the ancient discoveries, may perhaps lose their reputation, and much of the ardor, with which men study the abstruse and fine geometry be abated, if what to me, and those few to whom I have imparted it, seems evidently true, should really prove to be so. (*Theory*, §160, first two editions only)

The first sentence of this closing passage was retained in the third edition (1732) and incorporated into the preceding section, but the second sentence with its clear allusion to Berkeley's radical geometry of the minimum sensible was deleted. There can be no reasonable doubt that, when Berkeley refers to a conception of geometry so "far out of the common road" as to threaten both modern and classical geometry, he means the radical program set forth in the *Commentaries*. Thus, if we take this passage seriously, it would seem that Berkeley was still prepared to abandon essentially all of classical geometry when he wrote the *New Theory of Vision*.[34]

Although a case can be made for reading the *New Theory of Vision* as continuing much of Berkeley's early hostility toward traditional geometry, I prefer not to assume that his views on the subject were completely settled. Indeed, the accepted view of Berkeley's early work is that his thought underwent a significant development while he wrote the *Commentaries*,[35] and it is by no means implausible to think that there were unresolved issues concerning the nature of geometry in Berkeley's mind as he wrote the *New Theory of Vision*.

---

34. This passage drew the attention of Leclerc, who remarked that "Our author is not the only one who is not convinced that everything in nature operates mathematically, as many these days think. Perhaps there is some principle in bodies themselves, of which we have no idea, and which is the cause of their principal effects. We will view with pleasure, that which he has discovered, contrary to the common sentiments of mathematicians" (Leclerc 1711, 88). Because he did not share Berkeley's hostility to the doctrine of infinite divisibility, it is unlikely indeed that Leclerc would have been favorably impressed by Berkeley's "great discovery."

35. There are many works which deal with the development in Berkeley's thought in the *Commentaries* and other early works, the foremost being Luce (1963). See also Belfrage (1986) and McKim (1986) for a discussion of a change in Berkeley's theory of meaning in the *Commentaries*.

Berkeley's New Foundations for Geometry

In fact, several entries in the *Commentaries* themselves show that elements of the account of geometry in the *Principles* had already occurred to Berkeley at the time he was filling in his notebooks:

> Mem: a great difference between considering length wthout breadth, and having an idea of or imagining length without breadth.
>
> Suppose an inch represent a mile. 1/1000 of an inch is nothing, but 1/1000 of ye mile represented is something therefore 1/1000 of an inch tho' nothing is not to be neglected, because it represents something *i.e.* 1/1000 of a mile.
>
> Particular Determin'd lines are not divisible ad infinitum, but lines as us'd by Geometers are so they not being determin'd to any particular finite number of points. Yet a Geometer (He knows not why) will very readily say he can demonstrable an inch is divisible ad infinitum. (*Commentaries*, 254, 260–61)

These entries do not suggest that the full doctrine of the *Principles* had occurred to Berkeley by the time he wrote them. They are too vague and incomplete to be of much use in constructing the later account. Moreover, they are isolated exceptions in a program otherwise directed to the revision of classical geometry. Nevertheless, it seems safe to conclude that these entries represent the beginnings of the account of geometry which is given full expression in the *Principles*. If this is correct, then the *New Theory of Vision* would have been completed at a time when Berkeley's views on the nature of geometry were in a process of change, and we need not expect a completely unified treatment of geometry from it.

Geometry and Abstraction in the Later Works

A further evolution of Berkeley's philosophy of geometry can be traced by noticing the place accorded geometry in two later works, *Alciphron* (1732) and the *Analyst* (1734). There are no fundamental changes in view between the *Principles* and these later works—certainly nothing to compare with the shift between the *Commentaries* and the *Principles*—but there are some developments worth documenting.

Alciphron contains no extended treatment of the philosophy of geometry, but in dialogue 7 Berkeley considers the importance of geometry in human affairs. He disputes the claim that abstractions are the object of geometric knowledge and gives a succinct statement of a kind of geometric instrumentalism which has strong affinities with the doctrine set forth in the *Principles*:

Be the science or subject what it will, whensoever men quit particulars for generalities, things concrete for abstractions, when they forsake practical views, and the useful purposes of knowledge for barren speculation, considering means and instruments as ultimate ends, and labouring to obtain precise ideas which they suppose indiscriminately annexed to all terms, they will be sure to embarrass themselves with difficulties and disputes. Such are those which have sprung up in geometry about the nature of the angle of contact, the doctrine of proportions, of indivisibles, infinitesimals, and diverse other points; notwithstanding all which, that science is very rightly esteemed an excellent and useful one, and is really found to be so in many occasions of human life, wherein it governs and directs the actions of men, so that by the aid or influence thereof those operations become just and accurate which would otherwise be faulty and uncertain. (*Alciphron*, dialogue 7, §15)

Given what Berkeley says here about the excellence and usefulness of geometry, it is no surprise to find that the third edition of the *New Theory of Vision* (printed in 1732 and bound together with the first edition of *Alciphron*) deletes §160, where Berkeley had expressed his reservations about the whole of traditional geometry and suggested that the entire subject was in need of revision. Berkeley's decision to make this deletion verifies that the approach to geometry expressed in *Alciphron* was, by 1732, his considered view and that he had repudiated his earlier project for a complete overhaul of the subject.

By 1734, in the *Analyst*, Berkeley invokes results of classical geometry in attacking the new methods introduced in the Newtonian calculus of fluxions and the differential calculus of Leibniz. In setting out his case against Newton and Leibniz he relies upon classical results in the theory of conic sections and presents himself as a defender of classical standards of rigor against the infinitesimal methods of the seventeenth and eighteenth centuries. A comprehensive treatment of the *Analyst* will follow in chapter 6, but here I will conclude my discussion of the Berkeleyan philosophy of geometry by considering the role of classical geometry in the argument of the *Analyst*.

Berkeley makes it clear very early in the *Analyst* that he regards classical geometry favorably and that his attack is directed against those who have introduced certain new methods into the subject:

It hath been an old remark, that Geometry is an excellent Logic. And it must be owned that when the Definitions are clear; when the Postulata cannot be refused, nor the Axioms denied; when from the distinct Contemplation and Comparison of Figures, their Properties are derived, by a perpetual well-connected chain of Consequences, the Objects being still kept in view, and the attention ever fixed upon them; there is acquired an habit of Reasoning, close and exact and methodical: which habit strengthens the Mind, and being transferred to other Subjects is of general use in the inquiry after Truth. But how far this is the case of our Geometrical Analysts, it may be worth while to consider. (*Analyst*, §2)

Berkeley's remarks echo Malebranche's comments on mathematics (and particularly geometry) in *Recherche de la Vérité* where he discusses the general method necessary to find truth (Malebranche 1963, 2:369–91). To find such pronouncements coming from the man who once complained that "sense rather than reason and demonstration ought to be employed about lines and figures" is further testimony to the extent to which the radical program of the *Commentaries* has been abandoned. The requirements enumerated here for clarity and cogency in geometric reasoning amount to a methodological manifesto, and in the course of the *Analyst* Berkeley shows that he thinks that the entire body of classical geometry satisfies his criteria for proper geometric demonstration.

The best way to see this is to consider Berkeley's procedures in §§21–29 of the *Analyst,* where he purports to show that the results obtained by the (illegitimate) infinitesimal methods of Newton and Leibniz are nevertheless true, because the procedures which yield these results contain two cancelling or self-rectifying errors. The details of the compensation of errors thesis need not concern us here, except to note that in attempting to show that the procedures of the calculus contain a double error, Berkeley carries out *purely classical* geometric arguments, even to the point of explicitly invoking such results as a theorem of Apollonius.[36] Berkeley's willingness to rely upon such classical results in an effort to explain the effectiveness of the calculus testifies convincingly to his acceptance of traditional geometry and underscores how dramatically his views changed between 1707 and 1734.

But although he accepts essentially all of classical geometry in the

36. The theorem of Apollonius, known as the tangent-axis theorem, states that the subtangent to a parabola is bisected at the vertex (Apollonius 1952, 640). Berkeley appeals to it in §22 of the *Analyst* in arguing that the calculus-based formula for the subtangent is derived by two compensating errors. The Apollonian result is crucial to Berkeley's argument and he seems to have no qualms about relying on it. Moreover, the amount of classical geometry required to prove the tangent-axis theorem is sufficient to warrant the conclusion that Berkeley's acceptance of the theorem amounts to his acceptance of all of classical geometry.

Analyst, his philosophical interpretation of geometry in this work is not fundamentally different from the view put forward in the *Principles*. The similarities between the *Principles* and the *Analyst* in regard to the philosophy of geometry can be seen in the list of queries which appear at the end of the *Analyst:*

> *Qu.* 1 Whether the Object of Geometry be not the Proportions of assignable Extensions? And whether, there be any need of considering Quantities infinitely great or infinitely small?
>
> *Qu.* 2 Whether the end of Geometry be not to measure assignable finite Extension? and whether this practical View did not first put Men on the study of Geometry? . . .
>
> *Qu.* 5 Whether it doth not suffice, that every assignable number of Parts may be contained in some assignable Magnitude? And whether it be not unnecessary, as well as absurd, to suppose that finite Extension is infinitely divisible?
>
> *Qu.* 6 Whether the Diagrams in a Geometrical Demonstration are not to be considered as Signs, of all possible finite Figures, of all sensible and imaginable Extensions or Magnitudes of the same kind?
>
> *Qu.* 7 Whether it be possible to free Geometry from insuperable Difficulties and Absurdities, so long as either the abstract Idea of Extension, or absolute external Extension be supposed its true Object? . . .
>
> *Qu.* 18 Whether from Geometrical Propositions being general, and the Lines in Diagrams being therefore general Substitutes or Representatives, it doth not follow that we may not limit or consider the number of Parts, into which such particular Lines are divisible? . . .
>
> *Qu.* 20 Whether all Arguments for the infinite Divisibility of finite Extension do not suppose and imply, either general abstract Ideas or absolute external Extension to be the Object of Geometry? . . .
>
> *Qu.* 53 Whether, if the end of Geometry be Practice, and this Practice be Measuring, and we measure only assignable Extensions, it will not follow that unlimited Approximations compleatly answer the Intention of Geometry? (*Analyst*, queries 1, 2, 5–7, 18, 20, 53; *Works* 4:96–101)

These queries show that Berkeley's conception of geometry in the *Analyst* retains all of the essentials of the view put forward in the *Principles*. In particular, Berkeley uses the theory of representative generalization to provide an account of geometric demonstration consistent with his epistemological principles, and he insists that an understanding of the role of representative generalization will suffice to rid geometry of the absurd thesis of infinite divisibility. Further, his earlier concern with the practical applications of geometry is still intact, and he reiterates his insistence that the traditional conception of geometry is flawed by the doctrine of abstract ideas. The differences in doctrine between the *Principles* and the *Analyst* are relatively minor and concern the question of how much of classical geometry can pass muster in the Berkeleyan scheme. As we have seen, Berkeley's views on this point have undergone an implicit liberalization. In the *Principles* he was prepared to jettison some of "the more abstruse and subtle parts of speculative geometry," but in the *Analyst* he is prepared to accept the whole of classical geometry, concentrating his attack on the more recent methods of the infinitesimal calculus.

This conception of geometry is considerably more plausible and attractive than Berkeley's earlier geometry of minima. Indeed, given his opposition to the theory of abstract ideas and his insistence that the object of geometry is perceived extension, it is not clear that Berkeley could have developed a better philosophical account of the subject. I will return to some of the issues raised in this chapter in chapter 6 when I consider Berkeley's critique of the calculus. In particular, the question of Berkeley's attitude toward classical geometry and his account of geometric truth will be important when I examine his attack on the use of infinitesimal magnitudes in the calculus. But I will first consider Berkeley's approach to arithmetic, since his views on that subject provide an interesting contrast to his philosophy of geometry.

CHAPTER THREE

Berkeley's New Foundations for Arithmetic

Thus far, I have shown how Berkeley's rejection of the received view in the philosophy of mathematics leads him to articulate a theory of geometry which departs in important ways from traditional treatments of the subject. He also opposes the standard account of arithmetic, but his approach to this branch of mathematics develops along completely different lines, and his views on the nature of arithmetic contrast in interesting ways with his account of geometry. My purpose here is to set forth the principal features of Berkeley's philosophy of arithmetic and to show that his work here is intended to solve important philosophical disputes about the nature of numbers and the status of arithmetical theories. As with his philosophy of geometry, Berkeley's guiding principle is the rejection of abstract ideas; but the consequences he draws from this principle lead him to articulate a strongly nominalistic account of arithmetic which has important connections with contemporary philosophies of mathematics grouped under the rubric of *formalism*.

This chapter will comprise a discussion of formalism and of the extent to which Berkeley's views are formalistic, but I think it proper to begin with an account of the philosophical context in which Berkeley's treatment of arithmetic developed. After an overview of seventeenth-century philosophical debates on the status of arithmetic, I will turn to an exposition of Berkeley's principal theses on the nature of numbers, followed by an account of his views on algebra, which turn out to be a straightforward extension of his interpretation of arithmetic. Finally, I will try to assess the novelty and plausibility of Berkeley's philosophy of arithmetic and algebra.

Geometry versus Arithmetic

A significant dispute in seventeenth-century philosophy of mathematics concerned the relationship between geometry and arithmetic, with attention focusing on the question of which was the more fundamental branch of mathematics. Earlier philosophers of mathematics, working in the Euclidean tradition, had distinguished arithmetic from geometry by assigning each a different object. On this view, the general object of mathematical investigation, quantity, has two specific kinds: continuous quantity ("magnitude") and discrete quantity ("multitude" or "number"). Geometry is conversant about the former, arithmetic about the latter. This approach avoids the question of whether one or the other science is the true foundation of mathematics, since each has its particular objects and can develop autonomously. Euclid, for example, develops the theory of proportions twice: in book V as applied to magnitudes, and again in book VII with reference to numbers.

The development of algebra and its application to geometric problems is one of the key episodes of seventeenth-century mathematics, but it reopened questions about the classification of the mathematical sciences. It was not obvious whether algebra has an appropriate object, since the theory of equations can be applied to all quantities. And yet, the fundamental algebraic operations of addition, subtraction, multiplication, and division seem to derive from arithmetic. Many took this to mean that algebra was ultimately a kind of "arithmetic with letters" in which symbols (or "species") representing arbitrary quantities were used to develop the "arithmetic of species." Joseph Raphson's *Mathematical Dictionary* contains an instructive account of the understanding of algebra in Berkeley's day, complete with a fanciful etymology for the term:

> *Algebra,* from *Al* in Arabic which signifies Excellent, and *Geber* the Name of the supposed Inventor of it, is a Science of Quantity in general, whence it has also got the name among some of *Mathesis Universalis,* & is chiefly conversant in finding Equations, by comparing of unknown and known Quantities together whence also by some it is called the Art of Equation, and is distinguished into
>
> [Algebra] *Numeral,* which is the more ancient and serves for the Resolution of Arithmetical Problems: For these see *Diophantus.*
>
> [Algebra] Specious, or the new *Algebra,* which is also called *Logistica Speciosa,* and is conversant about Quantity denoted by General or Universal Symbols, which are commonly the

Chapter Three

letters of the Alphabet; and serves indifferently for the Solution of all Mathematical Problems, whether Arithmetical or Geometrical. (Raphson 1702, 2–3)

With the advent of analytic geometry, we thus have a fusion of geometric and arithmetical theories which seems to break down the classical distinction between arithmetic and geometry. Some philosophically-minded mathematicians of the period thus asked whether arithmetic, geometry, or algebra was the truly foundational mathematical science. As might be expected, there were three different answers.[1]

Wallis argued that arithmetic was the foundation upon which all of mathematics ultimately rested. His commitment to this thesis is plain even in the full title of his *Mathesis Universalis*.[2] This, his most philosophical work, promises to treat of universal mathematics, in his view arithmetic, including the symbolic arithmetic of species. Klein (1968, 211–24) has identified Wallis's account of number as the "final act" in the introduction of a new conception of numbers which displaces the classical division between arithmetic and geometry. On Wallis's account of the matter, geometrical results ultimately depend upon more general arithmetical considerations:

> [I]f someone should affirm that a line of three feet, added to a line of two feet, makes a line five feet long, this is because the numbers two and three added together make five; this is not a geometrical calculation, but clearly an arithmetical one, although it is used in geometric measurement. But this assertion of the equality of the number five with the numbers two and three taken together is a general assertion, applicable to any other kinds of things whatever, no less than to geometrical objects. For also two angels and three angels make five angels. And the very same reasoning holds for all arithmetical and especially algebraic operations, which proceed from principles more general than those in geometry, which are restricted to measure. (Wallis, 1693–99, 1:56).

---

1. Pycior (1987) surveys this dispute and Berkeley's relation to it. Unfortunately, she does not take his *Arithmetica absque algebra* into account.
2. The ponderous title of this work reads: *Mathesis Universalis; sive, Arithmeticam Opus Integrum, Tum Philologice, tum Mathematice traditum, Arithmeticam tum numerosam, tum Speciosam sive Symbolicam complectens, sive Calculum Geometricum; tum etiam Rationum Proportionumve traditionem; Logarithmorum item Doctrinam; aliaque, quæ Capitum Syllabus indicabit*. Both Berkeley and Barrow cite it as the *Arithmetica*, which is fitting for a book which raises arithmetic to the level of universal mathematics.

Berkeley's New Foundations for Arithmetic

In Wallis's scheme, then, arithmetic is the foundation of mathematics. Algebra, although a very general science applicable to all quantities, is ultimately rooted in arithmetical considerations, and geometry is a subordinate science whose principles are less general than those of arithmetic.

Barrow took quite a different view of the matter and insisted that Wallis had ignored the fundamental role of geometry. On Barrow's account, geometry is the foundation of all other mathematics. He develops this line of thought in the third of his *Lectiones Mathematicæ*, where he argues that arithmetic and geometry are not conversant about different objects, and that the science of number must be seen as part of geometry:

> Nevertheless I can attest that I by no means remove or exclude from mathematics such a beautiful and useful science as that of number. This is far from my intention. In truth I intend rather to return it to its rightful place, as having been removed from its proper seat, and to graft it and reunite it again to the stock of its native geometry, from which it has been plucked. . . . For I am convinced that number (at least that which the mathematician contemplates) does not differ in the least from that quantity which is called continuous, but is formed wholly to express and declare it. And neither are arithmetic and geometry conversant about diverse matters, but equally demonstrate properties common to one and the same subject, and from this it will follow that many and great advantages derive to the republic of mathematics. (*Lectiones*, 3:46–47)

Barrow responds to Wallis's arguments for the primacy of arithmetic over geometry by insisting that numbers have no determinate existence apart from the magnitudes which they are taken to measure. In a passage which finds an echo in some of Berkeley's remarks that number is a "creature of the mind," Barrow declares that

> I observe that no number of itself signifies any thing distinctly, or agrees to any determinate subject, or certainly denominates any thing. For any number at all may with equal right denote and denominate any quantity. And likewise any one number may be attributed to every quantity. For instance, any line *A* may be indifferently called unit, two, three, four, or any other number . . . as it remains undivided, may be cut into, or compounded of, two, three, four, or any other number of parts. . . . Whence it appears that no number can des-

ignate any thing certain and absolute, but may be applied to any quantity at pleasure. (*Lectiones*, 3:51)

Barrow eventually adopts an interpretation of arithmetic in which numbers become simply signs for the designation of (geometric) magnitudes:

> I say that mathematical number is not some thing having existence proper to itself, and really distinct from the magnitude which it denominates, but is only a kind of note or sign of magnitude considered in a certain manner; so far as the magnitude is considered as simply incomposite, or as composed out of certain homogeneous equal parts. . . . For in order to expound and declare our conception of a magnitude, we designate it by the name or character of a certain number, which consequently is nothing other than the note or symbol of such magnitude so taken. This is the general nature, meaning, and account of a mathematical number. (*Lectiones*, 3:56)

As we will see, Berkeley accepts the proposition that numbers are not self-subsistent objects, but he goes beyond Barrow to embrace a wholehearted nominalism in regard to arithmetic.

A third response to the philosophical problems posed by the development of analytic geometry was to take algebra as the true foundation of all mathematics. On this view, algebra is a *mathesis universalis*, or a generalized science of quantity whose principles apply to all mathematical objects. In this scheme arithmetic is a special case of algebra, and geometry is ultimately dependent upon algebraic operations. Proclus's commentary on Euclid's *Elements* is the root of this idea of a universal mathematical science, although Proclus does not declare algebra to fill this role.[3] Italian commentators of the sixteenth century developed the notion further, so that the idea that there should be a *mathesis universalis* was a familiar philosophical doctrine by the beginning of the seventeenth century (Crapulli 1969). The identification of *mathesis universalis* with algebraic methods emerged in the work of François Viéte and in interpretations of Descartes's analytic geometry (Klein 1968, chapter 11). André Tacquet tended to take this view, as is evident from the presentation of arithmetic in his *Arithmeticæ Theoria et Praxis* (1665). His treatment contains a number of general laws such as the associativity of addition and the distribu-

---

3. See Proclus (1970, chapters 11–13) for the classical discussion, which considers various possible classifications of the mathematical sciences.

tion of multiplication over addition, which he regards as the algebraic foundation of arithmetic.[4]

Berkeley was familiar with these philosophical discussions, and his first publication includes a treatise on arithmetic which was intended to resolve this dispute by showing arithmetic to be an autonomous science, demonstrable without recourse to geometry or algebra. The title of the work, *Arithmetica absque Algebra aut Euclide demonstrata*, indicates this intention well enough: arithmetic is to be demonstrated without algebra or Euclid, so that the science can be shown to be independent of algebraic or geometrical considerations.[5] This intention becomes clearer in the preface, when Berkeley denounces Tacquet's *Arithmetica* as a work tainted by an overreliance on algebra:

> In truth if we read Tacquet's *Arithmetic*, which no one thoroughly understands who has not already learned algebra, hence it happens, that most who apply themselves energetically in mathematics, while they studiously and successfully read the demonstrations of theorems of inferior value, yet they go forth and leave untouched the principles and reasonings of arithmetical operations, the power and excellence of which is such that they serve most excellently, not only other branches of mathematics, but even all manner of applications in human life. Thus if someone, after a long mathematical course, would turn his eyes back to the aforementioned work of Tacquet, he will find many things demonstrated in an obscure manner, and not so much to enlighten as force conviction on the mind, being surrounded with a horrible array of porisms and theorems. (*Arithmetica*, Preface)

Berkeley's promise that he can demonstrate the "principles and reasonings of arithmetical operations" independently of algebra is matched by an insistence that nothing from geometry is needed for his project, that arithmetic can be demonstrated and the theory of roots developed without relying on the *Elements*.

The curious feature of Berkeley's *Arithmetica* is that its "demonstrations" of the fundamental principles of arithmetic come to nothing more than a statement of the computational algorithms for finding

---

4. See Bosmans (1927) for a study of Tacquet's *Arithmetica* in the context of Jesuit education in the seventeenth century.

5. Pycior (1987) neglects this text, and eventually concludes that Berkeley ranked arithmetic and algebra as less worthy sciences than geometry. This conclusion seems rather hasty, however. There is nothing in Berkeley's works to indicate that he held algebra or arithmetic in particularly low esteem, or that he was more kindly disposed toward geometry.

sums, differences, quotients, or extracting roots. The basic theme of the *Arithmetica* is that arithmetical operations stand in no need of any special justification, but can simply be stated and seen to follow from the nature of the notational system. Thus, in a sense this early work shows Berkeley willing to adopt a rather extreme nominalism in regard to arithmetic, with the basic arithmetical operations being little more than rules for the manipulation of symbols. For Berkeley, "nature teaches us" certain basic arithmetical facts (namely those which can be done by working sums on the fingers), but arithmetic is an extension of this primitive counting which allows us to extend our knowledge to numbers which can not be treated in the primitive system.

Berkeley apparently felt that once the notational scheme had been mastered, the whole of arithmetic could be developed almost effortlessly by simply stating rules for the combination of numerical signs. Thus, at the end of the first chapter, "On the notation and enunciation of numbers," he declares:

> It will be easy to write and express numbers however large, if what has been said is carefully observed, knowledge of which will be of greatest importance in what follows. For indeed nature itself teaches us how the operations of arithmetic are performed on the fingers; but there is need of art to employ them accurately with greater numbers, all of which turns on this, that as the operation cannot be done all at once because of the limitations of the human mind, we divide it into many smaller operations, inquiring separately into the aggregate, difference, product, etc. of the digits, and then so combining these as to exhibit the whole sum, remainder, or product, etc. The whole reason and artifice of which derives from the simple progression of places, and is ultimately founded on it. (*Arithmetica*, chap. 1)

Typical of the "demonstrations" of the principles of arithmetic in the *Arithmetica* is the following discussion of multiplication:

> In multiplication the multiplicand is taken as often as the multiplier demands, or a number is sought which bears the same ratio to the multiplicand as the multiplier bears to unity. And this number is called the product or rectangle; the factors or sides of which are called respectively, the multiplicand and the number by which it is multiplied.
>
> When we find the product of two numbers, the multiplying number being written under the multiplicand, the latter should be multiplied by each sign of the former, beginning

from the right hand: the first sign of the product should be written directly under the multiplier's sign and the rest in order towards the left.

The multiplication being finished, the particular products should be collected into one sum, in order to have the total product, in which as many places are to be assigned to the parts as there are in both factors. (*Arithmetica*, chap. 4)

The formalistic element in the *Arithmetica* should be fairly clear, although Berkeley does not present arithmetic as a purely formal system in the style of modern formalists. It is evident that his attempt to show how arithmetic can be demonstrated without recourse to geometry or algebra depends upon taking arithmetic to be a computational science in which rules are stated for the formation of numerals out of a vocabulary of primitive symbols and further rules are given for combining these numerals to form sums, products, quotients, etc. On Berkeley's view, these rules require no special justification, but can be seen to follow from the nature of the notational system itself.

Although the *Arithmetica* is interesting for the light it sheds on Berkeley's early approach to arithmetic, it is not itself a treatise of great philosophical significance.[6] We must look to Berkeley's subsequent works to find a more thoughtful elaboration of his formalistic philosophy of arithmetic. In the following sections, I will be concerned with the exposition of Berkeley's views on arithmetic as expressed in the *Commentaries*, the *Principles*, and *Alciphron*. In treating these works we will see how the conception of arithmetic as a science of signs emerges much more clearly in the later stages of Berkeley's philosophical career. In presenting Berkeley's more considered views on the nature of arithmetic, I begin with a brief discussion of his claim that numbers are "creatures of the mind."

Numbers as Creatures of the Mind

The basic point behind Berkeley's slogan "number is a creature of the mind" is that numbers do not exist in the world independent of any thought. Phrased in this way, it is obviously tied to the principle *esse* is *percipi*. Numbers, however, are not so much perceived as *conceived* on Berkeley's view. He finds numbers to be more radically mind-dependent than ordinary objects of perception, at least with respect to human minds. Essentially, the difference here is that we are passive

---

6. Breidert (1989, 84) shares this view, noting that Berkeley's account of arithmetic here provides only computational rules, without justification, and never attempts to explain what a number is.

in our perception of objects, but active in our assigning numbers to collections. Thus, while everything is a "creature of the mind" when we consider it as a product of the divine mind, numbers turn out to be things of our own making. The strategy which Berkeley uses to show that numbers are mind-dependent is to argue that the number properly attributed to a given idea (or collection of ideas) depends upon the activity of some conceiving mind. In particular, the mind must make a choice of unit before a unique number can be properly attributed to any collection.

In the *Commentaries,* this kind of argument is advanced in several entries:

> Twelve inches not the same Idea with a foot. Because a Man may perfectly conceive a foot who never thought of an inch. (*Commentaries,* 557)[7]
>
> 2 Crowns are called ten shillings hence may appear the nature of Numbers.
>
> Complex ideas are the Creatures of the Mind, hence may appear the Nature of Numbers. this to be deeply discussed. (*Commentaries,* 759–60)

These entries clearly foreshadow a more complete statement of the thesis in the *New Theory of Vision:*

> [I]t ought to be considered that number (however some may reckon it amongst the primary qualities) is nothing fixed and settled, really existing in things themselves. It is intirely the creature of the mind, considering either an idea by it self, or any combination of ideas to which it gives one name, and so makes it pass for a unit. According as the mind variously combines its ideas the unit varies: and as the unit, so the number, which is only a collection of units, doth also vary. We call a window one, a chimney one, and yet a house in which there are many windows and many chimneys hath an equal right to be called one, and many houses go into the making of one

---

7. Berkeley makes an interesting concession on this point when he admits in the following entry, "A foot is equal to or the same with twelve inches in this respect viz. the[y] both contain the same number of points" (*Commentaries,* 558). Among other things, this observation shows how Berkeley's early claim that perceivable extension is composed of finite collections of minima turns out to provide an absolute intrinsic metrication of space: any extended object has a unique measure of its magnitude, independent of any other features of space. This note raises complex issues regarding Berkeley's conception of space, as well as questions about what it is for two ideas to be the same. But because Berkeley later abandoned the project of founding geometry on the doctrine of minima and his point can be made without reference to the doctrine, I will ignore the possible complexities introduced by this concession.

Berkeley's New Foundations for Arithmetic

city. In these and the like instances it is evident the unit constantly relates to the particular draughts the mind makes of its ideas, to which it affixes names, and wherein it includes more or less as best suits its own ends and purposes. Whatever, therefore, the mind considers as one, that is an unit. Every combination of ideas is considered as one thing by the mind, and in token thereof is marked by one name. Now, this naming and combining together of ideas is perfectly arbitrary, and done by the mind in such sort as experience shews it to be most convenient; Without which our ideas had never been collected into such sundry distinct combinations as they now are. (*Theory*, §109)

In the *Principles* we find another explicit statement of the same thesis:

That number is entirely the creature of the mind, even though the other qualities be allowed to exist without, will be evident to whoever considers, that the same thing bears a different denomination of number, as the mind views it with different respects. Thus, the same extension is one or three or thirty six, according as the mind considers it with reference to a yard, a foot, or an inch. Number is so visibly relative, and dependent on men's understanding, that it is strange to think how any one should give it an absolute existence without the mind. We say one book, one page, one line; all these are equally units though some contain several of the others. And in each instance it is plain, the unit relates to some particular combination of ideas arbitrarily put together by the mind. (*Principles*, §12)

Berkeley's observations here are chiefly intended to undermine the distinction between primary and secondary qualities. He admits as much with the parenthetical remark at the beginning of §109 of the *New Theory of Vision*, and the passage from the *Principles* appears in the context of an attempt to show that there is no meaningful distinction between the primary and secondary qualities. As I understand the distinction, the primary qualities are supposed to inhere in objects and remain unchanged regardless of the state of the observer, while secondary qualities such as color, taste, or smell are not inherent in objects in exactly the way we perceive them but are only powers of bodies to cause sensations in us. According to proponents of the distinction, nature is ultimately composed of material particles endowed only with such primary qualities as size, number, figure, and motion,[8]

---

8. Opinion was divided as to whether such physical properties as force should be considered primary qualities of bodies, and if so whether only one kind of force or

while the secondary qualities are irreducibly subjective and need not figure in a properly scientific account of nature. Berkeley notoriously opposed this distinction, arguing that there really are no Lockean primary qualities and that nothing can truly be said to exist "without the mind." It is important to read the claim that number is a creature of the mind as part of an attack on this distinction.

Number is prominent among the primary qualities, and Berkeley's insistence that numbers are creatures of the mind is designed to show that the number properly assigned to a given object is not invariant under changes in the state of the observer in the way that primary qualities are supposed to be. Berkeley insists that numbers are not properties that objects have independent of any observer, but rather properties of objects viewed in a certain way, i.e., properties of objects which depend upon the observer's specification of a unit, where the specification of the unit is expressed in terms of the mind freely selecting a certain collection of ideas to count as one.

Berkeley's arguments here are fatal to a naive theory which takes numbers to be properties of objects (or collections of objects) such that each object (or collection) has a unique number. But success against such a naive theory does not, I think, establish the kind of mind-dependence of number which he desires. It is possible to grant the Berkeleyan insight that a unit must be "freely chosen," without thereby being led to the conclusion that numbers are mental constructions. The first way to do this would be to reject his definition of a number as a collection of units. Berkeley uses this definition to move from the observation that the unit is mind-dependent to the claim that number itself is mind dependent. This move is made explicit in the passage from the *New Theory of Vision:* "According as the mind variously combines its ideas the unit varies: and as the unit, so the number, which is only a collection of units, doth also vary" (*Theory*, §109). I will consider the significance of this definition of number as a collection of units when I look into Berkeley's case for the non-abstract nature of numbers; for now it suffices to observe that he does rely upon this definition in order to argue that numbers are "creatures of the mind."

Another way to avoid the conclusion that numbers are mind-dependent entities is to deny that a number is properly understood

---

many different ones (*vis inertiæ*, gravitation, etc.) should be ranked among the primary qualities of bodies. The details of the primary-secondary quality distinction are unimportant for our purposes, however. What matters is that number was generally ranked among the primary qualities.

as a property of a collection at all. This strategy could admit that what number applies to a given collection depends upon choice of a unit, but would argue that numbers are simply not properties of collections at all, but rather self-subsistent objects (to use the Fregean terminology). Berkeley never considered either of these alternatives to his argument for the thesis that numbers are creatures of the mind, and I will not try to reconstruct a Berkeleyan line of argument to counter such alternative analyses of the concept of number. What is important for my purposes is to show that Berkeley was committed to the claims that a number is a collection of units and that a unit is a perceptual object regarded in a certain way. Both claims will become important as I investigate the Berkeleyan treatment of arithmetic.

In claiming that numbers are creatures of the mind Berkeley is not forced to adopt a radical relativism about arithmetical truth. From the fact that the number assigned to a collection varies as we change the unit, it does not follow that there is no determinate answer to numerical questions. Given a specification as to the unit we are considering, a question such as How many books are on the table? has a determinate and objective answer, and Berkeley has no interest in denying this. Similarly, there is no reason to think that Berkeley's analysis of number would leave such arithmetical truths as $3 + 4 = 7$ open to question. Thus, we should not take Berkeley's slogan "number is a creature of the mind" to be an expression of doubt about the objectivity of arithmetical truth.

The Nonabstract Nature of Numbers

From Berkeley's remarks about numbers being creatures of the mind I conclude that he construed numbers as collections of perceived objects. But a more important Berkeleyan claim concerning the nature of numbers is that we do not have (nor can we have) abstract ideas of number. As I will show, Berkeley develops an account in which arithmetic is concerned with the manipulation of symbols, and his chief motivation is to provide an alternative to the view that arithmetic is the science of abstract ideas of number.

Berkeley viewed the assertion that there are no abstract ideas of numbers as a consequence of his stronger claim that there are no abstract ideas *simpliciter*. He advanced this thesis in conscious opposition to the common seventeenth and eighteenth century view that the object of arithmetic is "number in the abstract." Thus he writes:

> *Arithmetic* hath been thought to have for its object abstract ideas of *number*. Of which to understand the properties and

mutual habitudes is supposed no mean part of speculative knowledge. The opinion of the pure and intellectual nature of numbers in abstract, hath made them in esteem with those philosophers, who seem to have affected an uncommon fineness and elevation of thought. (*Principles*, §119)

As I have indicated, the view that arithmetic is the science of abstract ideas was fairly widely held in Berkeley's day. Barrow, for example, distinguishes between pure and mixed mathematics in terms of the degree to which the mind abstracts from "all matter, material circumstance, and accidents." Pure arithmetic, on Barrow's account, is the science of abstracted number; mixed (or "concrete", or "applied") arithmetic is the study of the properties of particular finite collections of objects. As we will see, Berkeley is eager to deny that there is any such thing as pure arithmetic in this sense. He eventually characterizes all of arithmetic as either the counting of finite perceived collections or the manipulation of arithmetical symbols in accordance with specified rules. But there is no room in his philosophy for pure or abstract arithmetic in this sense.

Berkeley explicitly argues for the thesis that there is no abstract idea of number in both the *Principles* and *Alciphron*, although the differences between the two arguments are sufficient to warrant treating them separately. In the *Principles*, Berkeley takes a rather roundabout route in his attempt to show that the abstract idea of number is not the object of arithmetic:

> Unity in abstract we have before considered in §13, from which and what hath been said in the Introduction, it plainly follows there is not any such idea. But number being defined a *collection of units*, we may conclude that, if there be no such thing as unity or unit in abstract, there are no ideas of number in abstract denoted by the numerical names and figures. (*Principles*, §120)

This argument depends crucially upon the definition of number as a collection of units and on the claim in §13 of the *Principles* that the idea of unity (or "unity in the abstract") is incomprehensible. In effect, Berkeley employs a certain definition of the term "number" to reduce the task of showing the incomprehensibility of number in the abstract to the problem of showing the incomprehensibility of the abstract idea of unity. Since he is confident that this latter task has been achieved by the work done in the Introduction and §13, he declares that the abstract idea of number has been shown to be unintelligible.

It is evident that this argument will not persuade someone who rejects the definition of number as a collection of units but instead regards numbers as abstractions. Newton clearly held such a position, as he announces at the beginning of the *Universal Arithmetick:* "By *Number* We understand not so much a Multitude of Unities, as the abstracted Ratio of any Quantity to another Quantity of the same kind, which we take for Unity" (Newton 1964–67, 2:7). It is also clear that Barrow's insistence on the primacy of geometry over arithmetic led him to endorse a position similar to Newton's, according to which a number is a kind of ratio which holds between magnitudes.

This is not to say that the characterization of number as a collection of units was uncommon in Berkeley's day. Many standard works in the period distinguished geometry and arithmetic by declaring geometry to be the science of continuous quantity and arithmetic the science of discrete quantity. A natural consequence of this conception is the view that geometric numbers are proportions of continuous magnitudes while arithmetic numbers are collections of discrete units. Indeed, the definition of number as a collection of units can be found in almost all of the standard works on arithmetic in the seventeenth and early eighteenth centuries. The source of this definition is book VII of Euclid's *Elements,* where the theory of arithmetic is developed. Among his definitions, Euclid includes the following:

> 1. An *Unit* is that in virtue of which any of the things that exist is called one.
> 2. A *Number* is a multitude composed of units. (*Elements* VII, def. 1–2)

These definitions appear in many works of the period. The first part of Tacquet's *Arithmeticæ Theoria et Praxis* aims at a presentation of the classical doctrine and adopts the Euclidean definitions, while Wallis' *Mathesis Universalis* uses the same account of number.[9] Thus, even mathematicians who had departed from the classical tradition in some important respects were not prepared to abandon Euclid's account entirely.

Berkeley has an ulterior motive, apart from the wide acceptance, in his day, of the definition of number as a collection of units, for adopting this definition. The argument that there are no abstract ideas of numbers is much more effective if directed first as discrediting the

---

9. Wallis writes: "As it is the nature of number to designate a multitude of units, or (which is the same) to say how often a unit is taken, or how much the units signify, then with different collections of units there will arise differences of numbers. And with whatever change of the unit, the species of things numbered changes" (Wallis 1693–99, 1:28).

abstract idea of unity. Berkeley can give a reasonably good argument for why there can be no abstract idea of unity, but an attempt to apply the same kind of argument directly to the abstract idea of number would involve a wildly implausible assumption.

We may provisionally accept the Berkeleyan thesis that a number is a collection of units, but the argument in the *Principles* that there can be no abstract idea of unity is not immediately compelling. The incomprehensibility of the idea of unity is not so much demonstrated as simply asserted:

> Unity I know some will have to be a simple or uncompounded idea, accompanying all other ideas into the mind. That I have any such idea answering to the word *unity*, I do not find; and if I had, methinks I could not miss finding it; on the contrary it should be the most familiar to my understanding, since it is said to accompany all other ideas, and to be perceived by all the ways of sensation and reflexion. To say no more, it is an *abstract idea*. (*Principles*, §13)

Berkeley's dismissal of the idea of unity is clearly directed against Locke's claim that the idea of unity is a completely simple and uncompounded idea which is present to the mind with all of its perceptions. Locke's account of the genesis of our ideas of numbers takes the idea of unity as basic and then treats the ideas of all other numbers as constructions from it.[10] It is clear that Locke is a primary target of Berkeley's attack in §§13 and 120 of the *Principles* and that Berkeley felt that the characterization of arithmetic as "the science of number in the abstract" had been shown to be absurd by the arguments against abstract ideas in the Introduction. I would like to try to spell out in more detail the argument against the claim that there is an abstract idea of unity, since Berkeley felt it unnecessary to develop it fully in the *Principles*.

The argument from impossibility applied to the abstract idea of unity would then run as follows: (*a*) whatever is impossible is incon-

---

10. "Amongst all the *Ideas* we have, as there is none suggested to the Mind by more ways, so there is none more simple, than that of *Unity*, or One: it has no shadow of Variety or Composition in it: every Object our Senses are employed about; every *Idea* in our Understandings; every Thought of our Minds brings this *Idea* with it. And therefore it is the most intimate to our Thoughts, as well as it is, in its Agreement with all other things, the most universal *Idea* we have" (*Essay* II, xvi, 1). "By repeating this *Idea* in our Minds, and adding the Repetitions together, we come by the *complex* Ideas *of the Modes of it*. Thus by adding one to one, we have the complex *Idea* of a Couple; by putting twelve Unites together, we have the complex *Idea* of a dozen; and so of a Score, or a Million, or any other Number" (*Essay* II, xvi, 2).

ceivable; (*b*) it is impossible that unity exist apart from any single thing; (*c*) the abstract idea of unity is the idea of unity exclusive of any single thing; therefore, (*d*) the abstract idea of unity cannot be conceived. I think that this formulation is the kind of argument that Berkeley would have endorsed, and it applies reasonably well to Locke's account of abstraction and discussion of the idea of unity. It also makes a very strong case against the claim that the human mind is capable of framing such an abstract idea of unity. A Lockean abstract idea of unity would apparently have to be an idea of "bare oneness" as described in premise *c*, and it is hard to see how to give any content to the claim that the mind is capable of separating out the idea of unity. from any particular perception. If we grant that Berkeley discredits the claim that the mind can frame such an abstract idea of unity, we seem compelled to accept his conclusion that there can be no abstract idea of number, provided that the abstract idea of number is to be taken as a collection of abstract ideas of unity.[11]

Note again the important role which the definition of number as a collection of units plays. If we attempt to apply a Berkeleyan argument against the abstract idea of number without first reducing the task to that of attacking the abstract idea of unity, the argument is substantially weaker. Consider the following argument against the abstract idea of number: (*a*) whatever is impossible is inconceivable; (*b*) it is impossible that a number exist apart from any collection of which it is the number; (*c*) the abstract idea of number is the idea of a number apart from any collection of which it is the number; therefore, (*d*) the abstract idea of number cannot be conceived. The most obvious difficulty with this argument is premise *b*, which seems to beg the question at issue and is certainly not likely to be accepted by anyone who defends an abstractionist account of numbers. Whereas such

---

11. This is Frege's conclusion as well. In discussing the Weierstrassian definition of number he complains that the definition of a number as a collection of abstract units is simply incomprehensible. The tone of his complaint is strongly reminiscent of Berkeley's polemics against abstraction: "According to [this view] a numerical sign designates a numerical magnitude itself, but not one composed of concrete units. Rather, it designates a numerical magnitude composed of abstract units, or even of a single unit, namely the unique abstract unit or one. Thus, as a row of books consists of books, the number three consists of abstract units, or even better it consists of the—of course repeatedly occurring—one. Whatever this might be, we are not told. It is probably so abstract that in order to conceive of it, one must not conceive of anything at all. It is somewhat less difficult to conceive of the abstract cow, from which a cow herd can be constructed. As a cow it is surely not yet abstract enough to make the sign '50' appear fitting for a herd composed of it. For that a yet greater abstraction would be necessary" (Frege [1893–1903] 1962, 2, §153).

claims as the impossibility of color without extension or the impossibility of there being a unit without some unitary thing all have a fair degree of plausibility, there is no obvious reason for thinking that premise *b* or anything like it is true in this case. There is presumably no collection of objects in the universe which corresponds to the number $(337^{337})^{337}$, but we make a very strong assumption indeed if we take this to mean that there is literally no such number.

I consider that Berkeley would have to give some reason for accepting premise *b* if he wanted to mount a direct argument against the claim that it is possible to frame abstract ideas of number, but it is not clear what kind of reason could be given. The difficulty of mounting a direct argument may account for his roundabout attack on the abstract idea of number in the *Principles*. By relying on the definition of number as a collection of units, Berkeley avoids the less plausible assumption needed to argue directly against the abstract idea of number.

The argument in the *Principles* against the abstract idea of number is thus effective only against a conception of number in which the abstract idea of number is the idea of a collection of abstract units. But this does not suffice to show that there can be no abstract idea of number at all, since an abstractionist treatment of number which does not treat numbers as collections of units can apparently survive Berkeley's attack. Thus, Berkeley's claim, to be completely persuasive, must be supported either with a demonstration that "number" can only be legitimately defined as a collection of units or with a more direct, general argument against the abstract idea of number. Although I cannot say whether Berkeley was aware of the shortcomings of his argument for the impossibility of the abstract idea of number in the *Principles*, it is interesting to note that in *Alciphron* he follows a completely different line of argument.

Alciphron includes the following exchange between Berkeley's spokesman Euphranor and the "minute philosopher" Alciphron:

> EUPHRANOR. I do not deny [the mind] may abstract in a certain sense, inasmuch as those things that can really exist, or be really perceived asunder, may be conceived asunder, or abstracted one from the other; for instance, a man's head from his body, colour from motion, figure from weight. But it will not thence follow that the mind can frame abstract general ideas, which appear to be impossible.
> ALCIPHRON. And yet it is a current opinion that every substantive name marks out and exhibits to the mind one distinct idea separate from all others.

EUPHRANOR. Pray, Alciphron, is not the word *number* such a substantive name?
ALCIPHRON. It is.
EUPHRANOR. Do but try now whether you can frame an idea of number in abstract, exclusive of all signs, words, and things numbered. I profess for my own part I cannot.
ALCIPHRON. Can it be so hard a matter to form a simple idea of number, the object of a most evident demonstrable science? Hold, let me see if I can't abstract the idea of number from the numerical names and characters, and all particular numerable things—Upon which Alciphron paused a while, and then said, To confess the truth I do not find that I can. (*Alciphron*, dialogue 7, §5)

Admittedly, this argument turns out to be no more than an appeal to introspection (or rather a report of Alciphron's inability to frame the abstract idea of number) but, significantly, it does not employ the definition of number as a collection of units and thus appears to be intended as a more general criticism of the thesis that the mind can frame abstract ideas of number.

Even more importantly, Berkeley's demand here that the abstract idea of number be "exclusive of all signs, words, and things numbered" provides an insight into his antiabstractionist treatment of numbers. Whereas I earlier considered the abstraction which is supposed to yield ideas of numbers simply as abstraction from any collection of things numbered, this statement suggests that Berkeley conceived of mathematical abstraction as going beyond the abstraction from particular numbered collections. Not only must the abstract idea of number be exclusive of all "things numbered" but also of "all signs and words." This, I think, makes Berkeley's case against the abstract idea of number considerably stronger.

Berkeley clearly thinks that it is impossible to have an idea of number that is not the idea of a sign, word, or collection, so that his denial that there can be abstract ideas of number can be rephrased as the positive thesis that all arithmetical thought is concerned either with signs, words, or counted collections. Rephrasing Berkeley's position thus overcomes the difficulty that a straightforward Berkeleyan argument against the abstract idea of number faces because of its dependence on the assumption that a number can only be the number of some collection. Berkeley is not actually committed to this absurd thesis; he holds that numbers can enjoy a two-fold existence, either as properties of collections of objects or as signs which denote no actual collection.

Berkeley's Arithmetical Formalism

I have shown that Berkeley's analysis of the concept of number is strongly influenced by his opposition to the theory of abstract ideas and that in this early work *Arithmetica* he is inclined to treat arithmetic as a science concerned purely with computational procedures. In this section I want to give a more detailed presentation of the theory of arithmetic that grows out of this analysis of number. At the risk of anachronism, I think that Berkeley's approach to arithmetic can best be characterized as a kind of formalism.[12] Berkeley's statements about the nature of arithmetic will more fully justify this characterization, but beginning with a rough description of the formalistic point of view will clarify what is involved in my claim that Berkeley presents a formalistic treatment of arithmetic.

Formalism, as I understand it, is the doctrine that the subject matter of mathematics is the mathematical symbols themselves and not some external realm of objects to which these symbols refer. Of utmost importance in the formalistic treatment of mathematics is the concept of a formal system. For our purposes, a formal system will be taken to consist of a vocabulary of primitive symbols, rules for combining these symbols into formulas, a privileged class of formulas (which we might tendentiously call "axioms"), and rules for deriving additional formulas from the axioms. In the case we are considering, namely arithmetic, the formalist identifies the truth of an arithmetical formula with its derivability from the axioms via the rules.

David Hilbert is the most famous of the formalist philosophers of mathematics, but I do not want to identify formalism too closely with the views of Hilbert. There is a formalist tradition which antedates Hilbert, and I see Berkeley as the first in this line of formalists. Before Hilbert, we can find a recognizably formalistic account of arithmetic in the writings of Eduard Heine:

> In matters of definition I adopt the purely formal standpoint, *in which I call certain tangible signs numbers* so that the existence of these numbers thus does not come into question. *The main emphasis is to be put on the computational operations;* and the numerical sign must be so chosen or equipped with such an apparatus as lends support for the definitions and operations. *Computational operations are called rules, according to which two numbers which are combined by an operation sign can be replaced by a single number.* (Heine 1871, 173)

---

12. I am not alone in reading Berkeley's treatment of arithmetic as a species of formalism. Baum (1972) and Brook (1973) make similar claims.

It is clear from Heine's writings that the formalistic treatment of arithmetic had a life before Hilbert, but the similarity to much of what Berkeley had to say about arithmetic is, for my purposes, more interesting.

Although there are many varieties of formalism, all share a basic commitment to the thesis that mathematical theories do not describe a separate "Platonic" world of mathematical objects. The formalism of Heine has been described as "game formalism," since it treats mathematics as a game of symbol-manipulation analogous to chess.[13] Heine's contemporary Johannes Thomae was explicit in claiming that mathematics is simply a game of symbol manipulation:

> In the formalist conception arithmetic is a game with signs which are called empty, by which is meant that they have no other content (in the calculating game) than that which they are assigned in relation to their behavior under certain combination rules (rules of the game). The chess player makes similar use of his pieces, assigning them certain characteristics which determine their behavior in the game, and the figures are only the outer signs of this behavior. Of course there are important differences between chess and arithmetic. The rules of chess are arbitrary, but the system of rules for arithmetic is such that the numbers can be referred to perceptual manifolds by means of simple axioms and thereby furnish us with important contributions in the knowledge of nature. (Thomae 1898, 7–8)

Berkeley's nominalistic treatment of arithmetic can best be seen as a variety of game formalism, although he never explicitly describes arithmetic as a game.[14] The best way to see this is to begin with an account of a special kind of problem which large numbers pose for Berkeleyan epistemology and to indicate how his formalistic approach to arithmetic can be seen as a response to this problem.

In discussing Berkeley's claim that numbers are creatures of the mind, I noted that he tentatively adopts the view that numbers are

---

13. I take the term "game formalism" from Resnik (1980, chap. 2). Resnik contrasts game formalism with "theory formalism," which views mathematics as a theory of formal systems, and "finitism," which takes mathematics to be partly a meaningful theory of symbolic objects and partly an instrumentalistic extension of this theory. Since theory formalism and finitism tend to be concerned with proof-theoretic considerations such as consistency and conservative extension, there is little in Berkeley which fits neatly into these versions of formalism. As I will show, Berkeley's formalism is elaborated as a form of game formalism.

14. He does, however, characterize algebra as a game in *De Ludo Algebraico*, which we will consider in the next section.

properties of collections of objects. Of course, in the Berkeleyan idiom, such collections can only be collections of ideas. But this account of the nature of numbers breaks down when we try to relate any but very small natural numbers to collections of ideas. It is abundantly clear that Berkeley recognized the problems inherent in a philosophy of arithmetic which treats numbers as collections of ideas, and it is in his solution to this problem that he took the first steps toward formalism.

Briefly stated, the problem which undermines the conception of numbers as collections of ideas is that it does not allow us to distinguish between large numbers. To take an example, suppose we try to frame an idea of a group of twenty-six thousand people and contrast it with the idea of a group of twenty-six thousand and one people. I take it that we simply cannot frame any distinct idea of twenty-six thousand people, and that any mental image that might be an image of twenty-six thousand people cannot be easily distinguished from an idea of twenty-six thousand and one people. We surely do not want to claim that there is no meaningful distinction between the numbers 26,000 and 26,001, but we seem to be forced to this conclusion if we stick to the definition of a number as a collection of ideas.

Berkeley was acutely aware of this problem. In the *Commentaries* he writes: "Qu: whether we have clear ideas of large numbers themselves, or only of their relations" (*Commentaries*, 77). "We seem to have clear & distinct ideas of large numbers v.g. 1000 no otherwise than by considering 'em as form'd by the multiplying of small numbers" (*Commentaries*, 217). The Berkeleyan solution to the problem of large numbers is to deny that numbers must always be collections of ideas. Indeed, Berkeley claims that arithmetic is a case of reasoning in which we do not have distinct ideas annexed to our words, and it is the problem of large numbers that leads him to claim that arithmetic is really only concerned with signs. In a letter to Samuel Molyneux from 8 December 1709, Berkeley writes:

> We may very well, and in my Opinion often do, reason without Ideas but only the Words us'd, being us'd for the most parts as Letters in Algebra, which tho they denote particular Quantities, Yet every step do not suggest them to our Thoughts, and for all that We may reason or perform Operations intirly about them. Numbers We can frame no Notion of beyond a certain Degree, and yet We can reason as well about a Thousand as about five, the Truth on't is Numbers are nothing but Names. (*Works* 8:25)

The letter shows the clear link between the problem of large numbers and Berkeley's doctrine that numbers are "nothing but names," and it is this doctrine which is the main conclusion of Berkeley's discussion of arithmetic in the *Commentaries*, as seen in the following sequence of entries:[15]

> I am better inform'd & shall know more by telling me there are 10000 men than by shewing me them all drawn up. I shall better be able to Judge of the Bargain you'd have me make when you tell me how much (i.e. the name of y^e) money lies on ye Table than by offering & shewing it without Naming. In short I regard not the Idea the looks but the Names. Hence may appear the Nature of Numbers. (*Commentaries*, 761)

This doctrine is elaborated in interesting ways in the entries which follow:

> Children are unacquainted with Numbers till they have made some Progress in language. This could not be if they were Ideas suggested by all the senses.
> Numbers are nothing but Names, meer Words. (*Commentaries*, 762–63)
> In Arithmetical Problemes Men seek not any Idea of Number. They onely seek a Denomination. this is all that can be of use to them.
> Take away the signs from Arithmetic & Algebra, & pray w^t remains?
> Those are sciences purely Verbal, & entirely useless but for Practise in Societys of Men. No Speculative knowledge, no comparing of ideas in them. (*Commentaries*, 766–68)
> The great use of the Indian figures above the Roman shews Arithmetic to be about Signs not Ideas, or not Ideas different from the Characters themselves. (*Commentaries*, 803)

From what these entries reveal, I conclude that Berkeley, by the time he finished writing the *Commentaries*, has adopted the position that numbers are nothing but words. The placement of the entries on arithmetic at the end suggests that Berkeley turned his attention to the consideration of arithmetic at a rather late stage in the writing of the *Commentaries* and some time after developing his early antiabstractionist account of geometry.

The *Commentaries* do not tell the whole story of Berkeley's views on arithmetic, however. A much more detailed account of his brand of arithmetical formalism is presented in the *Principles* (§§119–22) and *Alciphron* (dialogue 7). A comparison of the *Commentaries* with the *Principles* and *Alciphron* does not reveal the sharp change in doctrine which

---

15. This sequence of entries actually begins with entries 759 and 760, where the thesis that numbers are creatures of the mind is first proclaimed.

I demonstrated in Berkeley's views on geometry. Indeed, Berkeley's approach to arithmetic remained essentially the same throughout his career, and some of what he has to say in the later works is strongly reminiscent of the *Arithmetica*. What is new in the *Principles* and *Alciphron* is the more detailed exposition of the claim that arithmetic is a science of signs and the application of the theory of representative generalization to strengthen the claim that numerical signs do not denote abstract ideas.

In the *Principles* Berkeley elaborates the thesis that arithmetic is a science of signs by "taking a view of arithmetic in its infancy." He suggests that the first arithmetical notation was a simple system of tally strokes, each taken to designate a unit, where a unit is "some one thing of whatever kind [men] had occasion to reckon." This system was developed to aid in counting finite collections, an activity which Berkeley maintains is the sole proper use of arithmetic. In the course of time, more concise forms of notation were adopted, culminating in the Hindu-Arabic numerals which Berkeley describes (much as in the *Arithmetica*) as a system "wherein by the repetition of a few characters or figures, and varying the signification of each figure according to the place it obtains, all numbers may be most aptly expressed" (*Principles*, §121).

More important than the decimal notation system are the rules of computation, characterized by Berkeley as "methods of finding from the given figures or marks of the parts, what figures and how placed, are proper to denote the whole or *vice versa*." This system of rules is a special kind of language which is of utmost utility because it allows us to accomplish by the manipulation of symbols many things which would be impossible (or at least highly impractical) if we were to attempt to work directly with those objects which are our ultimate concern:

> For these signs being known, we can by the operations of arithmetic, know the signs of any part of the particular sums signified by them; and thus computing in signs (because of the connexion established betwixt them and the distinct multitudes of things, where of one is taken for an unit), we may be able rightly to sum up, divide, and proportion the things themselves that we intend to number. (*Principles*, §121)

This philosophy of arithmetic has obvious affinities with formalism and shows how Berkeley accounts for the application of arithmetic. In applied arithmetic, we begin with a collection (or collections) of objects in the world and assign numbers to the collections in accordance with our specification of a unit. We then perform operations on the numerical signs, obtaining as a result a new sign which can be interpreted "back into the world" and used to guide our practice: "In *arithmetic* therefore we regard not the *things* but the *signs*, which nevertheless are not regarded for their own sake, but because they direct us how to act with relation to things, and dispose rightly of them" (*Principles*, §122). Berkeley concludes this exposition of his views on arithmetic with a dismissal of the claim that arithmetic is a science whose object is the abstract idea of number:

> Now agreeably to what we have before observed, of words in general (*Sect* 19 *Introd.*) it happens here likewise, that abstract ideas are thought to be signified by numeral names or characters, while they do not suggest ideas of particular things to our minds. I shall not at present enter into a more particular dissertation on this subject; but only observe that it is evident from what hath been said, those things which pass for abstract truths and theorems concerning numbers, are, in reality, conversant about no object distinct from particular numerable things, except only names and characters; which originally came to be considered, on no other account but their being *signs*, or capable to represent aptly, whatever particular things men had need to compute. (*Principles*, §122)

In dialogue 7 of *Alciphron*, Berkeley does "enter into a more particular dissertation on this subject," presenting a long discussion of the importance of signs in all branches of learning. Berkeley's chief point is that "all sciences, so far as they are universal and demonstrable by human reason, will be found conversant about signs as their immediate object, though these in the application are referred to things" (*Alciphron*, dialogue 7, §13). Failure to appreciate the role of signs in the sciences has, on Berkeley's account, misled people into holding that general terms in the sciences (such as "force," "number," or "triangle") must refer to abstract general ideas. One of the central themes of dialogue 7 is Berkeley's claim that all of the explanatory roles assigned to the theory of abstract general ideas can be filled by the theory of signs:

> It is not, therefore, by mere contemplation of particular things, and much less of their abstract general ideas, that the mind makes her progress, but by an apposite choice and skillful management of signs: for instance, force and number, taken in concrete, with their adjuncts, subjects, and signs, are what every one knows; and considered in abstract, so as making precise ideas of themselves, they are what nobody can

comprehend. That their abstract nature, therefore, is not the foundation of science is plain: and that barely considering their ideas in concrete is not the method to advance in the respective sciences is what everyone that reflects may see; nothing being more evident than that one who can neither write nor read, in common use understands the meaning of numeral words as well as the best philosopher or mathematician. (*Alciphron,* dialogue 7, §11)

Berkeley claims that the distinction between the illiterate common man and the educated philosopher or mathematician (at least with respect to their understanding of arithmetic) lies in the ability of the educated man to "express briefly and distinctly all the variety and degrees of number, and to perform with ease and dispatch several arithmetical operations by the help of general rules." But this ability depends only upon the educated man's familiarity with arithmetical symbolism and does not derive from his superior grasp of the "abstract, intellectual, general idea of number" (*Alciphron,* dialogue 7, §11).

He then characterizes arithmetic in terms which recall the doctrine of the *Principles:*

> I imagine one need not think much to be convinced that the science of arithmetic, in its rise, operations, rules, and theorems, is altogether conversant about the artificial use of signs, names, and characters. These names and characters are universal, inasmuch as they are signs. The names are referred to things, the characters to names, and both to operation. The names being few, and proceeding by a certain analogy, the characters will be more useful, the simpler they are, and the more aptly they express this analogy. Hence the old notation by letters was more useful than words written at length; and the modern notation by figures, expressing the progression or analogy of the names by their simple places, is much preferable to that, of ease and expedition, as the invention of algebraical symbols is to this, for extensive general use. (*Alciphron,* dialogue 7, §12)

It is not difficult to see this conception of arithmetic as a variety of game formalism. On Berkeley's view, arithmetic is a "purely nominal" science concerned with the manipulation of symbols, although we can apply this purely symbolic reasoning to the world by interpreting the formalism to represent collections of objects. Arithmetical theorems are demonstrable precisely because the demonstrations are concerned only with the manipulation of symbols in accordance with specified rules. Thus, the truths of arithmetic concern only what combinations of symbols are constructible by using the rules, and are essentially true by definition. This means, among other things, that Berkeley's account of arithmetical truth does not face the difficulties we observed in his treatment of geometry. Berkeley has difficulty accommodating geometric truth within his philosophy of geometry, and eventually treats geometric truth as a kind of "truth in the limiting case." But his formalistic account of arithmetic can accept the literal truth of arithmetical theorems by construing them as truths about the symbols employed in a demonstration. This is not to say that Berkeley's treatment of arithmetic is altogether free from difficulty; I shall attend to some of its problems at the end of this chapter.

There is another difference between Berkeley's treatment of arithmetic and geometry which is worth noting here. Berkeley regards arithmetic as a purely verbal science in which numerical signs are manipulated in accordance with specified rules. However, he does not conceive geometry as a purely verbal activity, but rather as the study of perceived extension in which properties of perceived objects are generalized so as to hold of all similar figures. The difference here lies in what we might cautiously call the degree of abstraction involved in arithmetic as opposed to geometry. The theorems of arithmetic (although ultimately about the objects in the world) are immediately about what combinations of signs can be produced by following computational rules. The immediate object of a geometrical demonstration, however, is a perceived figure which represents a class of other figures which resemble it. This resemblance between the immediate and ultimate objects of geometry is lacking in arithmetic. Because arithmetical signs do not resemble the things they represent, there is an arbitrary or conventional element in the choice of arithmetical signs that is not matched by a similar arbitrariness or conventionality in geometry.

The difference is perhaps best brought out by considering that, on Berkeley's account, we can choose whatever signs we want for numbers, while there is no similar freedom of choice in geometry. Berkeley was clearly aware of this difference between arithmetic and geometry in his system:

> Qu: whether Geometry may not be properly reckon'd among the Mixt Mathematics. Arithmetic and Algebra being the Only abstracted pure i.e. entirely Nominal. Geometry being an application of these to Points. (*Commentaries*, 770)

In another entry he connects the arbitrariness of our selection of signs to the nature of demonstration, arguing that it is by virtue of the fact that we create and choose such signs that we are able to carry out demonstrations: "The reason why we can demonstrate So well about signs is that they are perfectly arbitrary & in our power, made at pleasure" (*Commentaries,* 732). Although there is a clear difference between Berkeley's treatments of arithmetic and geometry, the similarities are also significant. In each case Berkeley is concerned to develop an account of mathematics which does not rely upon the doctrine of abstract ideas, and in each case he avoids invoking abstract ideas by treating the objects of mathematics as signs. In his account of geometry, Berkeley treats geometric figures as signs for classes of figures which they resemble, and thus characterizes geometry as a science whose immediate objects are signs. Similarly, arithmetic is treated as a science whose immediate objects are signs, but in this case the signs are arbitrarily selected symbols which can stand for collections of objects. Although it is true that "all sciences, so far as they are universal and demonstrable by human reason, will be found conversant about signs as their immediate object" (*Alciphron,* dialogue 7, §13), this does not mean that arithmetic and geometry must be treated in exactly the same way. The signs used for geometry must be geometric objects that represent other geometric objects, while the signs used in arithmetic or algebra are arbitrarily chosen symbols.

Algebra as an Extension of Arithmetic

Although Berkeley did not devote a great deal of discussion to algebra, it is clear that he favored a formalistic interpretation along the same lines as his treatment of arithmetic. In the Berkeleyan scheme, algebra is a science of signs with an even higher level of generality than those of arithmetic. Berkeley claims that algebraic letters are signs for numerals, which in turn are signs for collections of objects, making algebra a kind of "meta-arithmetic": "Algebraic Species or letters are denominations of Denominations. Therefore Arithmetic to be treated of before Algebra" (*Commentaries,* 758). In this formalistic scheme, the truths of algebra are to be interpreted as general statements about the relations between numerical signs in arithmetic. For example, the Berkeleyan interpretation of the algebraic identity $(x + y)^2 = x^2 + 2xy + y^2$ would be that the theorem is a general truth which holds for all numerical signs substituted for x and y in the algebraic formula.

Two of Berkeley's early works in the *Miscellanea Mathematica, De Ludo Algebraico* and *De Radicibis Surdis,* bring out his strongly formalistic treatment of algebra. In *De Ludo Algebraico* he introduces an "algebraic game" designed to help students master techniques for solving equations. The point of the algebraic game is to randomly

Berkeley's New Foundations for Arithmetic

assign a system of equations and then find a solution for the system. The most striking feature of the game is that Berkeley treats it as a worthy alternative to chess.[16] In drawing the link between algebra and such games as chess, Berkeley's commitment to game formalism could hardly be made more obvious.

The paper *De Radicibis Surdis* contains a suggestion for improving algebra by altering the standard notation for roots. Berkeley proposes eliminating the radical sign from expressions involving roots of unknown quantities in favor of Greek letters. The idea here is that an expression such as $\sqrt{a}$ should be replaced by a single sign α. The advantage that Berkeley supposes to derive from this notational convention is that computations involving roots can be more easily assimilated into ordinary algebraic operations. For example, the square root of the product bc is normally written $\sqrt{bc}$, and the product of the roots $\sqrt{b}$ and $\sqrt{c}$ is written $\sqrt{b}\sqrt{c}$. This introduces a notational difference which corresponds to no real difference, since the root of the product is the product of the roots. Berkeley's suggestion would eliminate the notational ambiguity; replacing $\sqrt{bc}$ with $\beta\kappa$, we have $\sqrt{bc} = \sqrt{b}\sqrt{c} = \beta\kappa$ with one expression in Greek letters replacing two equivalent expressions using the radical sign. This suggestion is not developed at length and Berkeley recognizes some drawbacks (most notably, that it cannot express cube or higher roots without either ambiguity or new notation), but the interesting feature is his emphasis on the importance of notation. This is quite in keeping with Berkeley's formalistic conception of arithmetic and his work in the *Arithmetica* and underscores the extent to which he treated both algebra and arithmetic as sciences whose principal object is signs.

There is no discussion of algebra in the *Principles,* except for Berkeley's remark in the Introduction that algebra is a paradigm case of purely verbal reasoning in which signs (in this case, algebraic letters) can be significant even though they do not stand for particular determinate ideas:

> And a little attention will discover, that it is not necessary (even in the strictest reasonings) significant names which stand for ideas should, every time they are used, excite in the understanding the ideas they are made to stand for: in reading and

---

16. As a whole, *De Ludo Algebraico* treats algebra as a game, and Berkeley is fairly explicit on this point in two passages. In the Introduction he asks, "But is mathematics a game?" and strongly suggests an affirmative answer. Again, after presenting the rules of his game, he exclaims, "You see what a mere game algebra is, and that both chance and science have a place in it. Why not, therefore, come to the gaming table?" (*Works* 4:214, 218).

discoursing, names being for the most part used as letters are in *algebra,* in which though a particular quantity be marked by each letter, yet to proceed right it is not requisite that in every step each letter suggest to your thoughts, that particular quantity it was appointed to stand for. (Introduction, §19)

This is essentially the same line of thought which Berkeley pursues in his account of arithmetic. He insists that in arithmetical calculations, we can simply manipulate the signs without thinking of any particular collections which might correspond to them. Similarly, in an algebraic calculation, we can work with symbols x, y, or ξ without having particular numbers in mind.

Berkeley's most complete statement of his formalistic account of algebra follows a statement of his views on arithmetic in *Alciphron:*

For instance, the algebraic mark, which denotes the root of a negative square, hath its use in logistic operations, although it be impossible to form an idea of any such quantity. And what is true of algebraic signs is also true of words or language, modern algebra being in fact a more short, apposite, and artificial language, and it being possible to express by words at length, though less conveniently, all the steps of an algebraical process. (*Alciphron,* dialogue 7, §14)

There is an important admission here which will be of interest later when I investigate Berkeley's critique of the calculus. Berkeley admits that algebra can employ symbols such as $\sqrt{-1}$ even though no quantity can be imagined to correspond to it. But in the case of the calculus, Berkeley continually complains that the proponents of "the modern analysis" have introduced a notation for fluxions and infinitesimals without being able to frame any ideas of the objects denoted by these symbols.

This is further evidence of the important distinction between geometry and arithmetic in Berkeley's philosophy of mathematics. For Berkeley, the calculus is fundamentally a geometric theory, whose proper object is perceivable extension. Thus, the key terms in the calculus must be interpretable in terms of perceivable extension, i.e., we must be able to frame ideas corresponding to these terms. But algebra, being an extension of arithmetic, is a purely nominal science which can be justified on formalistic grounds. The algebraic identity $i^4 = ((\sqrt{-1})^2)^2 = (-1)^2 = 1$ is true because the laws of exponentiation and multiplication dictate it, but these laws concern only the manipulation of symbols, and we need have no idea (other than the symbols themselves) corresponding to our notation. But a theorem of the calculus (such as the determination of the arc-length of a curve) concerns extended objects and cannot be legitimately obtained unless each step in the derivation has the appropriate ideas corresponding to it.

The Primacy of Practice over Theory

We saw when we considered Berkeley's treatment of geometry that he held a very low opinion of "speculative geometry" and thought it a subject unworthy of serious attention. This hostility toward pure mathematics is just as strong in the case of arithmetic, where Berkeley's formalism leads him to advocate the elimination of pure arithmetic. Because he regards arithmetic as a science of signs, he insists that there is no reason to study the signs themselves. Although Berkeley regards an understanding of the role of signs in human reasoning as a matter of supreme importance, he repeatedly asserts that such purely nominal sciences as arithmetic and algebra can be justified only in so far as they can be used to help solve practical problems in daily life.

This emphasis on the importance of practical mathematics is expressed in the *Commentaries* where, having styled the mathematicians as "Nihilarians," Berkeley complains, "If the wit & industry of the Nihilarians were employ'd about the useful & practical Mathematiques, w^t advantage had it brought to Mankind?" (*Commentaries*, 471). And at the end of the *Commentaries*, "Speculative Math: as if a Man was all day making hard knots on purpose to unty them again" (*Commentaries*, 868). This attitude should not be surprising, given Berkeley's formalistic conception of arithmetic. Because the immediate objects of arithmetical discourse are arbitrarily chosen signs and arithmetical theorems simply claims that certain combinations of signs can be produced by following given rules, the study of pure arithmetic is of little intrinsic interest. What matters, for Berkeley, is the way in which arithmetical theorems can be used to help solve practical problems, but pure arithmetic ignores this practical application altogether.

The most concise statement of Berkeley's opposition to pure arithmetic can be found in the *Principles,* when he begins his discussion of arithmetic:

> The opinion of the pure and intellectual nature of numbers in abstract has made them in esteem with those philosophers who seem to have affected an uncommon fineness and elevation of thought. It hath set a price on the most trifling numerical speculations, which in practice are of no use, but

> serve only for amusement; and hath heretofore so far infected the minds of some, that they have dreamed of mighty mysteries involved in numbers, and attempted the explication of natural things by them. But, if we narrowly inquire into our own thoughts, and consider what has been premised, we may perhaps entertain a low opinion of those high flights and abstractions, and look on all inquiries about numbers only as so many *difficiles nugae,* so far as they are not subservient to practice, and promote the benefit of life. (*Principles,* §119)

Similarly, he complains that pure arithmetic amounts to little more than empty verbal exercises, since arithmetic is nothing more than a handy symbolic language which can be used to abbreviate calculation:

> Whence, it follows, that to study [numbers] for their own sake, would be just as wise, and to as good a purpose, as if a man neglecting the true use or original intention and subserviency of language, should spend his time in impertinent criticisms upon words, or reasonings and controversies purely verbal. (*Principles,* §122)[17]

This rather disheartening hostility toward pure mathematics is certainly not evident in Berkeley's earlier *De Ludo Algebraico* where he praises algebra in particular as an excellent and important science. But it remains intact through all editions of the *Principles* and seems to have been Berkeley's considered view throughout most of his philosophical career.

Berkeley's Formalism Evaluated

Granted that Berkeley is a proponent of game formalism, I think we must acknowledge the historical and philosophical significance of his account of arithmetic and algebra, simply because he is the first proponent of the formalistic point of view. Certainly, there were nominalists of many varieties before Berkeley and some who interpreted the "mysterious" number $\sqrt{-1}$ as a mere symbol of use in calculation. But Berkeley was the first to apply his nominalistic

---

17. The manuscript version of this passage contains a legible deletion which is even harsher: "such as not a few philosophers are accused to have done. In which, perhaps, there may be a greater agreement between Schoolmen & some Speculative Mathematicians than one wou'd first imagine" (British Library Additional MS. 39304, fol. 84^r–85^r). The reference to "Schoolmen" recalls another statement: "We have learn'd from Mr. Locke that there may be and that there are several glib, coherent, methodical Discourses wch nevertheless amount to just nothing. this by him intimated with relation to the Schoolmen. We may apply it to the Mathematicians" (*Commentaries,* 492).

Berkeley's New Foundations for Arithmetic

scruples thoroughly enough to conclude, with no reference to abstract numbers or other mathematical entities, that all of arithmetic and algebra are essentially symbolic calculi.

Not everyone acknowledges Berkeley as the originator of formalism. Resnik maintains that:

> Formalism has a respectable mathematical history dating back at least to the introduction of imaginary numbers, by Bombelli in the sixteenth century, to furnish numerical roots for previously unsolvable equations such as "$x^2 = -1$." Because Bombelli believed merely that he was introducing symbols, treating them as if they designated numbers and subjecting them to algebraic operations without actually producing the numbers themselves, he reflected his many reservations about this feat by calling his numbers *imaginary numbers*. (Resnik 1980, 55)

But this misses much of the point. Bombelli and others of his era contrasted the "imaginary" numbers which they took to be mere symbols with the "real" numbers which denoted genuine magnitudes. Yet any formalism worthy of the name should treat all numbers as merely symbols, so there is really nothing terribly formalistic about these early treatments of complex numbers. Moreover, a well-conceived formalism requires an account of the relationship between the formal theory and its applications, and this is entirely lacking in the works of sixteenth-century algebraists. They introduce complex numbers to solve equations, but there is nothing to parallel Berkeley's account of how the numerical system can be interpreted in ordinary practice and applied to the solution of problems.

Of the nominalists before Berkeley, none seems to have been especially keen to apply his theory of mathematics. Medieval discussions of nominalism concentrate on the question of whether there are real universals corresponding to general terms, but make no explicit mention of arithmetic.[18] Even the notorious nominalist Hobbes avoided any trace of a formalistic conception of arithmetic or algebra.[19] In-

---

18. This is not to say that I have surveyed all of the writings of medieval nominalists, but I have certainly encountered no nominalistic treatment of arithmetic in the literature. To the extent that numbers were regarded as universals, a nominalist about universals would presumably be a nominalist about arithmetic, but I know of no well-developed mathematical nominalism from the period. See Mahoney (1978) and Masi (1983) for more on medieval accounts of arithmetic.

19. Hobbes followed Barrow in regarding geometry as the fundamental mathematical science. In his *Examinatio et emendatio mathematicæ hodiernæ* he holds that "geometry is the science of determining magnitudes, either of bodies, or times, or of any

deed, Hobbes is famous for his rejection of the methods of symbolic algebra as a "scab of symbols" which deface geometric demonstrations. It thus seems that Berkeley's exposition of a recognizably formalistic conception of arithmetic and algebra marks the beginning of mathematical formalism. Whether Berkeley's formalistic account was historically influential is difficult to judge and is a question I will leave open.

Whatever its historical status, Berkeley's formalistic philosophy of mathematics seems to offer him a way to accept arithmetic and algebra as genuine sciences without taking their object to be abstractions. Nevertheless, there are some notorious difficulties with formalism which I should discuss by way of closing out my investigation of Berkeley's account of arithmetic. To begin with, Berkeley's formalism, in its initial characterization, masks serious confusions between type and token. To declare that the truths of arithmetic are really truths about symbols is either to utter a patent falsehood or to confuse numerical inscriptions with inscription-types. For example, we would like to say that the sum of 11 and 12 is 23. If we express this as an identity, we would say "11 + 12 = 23." But we don't want to say that the actual inscription "11 + 12" is *identical* with the inscription "23." These are as distinct as two things can be. Moreover, we don't want to say that the inscription-type "11 + 12" is identical with the inscription-type "23." The former contains five occurrences of symbols (ignoring the blank characters) and the latter only two. The natural way of understanding arithmetic formalistically would be to take arithmetical equations as "rewrite rules" for inscription-types. On this reading, the theorem "11 + 12 = 23" is not an identity, but should instead be treated as the assertion that the inscription-type "11 + 12" can be replaced by the inscription-type "23." This makes reasonable sense of the thesis that the truths of arithmetic concern only symbols, but it leaves Berkeley in a somewhat embarrassing position. Having dispensed with numbers in the abstract, his theory can only be rescued by introducing inscription-types. But is an inscription-type not a paradigmatic abstraction?

The difficulty here is related to one which was passed over in my discussion of representative generalization in chapter 2 but which can

---

other nonmeasured magnitude, by comparison of it with one or another measured magnitudes," while "arithmetic is the science of determining multitudes of things that are not numerated by comparison of them with numbered multitudes." But because a continuous magnitude can be divided into any number of equal parts, with its ratio to any other magnitude remaining unchanged, Hobbes concludes that "it is manifest that arithmetic is contained in geometry." (Hobbes [1845] 1966a, 4:27–28).

Berkeley's New Foundations for Arithmetic

be raised here. It seems that Berkeley's account of demonstration (both for geometry and arithmetic) may not be as free from abstractions as he had imagined. To generalize a geometric demonstration we must be able to apply it to all figures which resemble the figure initially used in the proof (which we may call the resemblance-class of the figure), but resemblance seems a suspiciously abstract property. Similarly, to interpret arithmetic formalistically we must read arithmetical demonstrations as applying to all inscriptions of a given type, but this seems to require the suspiciously abstract inscription-type. The only way out of this difficulty is to define both resemblance-class and inscription-type in terms which are admissible within the Berkeleyan framework, but it is not immediately obvious how this might be done. One suggestion would be to take resemblance to be a matter of sharing a property, so that we could say x resembles y (with respect to a certain property P) if and only if there is a property P such that Px and Py. Similarly, we could try to define inscription-types in terms of the "structural" properties of inscriptions: two inscriptions x and y belong to the same inscription-type if and only if there is a structural property P_s such that $P_s x$ and $P_s y$. This strategy assumes that we can take for granted what it is for an object to have a property, and that quantification over properties is unproblematic. More dubiously, it assumes that there is a special class of "structural properties" which define what it is for two inscriptions to be of the same type. I assume that Berkeley could simply leave both concepts unanalyzed and proceed. Indeed, this seems the only sane course, since it is difficult to find anything enlightening to say on either issue and yet we all know what it is for two things to share a property or to have a similar structure. Nevertheless, it is clear that Berkeley's treatment of demonstrations is not quite as free from metaphysical baggage as he may have thought.

A second difficulty with Berkeleyan formalism is that it must be phrased in modal terms in order to be even moderately plausible. Since only a small finite number of inscriptions have ever been written down, any attempt at formalism which countenances only actually existing inscriptions makes a complete hash of arithmetic. Such obvious arithmetical truths as the claim that every number has a successor must be rejected if we take arithmetic to be a formalistic theory about only inscriptions that have actually been effected. The obvious way to avoid this problem would be to say that the truths of arithmetic depend upon the continued possibility of making inscriptions. Thus, the claim that every number has a successor would be rephrased as the assertion that, for any possible inscription, there could be another one succeeding it in accordance with the relevant notational rules. It is clear that Berkeley would have had no qualms about such a formulation, since he makes a similar move in his account of geometry. In the geometric case, Berkeley takes theorems to be about all possible lines or figures represented by those in a proof, and a modal interpretation of his arithmetical formalism is very much in this spirit. However, the restrictive epistemological tenets which underwrite Berkeley's critique of the doctrine of abstraction might well create problems for this modally-based philosophy of mathematics. His objection to abstract ideas is that they are impossible and therefore inconceivable. But there are plenty of things which are possible although they exceed our power to comprehend them. To get any reasonable theory of arithmetic off the ground, we need an infinity of numbers, which in this case corresponds to an infinite number of possible inscriptions. But plainly, Berkeley will allow that we can only comprehend finitely many inscriptions, so his theory of arithmetic seems to allow for the introduction of objects literally incomprehensible to us. There may be ways out of this difficulty, but it seems that a proper formulation of Berkeley's formalistic conception of arithmetic will require epistemological and metaphysical principles which Berkeley neither stated nor considered.

Despite these difficulties, Berkeley's formalistic approach to arithmetic and algebra is certainly worthy of our consideration. Many philosophers of mathematics have been attracted to formalism because of its apparent ability to sidestep thorny issues in the metaphysics and epistemology of mathematics, and Berkeley's formulation of the thesis is remarkably precise and consistent for an eighteenth-century philosopher of mathematics.

CHAPTER FOUR

Berkeley and the Calculus: The Background

Berkeley's most famous contribution to the philosophy of mathematics is his 1734 treatise, *The Analyst*. However, an appreciation of this Berkeleyan critique of the calculus requires that we first spend some time investigating its historical background. The level of mathematical sophistication required for this task is quite modest, but the differences between the calculus as practiced in Berkeley's day and current presentations of the subject are sufficient to warrant this brief excursion into the history of mathematics.

Before we begin, some terminological clarifications are in order. I use the term *analysis* to refer to a body of mathematical techniques and results concerned with the problems of finding tangents to curves, finding the area enclosed by a curve (known as "quadrature" in the parlance of the period), and finding the arc-length of a curve (known as "rectification"). I use the term *calculus* for the algorithmic procedures developed by Newton and Leibniz for constructing tangents and finding quadratures. The Newtonian method of fluxions and the Leibnizian *calculus differentialis* will be our principal concern, but there are other authors, including Euclid, Archimedes, Cavalieri, Wallis, and Barrow, whose work is relevant here.

A number of valuable and accessible works deal with the history of the calculus in broader scope and greater detail than I can here, and the reader is referred to them for a more comprehensive treatment of the historical background to the calculus.[1] My aim here is to present

---

1. See Boyer ([1949] 1959), Boyer ([1968] 1989, chaps. 16–20), Edwards (1979), and Fauvel and Gray (1987) for accessible treatments of the relevant history.

enough of the material to make Berkeley's criticisms of the calculus intelligible to the modern reader. I begin by discussing the classical approach to problems in analysis, with special attention to the method of proof by exhaustion, then examine the infinitesimal methods developed in the seventeenth and eighteenth centuries, with particular attention to the "method of indivisibles," the Newtonian calculus of fluxions, and the Leibnizian differential calculus.

Classical Geometry and the Proof by Exhaustion

As we saw in chapters 1 and 2, classical geometry was conceived of as a theory which investigated the properties of abstract magnitudes such as lines, angles, surfaces, and solids. This account of the object of geometry was paired with a methodological requirement that theorems be proved and problems solved by constructions in accordance with the fundamental axioms. For the investigation of problems in analysis, this methodology dictates that solutions to problems rely heavily upon the theory of ratios and proportions as presented in book V of Euclid's *Elements*. In essence, a problem of quadrature or rectification is solved classically by taking a given magnitude and constructing another magnitude which bears the desired ratio to it.

The Euclidean presentation of the theory of ratios has a strongly finitistic character which bars the use of infinitesimal magnitudes in the solution of problems in analysis. This is most clearly evident in the key definitions from book V of the *Elements*, which read as follows:

> 3. A *ratio* is a sort of relation in respect of size between two magnitudes of the same kind.
> 4. Magnitudes are said to *have a ratio* to one another which are capable, when multiplied, of exceeding one another.
> 5. Magnitudes are said to *be in the same ratio*, the first to the second and the third to the fourth, when, if any equimultiples whatever be taken of the first and third, and any multiples whatever of the second and fourth, the former equimultiples alike exceed, are alike equal to, or alike fall short of, the latter equimultiples respectively taken in corresponding order.
> 6. Let magnitudes which have the same ratio be called *proportional*. (*Elements* V, def. 3–6)

These definitions are significant in that they provide the means for comparing magnitudes within each species by the formation of ratios and for comparing ratios across species of magnitudes by constructing proportions.

The finitistic character of the classical theory is apparent when it is

Berkeley and the Calculus: The Background

understood that the multiplications in definitions 4 and 5 are finite multiplications. To compare two magnitudes α and β in a ratio $\alpha : \beta$, it is necessary that the continued multiplication of one will make it exceed the other. This explicitly bars infinitesimal elements and prevents ratios from being formed across species because there is no multiplication of a line which will allow it to exceed an angle or surface.[2] But proportions can be constructed from a pair of ratios whenever the criterion in definition 5 is satisfied, so it makes sense to say that the ratio between two lines is the same as that between two spheres, although a line and sphere cannot be directly compared with one another in a ratio.

The doctrine of ratios and proportions gives rise to a method of proof which is fundamental throughout all of classical geometry—the so-called method of exhaustion.[3] The fundamental idea in an exhaustion proof is that the unknown ratio between two magnitudes (or the proportion between two pairs of magnitudes) can be determined by considering sequences of known quantities which approximate the unknown to within any desired degree of accuracy. A characteristic feature of the exhaustion proof (especially as practiced by Archimedes) is the use of *compression*—the procedure of bounding the unknown above and below by monotonic sequences of known quantities which converge to a common limit. These sequences are generated by geometric constructions which provide successively better approximations to the unknown, typically in the form of figures inscribed within and circumscribed about the figure whose area is to be determined. When it can be shown that the unknown is compressed between sequences of inscribed and circumscribed figures, the proof is rounded off by a double *reductio ad absurdum* which shows that the unknown can be neither greater nor less than a specified amount.

The proposition at the foundation of this method follows immediately from definition 4 (cited above), and states a criterion by which a sequence of approximations can be brought arbitrarily close to a limit:

> Two unequal magnitudes being set out, if from the greater there be subtracted a magnitude greater than its half, and

2. In effect, definition 4 states the "Archimedean axiom" for the real numbers, which says that for any positive real numbers α and β there are finite positive integers m and n such that $m\alpha > \beta$ and $n\beta > \alpha$.

3. Dijksterhuis objects to the traditional term "exhaustion method" on the grounds that "for a mode of reasoning which has arisen from the conception of the inexhaustibility of the infinite, this is about the worst name that could have been devised" (Dijksterhuis 1987, 130). Nevertheless, the traditional name has stuck, and I will adopt it here.

from that which is left a magnitude greater than its half, and from that which is left a magnitude greater than its half, and if this process be repeated continually, there will be left some magnitude which will be less than the lesser magnitude set out. (*Elements* X, 1)

The general procedure for an exhaustion proof is to begin with upper and lower bounds for an unknown magnitude and then provide a method for systematically improving these bounds, after which the proof is rounded off by a double *reductio ad absurdum*. Both the rigor and the limitations of this approach can be seen by considering a version of the proof in Euclid that the ratio between the areas of two circles is as the square of their diameters (*Elements* XII, 2).

To begin, we need a principle which we can call the first compression lemma for the circle, as follows:

L_1. Given an area A less than that of a circle C, a regular polygon with area P can be inscribed within the circle, such that the inequality $A < P < C$ will hold.

The justification for this lemma lies in the observation that we can inscribe within any circle a square, whose area is a lower bound on the area of the circle. Doubling the sides of the square, we obtain an inscribed octagon area as a better approximation to that of the circle. By continuing to double the number of sides, we form a sequence of approximating regular polygons such that the difference between the area of the polygon and the area of the circle is reduced by more than half with each addition to the sequence. Applying *Elements* (X, 1) the difference between the area of the circle and the area of an inscribed regular polygon can be made as small as desired. Thus, the inequality in our first compression lemma can be constructed, for there will eventually be a polygon whose area falls between that of A and C. A second compression lemma for the circle is the claim that

L_2. Given any area A greater than that of a circle C, a regular polygon with area P can be circumscribed about the circle, such that the inequality $A > P > C$ will hold.

This can be proved in a manner exactly analogous to that of the first compression lemma by taking a square which circumscribes the circle and then forming a sequence of regular polygons by doubling the number of sides in the circumscribing figure.

In addition to these two compression lemmas for the circle, we need two further geometric results concerning the ratio of areas between a circle and a regular polygon:

Berkeley and the Calculus: The Background

1. Similar polygons inscribed in circles are to one another as the squares on their diameters.
2. Similar polygons circumscribed about circles are to one another as the squares on their diameters.

The first of these principles is *Elements* (XII, 1); the second can be proved in an analogous manner. With this machinery in place, we can work a complete exhaustion proof of the following proposition:

Proposition: Circles are to one another as the squares on their diameters.

Let circles C_1 and C_2 be given, and let A_1 and A_2 be their respective areas. Let D_1 and D_2 be the diameters of C_1 and C_2.

To show: $A_1 : A_2 = (D_1)^2 : (D_2)^2$.

Assume not. Then either $A_1 : A_2 < (D_1)^2 : (D_2)^2$ or $A_1 : A_2 > (D_1)^2 : (D_2)^2$.

Let C' be a circle with area A' such that $A' : A_2 = (D_1)^2 : (D_2)^2$. Then either $A' < A_1$ or $A' > A_1$. We show that neither case is possible by a double *reductio ad absurdum*.

Case 1. Assume $A' < A_1$.

If $A' < A_1$, then within C_1 we can inscribe a regular polygon P with area F, such that $A' < F < A_1$, by the first compression lemma for the circle (L_1 above).

Let P' be a similar polygon inscribed in C_2, and let F' be the area of P'.

Then $F : F' = (D_1)^2 : (D_2)^2$, by theorem 1 above.

But $(D_1)^2 : (D_2)^2 = A' : A_2$, by hypothesis. Thus, $F : F' = (D_1)^2 : (D_2)^2 = A' : A_2$.

But because $A' < F$, it follows that $A_2 < F'$.

Thus, the circle C_2 has an area less than that of an inscribed polygon, which is absurd.

Case 2. Assume $A_1 < A'$.

If $A_1 < A'$, then about the circle C_1 we can circumscribe a regular polygon P with area F such that $A_1 < F < A'$, by the second compression lemma for the circle (L_2 above).

Let P' be a similar polygon circumscribed about C_2 and let F' be the area of P'.

Then $F : F' = (D_1)^2 : (D_2)^2$ by theorem 2 above.

But $(D_1)^2 : (D_2)^2 = A' : A_2$, by hypothesis. Thus $F : F' = (D_1)^2 : (D_2)^2 = A' : A_2$.

But $F < A'$, so $F' < A_2$.

Thus, the circle C_2 has an area greater than that of a circumscribed polygon, which is absurd.

Thus, A' can be neither greater nor less than A_1, and the circles are to one another as the squares on their diameters. Q.E.D.

Two points should be clear from this elementary example. First, there is no need to consider infinitely small quantities in an exhaustion proof. Throughout the course of the argument, we refer only to the finite differences between finite magnitudes, and the key steps all involve the consideration of properties of finite ratios. Classical criteria for rigorous geometric demonstration require that a theorem be derived by steps which trace back to the fundamental axioms (in this case, the axioms governing ratios and proportions) and that the axioms themselves be clearly evident to reason. Because the infinite was generally thought to exceed the grasp of the human intellect, the classical criterion of rigor bars demonstrations which rely upon considerations of infinity.

Second, a fully worked out exhaustion proof is a cumbersome (not to say tortuous) chain of argument. A full proof requires specification of a limit toward which a sequence of approximations tend, as well as the compression lemmas which show that an appropriate sequence of approximations can be generated. In general, it is difficult to achieve this. Moreover, the proof requires a complex double *reductio ad absurdum* to fix the value of the result determinately.

These two features of the method of exhaustion were widely acknowledged by mathematicians of the seventeenth century, who agreed that exhaustion proofs were paradigmatically rigorous but complained that the technique was difficult to apply to any but the most simple cases. In particular, the required sequences of approximations can be difficult to generate. It is straightforward (if laborious) to prove the first two of our compression lemmas for the circle, and a similar situation obtains for other conic sections. Archimedes, the master of the exhaustion method, could exploit the known properties of conic sections to obtain quadrature results, as when he takes a series of inscribed triangles to approximate the area of a parabolic segment (Dijksterhuis 1987, chap. 11). But even here a great deal of work is required to show that compression is attained, and the situation is essentially hopeless if we try to apply the strict formulation of the traditional exhaustion proof to problems of quadrature or rectification for more complicated curves. Whiteside (1960–62) observes that the Archimedean approach can be rendered more general by adding a number of lemmas which effectively establish that the compression technique can be applied to any convex curve. Indeed, some seventeenth-century authors took an approach not unlike this and avoided having to establish particular compression lemmas for each problem being solved. Nevertheless, the great majority of mathema-

Berkeley and the Calculus: The Background

ticians of the period found the exhaustion method to be incapable of general application. The difficulties encountered in trying to apply exhaustion techniques to more complex problems eventually inspired the development of infinitesimal methods—one of the most philosophically interesting episodes in the history of mathematics.[4]

Infinitesimal Mathematics

Because theories of infinitesimal magnitudes are so important in the development of seventeenth-century mathematics, it makes sense to give a brief characterization of the infinitesimal before proceeding. A relatively simple problem—that of determining the area of the circle in terms of its radius and circumference—will illustrate the basic idea behind infinitesimal theories and the principal motivation for the introduction of infinitesimals into analysis. Take a circle with radius r and a circumference c and observe that the area of the circle can be interpreted as the sum of the sectors σ_1, σ_2, σ_3, and σ_4, as shown in the top half of figure 4.1. By rearranging the sectors, we can construct a "pseudoparallelogram," whose area can be estimated as $\frac{1}{2}rc$. If we then increase the number of sectors into which we decompose the circle (taking the sectors $\sigma_1 \ldots \sigma_8$), our pseudoparallelogram will more closely resemble a true parallelogram with area $\frac{1}{2}rc$. Now it seems natural to assume that if we could take an infinite number of sectors, we would get a true parallelogram (indeed, a rectangle) in the infinite case with an area of $\frac{1}{2}rc$ or πr^2. In doing this, we effectively treat the circle as consisting of an infinite number of isosceles triangles, each of which has an infinitely small, or infinitesimal, base.

One fundamental question to be asked here is What does it mean for a triangle to have an infinitesimal base? The answer is not at all obvious, since we want to deny that the base has zero length, but also want to deny that the base can be measured by any positive real number. The reason for denying that the base has a zero length is obvious

---

4. Whiteside (1960–62, §9) argues that the mathematicians of the seventeenth century were unduly dismissive in their evaluation of the exhaustion technique. He observes that a completely general proof procedure equivalent to Cauchy-Riemann integration can be extracted from the Archimedean treatment of exhaustion proofs and claims that the standard seventeenth-century evaluation of exhaustions as rigorous but prolix was based upon a failure to appreciate the real power of the method. It is beside my purpose here to determine whether there is a sound basis for the unfavorable evaluation of the exhaustion method in the seventeenth century. I want only to make clear that a typical exhaustion proof was regarded as rigorous, but the method was dismissed as both unnecessarily difficult and incapable of general application.

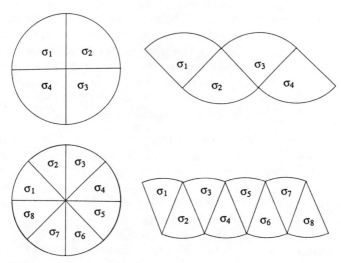

Fig. 4.1

enough: the admission of a zero length would force us to treat the circumference of the circle as an infinite sum of the form $0 + 0 + 0 + \ldots$, which is equal to zero. But to admit that the length of the base can be measured by any positive real number conflicts with our previous claim that the circle is composed of isosceles triangles since if the base of the triangle can be measured by any positive real number ρ, there will always be a tiny area ε left over from each sector (fig. 4.2).

Thus we are led to regard infinitesimals as quantities greater than zero but less than any positive real number. But although they are less than any positive real magnitude, infinitesimals can be greater or less than one another. This characterization may seem bizarre and inconsistent, arousing suspicion that talk of infinitesimals is simply incoherent. After all, to admit quantities which serve as an intermediary

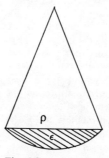

Fig. 4.2

Berkeley and the Calculus: The Background

between something and nothing offends rather grossly against any intuitive conception of magnitudes. As it turns out, contemporary model theory allows for the development of a consistent theory of infinitesimals.[5] The relevance of current accounts of the infinitesimal to issues in the seventeenth and eighteenth centuries is rather minimal, but it does show that it is possible to develop a consistent theory of infinitesimal magnitudes.

The mathematicians of the seventeenth and eighteenth centuries who spoke of taking "infinitely small" quantities in the course of solving problems often left the central concept unanalyzed and largely bereft of theoretical justification. Nevertheless, there are some salient features of infinitesimals that were generally accepted by those authors who employed infinitesimal methods. One way to characterize infinitesimals is as magnitudes which stand in the same ratio to a finite magnitude as any finite magnitude stands to an infinite magnitude. The intent of this definition is to capture the two properties we noted above as characterizing the infinitesimal—being greater than zero but less than any positive real number. Treated as ratios of positive finite numbers to an infinite number, infinitesimals can be treated as greater than zero (since they are ratios of *positive* magnitudes) and at the same time less than any positive real number (since any real number can be expressed as the ratio of *finite* magnitudes). The main difficulty is that the definition requires us to make sense of the notion of a ratio between finite and infinite: something quite explicitly barred by the classical account of magnitudes. A new theory of ratios which admits infinite magnitudes is by no means impossible, but there was no serious attempt made in the seventeenth and eighteenth centuries to work out the details of such a theory.

Infinitesimals were treated as quantities which could be disregarded when added to any finite magnitude. Thus, an infinitesimal ε was effectively governed by the law $\rho + \varepsilon = \rho$ for any finite real number ρ. Similarly, the product of an infinitesimal and a finite real number was regarded as an infinitesimal, which again could be disregarded when added to any finite magnitude. The multiplication of two infinitesimals was regarded by many as yielding another kind of infinitesimal altogether—the second-order infinitesimal.[6] Infinitesi-

---

5. The details of the theory can be found in Robinson (1965). See also Keisler (1976) for a presentation of the theory at the level of an introductory text in the calculus.

6. On those accounts which admit second-infinitesimals, the repeated multiplication of infinitesimals generates infinitesimals of the third and fourth order and so on in the obvious way. Modern presentations of nonstandard analysis depart from the traditional doctrine of infinitesimals in this respect. In present-day accounts, infinitesimals appear

mals of the second order were treated as greater than zero but less than any infinitesimal of the first order, in exact analogy with the relationship between first-order infinitesimals and finite magnitudes. Again, the rules for the combination of infinitesimals of different orders typically treated those of higher orders as negligible and they were routinely discarded when they appeared in sums which involved infinitesimals of a lower order.

Whatever conceptual problems they may present, infinitesimals were certainly of great importance in the advance of analysis. By exploiting infinitesimal techniques, mathematicians of the period obtained a great number of new results, even though the attitude toward infinitesimals was frequently one of ambivalence.[7] Indeed, some authors were quite happy to introduce this new kind of magnitude and regarded the theory of infinitesimal magnitudes as something that could be rigorously established. Others claimed that infinitesimal techniques can, at least in principle, be reduced to the classical method of exhaustion and regarded infinitesimals as a convenient abbreviation for the more complicated classical proofs. Still others sought to avoid infinitesimals altogether and presented their results in a form which, so they claimed, avoided the supposition of infinitely small magnitudes. Berkeley rejected the entire theory of infinitesimals, claiming that it violated principles of sound reasoning and produced numerous disputes and controversies. But before we turn to an investigation of Berkeley's critique of infinitesimal methods, we must outline the salient features of three important seventeenth-century approaches to analysis—the "method of indivisibles," the Leibnizian *calculus differentialis,* and the Newtonian calculus of fluxions. The remainder of this chapter will therefore be a brief exposition of these three targets of Berkeley's criticisms.

The Method of Indivisibles

As I have indicated, the advance of analysis in the seventeenth century was brought about largely by the search for alternatives to the method of exhaustion. One of these alternatives was the "method of indivisibles" developed by Cavalieri and extended by Wallis and others. Proponents of the new method saw it as a technique of far greater

---

as "hyperreal" numbers in certain nonstandard models of arithmetic, and contemporary accounts of hyperreal numbers define the product of two hyperreals as a hyperreal number.

7. See Jesseph (1989) for a study of some of these ambivalent attitudes.

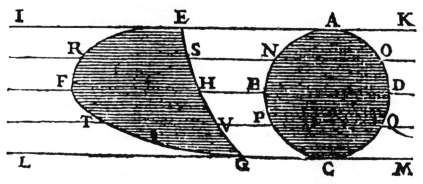

Fig. 4.3 From Cavalieri, *Exercitationes Geometricae Sex* (1647, 4).

scope and power than the exhaustion proof, and judicious application of the method yielded a large number of new and interesting results.

Cavalieri's major work, *Geometria indivisibilibus continuorum nova quadam ratione promota* (1635), promised to advance geometry by using the "indivisibles of the continuum" and laid the groundwork for later work in the infinitesimal calculus. The key to Cavalieri's treatment of quadratures is his concept of the indivisibles of a given figure, which he introduces in book 2 of the *Geometria*. If we take the figures *ABCD* and *EFGH* (see fig. 4.3), we can compare their areas by taking the line *LM* (called the *regula* in Cavalieri's terminology) and considering the transit of a line parallel to the *regula* through the two figures. The intersection of the moving line and the figures will produce "all the lines" in these figures or "the indivisibles of the figure."[8]

With the introduction of indivisibles, Cavalieri parted company with the classical tradition. No matter how we interpret the idea of the transit of a line producing "all the lines" of a figure, we are compelled to acknowledge that there are an infinity of lines produced by this transit and that each indivisible is infinitely small in comparison to the figure. Moreover, there is a strong temptation to regard the figure as composed of "all the lines," and such an approach seems to demand that the figure be seen as an infinite sum of infinitesimal

---

8. In an exactly analogous manner, the indivisibles of a line could be produced by passing a point through it, or the indivisibles of a solid can be generated by passing a plane through it. I will deal only with the two-dimensional case, however. Cavalieri also distinguished between the indivisibles formed by a transit of the *regula* parallel to the axis (called *recti transitis*) and by nonparallel transit (*obliqui transitis*). None of these distinctions will be of interest to us, although they appear in the statement of the key theorems. Studies of Cavalieri's work can be found in Andersen (1985), De Gandt (1991), and Giusti (1980).

elements. By stating his theory in the kinematic language of the transit of a line "producing" the indivisibles of the figure, Cavalieri seems intent on sidestepping questions concerning the nature of the infinite by casting his account in language that takes account of an intuitive understanding of continuous motion. Nevertheless, it is not obvious how to go about comparing the indivisibles of two figures or how such comparison can help to find quadratures without implicitly assuming that the figure is literally composed of an infinite number of indivisibles. Indeed, as we will see, this is exactly how Cavalieri's theory was interpreted.

Cavalieri was certainly aware of the conceptual problems posed by his nonclassical approach to analysis, but he hoped to resolve the difficulties by incorporating his theory within the classical theory of magnitudes. The fundamental claim in his *Geometria* is that the indivisibles of a figure comprise an entirely new species of magnitude, on a par with more familiar magnitudes such as lines, surfaces, and angles. True to this proposal, Cavalieri's principal line of attack in the solution of quadrature problems is to compare the ratios between the indivisibles of two figures and then to use these ratios of indivisibles as a means of establishing a proportion which will determine the ratio between the areas of the figures.[9] Thus, a fundamental part of Cavalieri's program is to state principles which will allow the indivisibles of a figure to be worked into the general theory of magnitudes in book V of Euclid's *Elements*.[10]

Cavalieri's attempt to expand the Euclidean treatment of magnitudes met with hostility from some quarters[11] but was enthusiastically

---

9. Indeed, in a corollary to theorem 3 of book 2, he declares: "It is clear from this that when we want to find what ratio two plane figures or two solids have to one another, it is sufficient for us to find what ratio all of the lines of the figure stand in (and in the case of solids, what ratio holds between all of the planes), relative to a given *regula*, which I lay as the great foundation of my new geometry" (Cavalieri 1635, 115).

10. Cavalieri declares, for example, in theorem 1 of book 2 that "All the lines *recti transitus* of any plane figures, and all the planes of any solids, are magnitudes having a ratio to one another" (Cavalieri 1635, 108). The proof begins with the postulate that the indivisibles of congruent figures are congruent, and then proceeds to consider noncongruent figures by considering remainders and forming sums. In this respect, the result is rather question-begging, since the entire procedure assumes that the indivisibles of figures can be compared with each other by a relation of equality, and that they can be ordered by a less-than relation. De Gandt (1991) and Andersen (1985) have some interesting observations on this aspect of Cavalieri's work and the extent to which it can be fit within the Euclidean model.

11. The most hostile reaction came from the Jesuit Paul Guldin, whose *Centrobaryca* contained an extended polemic against Cavalieri's methods. In rejecting Cavalieri's claim that the indivisibles of a figure are magnitudes (Cavalieri 1635, book 2, theo-

Berkeley and the Calculus: The Background

received by others. The most telling example of this acceptance is in the work of Wallis, who abandoned Cavalieri's cautious treatment of the infinite in favor of an unabashed acceptance of infinitesimal methods and regarded the foundational issues as having been settled by Cavalieri. Thus, in his 1670 *Mechanica: sive de Motu, Tractatus Geometricus*, he states as a definition, "It is understood that any continuum (according to the Geometry of Indivisibles of Cavalieri) consists of an infinite number of indivisibles" (Wallis 1693–99, 1:645). He then elaborates upon this definition and links the method of indivisibles with the classical method of exhaustion:

> As from an infinity of points, a line; a surface from an infinity of lines, and a solid from an infinite number of surfaces; so also from an infinity of temporal moments, time, etc.
> That is, . . . from homogeneous particles, infinitely small and infinite in number, and equal in at least one dimension. . . .
> This means (according to mathematical rigor) that there can at least be inscribed or circumscribed, or otherwise adapted, something composed of particles of this sort which will differ from the given thing by an infinitely small quantity, or a quantity less than any given quantity.
> For example: in the periphery of a circle, a curve will be composed of an infinite number of arcs which will coincide exactly with the periphery. But there can be inscribed within the same periphery a line composed of an infinite number of subtenses, or circumscribed about it a line composed of an infinite number of tangents, such that the inscribed will be less than the periphery, or the circumscribed line will be greater than the periphery by a difference which is less than any given amount. (Wallis 1693–99, 1:646)

Indeed, Wallis took infinitesimal methods to be essentially the same as the method of exhaustion but shorter and more readily applied. Thus, in his *Treatise of Algebra* he asserts:

> The Method of Exhaustions, (by Inscribing and Circumscribing Figures, till their Difference becomes less than any assignable,) is a little disguised, in (what hath been called) *Geometria Indivisibilium* . . . which is not, as to the Substance of it,

---

rem 1), Guldin exclaims: "This proposition is entirely false, and I oppose it with this argument: all the lines and all the planes of one another figure are infinite and infinite. But there is no proportion or ratio of an infinite to an infinite. Therefore, etc. Both the major and minor premises are clear to all geometers and so do not need many words. Therefore the conclusion of Cavalieri's proposition is false" (Guldin 1635–41, 4:341).

really different from the Method of Exhaustions, (used both by Ancients and Moderns,) but grounded on it, and demonstrable by it: But is only a shorter way of expressing the same Notion in other Terms. (Wallis 1685, 285)

Rather than follow Cavalieri's cautious lead, Wallis applied arithmetical results from the theory of infinite series to the solution of geometric problems, treating geometric objects as infinite sums of infinitesimal elements. For example, his quadrature of the cubic parabola in *Arithmetica Infinitorum* begins, in proposition 39, with arithmetical results, namely:

$$\frac{0+1}{1+1} = \frac{1}{2} = \frac{2}{4} = \frac{1}{4} + \frac{1}{4}$$

$$\frac{0+1+8}{8+8+8} = \frac{9}{24} = \frac{3}{8} = \frac{1}{4} + \frac{1}{8}$$

$$\frac{0+1+8+27}{27+27+27+27} = \frac{36}{108} = \frac{4}{24} = \frac{1}{4} + \frac{1}{12}$$

$$\frac{0+1+8+27+64}{64+64+64+64+64} = \frac{100}{320} = \frac{5}{16} = \frac{1}{4} + \frac{1}{16}$$

From these initial cases, Wallis concludes "by induction" that as the number of terms in the sum increases, the ratio approaches arbitrarily near to the ratio $1:4$. Proposition 41, which he takes to follow obviously from proposition 39, asserts that:

If an infinite series is taken of quantities in triplicate ratio to a continually increasing arithmetical progression, beginning with 0 (or, equivalently, if a series of cube numbers is taken) this will be to the series of numbers equal to the greatest and equal in number as one to four. (Wallis 1693–99, 1:382–83)

Given this result, Wallis turns to the quadrature of the cubic parabola (fig. 4.4), treating it as an infinite sum of lines forming a series of cubic quantities:

And indeed let *AOT* (with diameter *AT*, and corresponding ordinates *TO, TO, &c.*) be the complement of the cubic semi-parabola *AOD* (with diameter *AD* and corresponding ordinates *DO, DO, &c.*). Therefore, (by Proposition 45 of the *Treatise of Conic Sections*) the right lines *DO, DO, &c.* or their equals *AT, AT, &c.* are in subtriplicate ratio of the right lines *AD, AD, &c.* or their equals *TO, TO, &c.* And conversely these

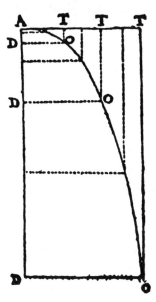

Fig. 4.4 From Wallis, *Arithmetica Infinitorum,* Proposition 42 (1693–99, 1:383).

TO, TO, &c. are in triplicate ratio of the right lines *AT, AT, &c.* Therefore the whole figure *AOT* (consisting of the infinity of right lines *TO, TO, &c.* in triplicate ratio of the arithmetically proportional right lines *AT, AT, &c.*) will be to the parallelogram *TD* (consisting of just as many lines all equal to *TO*) as one to four. Which was to be shown. And consequently the semiparabola *AOD* (the residuum of the parallelogram) is to the parallelogram itself as one to four. (Wallis 1693–99, 1:383)

What is interesting here is Wallis's conception of surfaces as composed of indivisibles and his application of algebraic and arithmetical methods for the summation of infinite series to problems in geometry. Note also that his method departs from the classical approach to geometry because it fails to observe the classical distinction between discrete and continuous magnitudes. By treating a continuous geometric figure as composed of sums of discrete points or lines, the method of indivisibles simply ignores the classical distinction. Moreover, the method's dependence upon infinite summations and the introduction of infinitely small magnitudes is a complete departure from the finistic viewpoint codified in the classical theory of ratios. A further point of interest is the almost casual way in which Wallis takes his key arithmetical results to have been established "by induction" after investigating only a few initial cases. Thus, as we saw above, Wallis asserted the result

$$\lim_{n\to\infty}\left(\frac{\Sigma_{k=0}^n k^3}{n^3(n+1)}\right) = \frac{1}{4}$$

(to use the modern notation) on the basis of an examination of the first five cases. Indeed, his technique for solving infinite sums amounts to a wholesale abandonment of the classical standard of rigor. In proposition 1 of his *Arithmetica Infinitorum* he declares:

> The easiest way of investigating this and some of the following problems is to consider a certain number of cases and observe the resulting ratios, and compare them with one another in order that the universal proposition can then be known by induction. (Wallis 1693–99, 1:365)

These inductive leaps to the infinite case are made without benefit of principles which guarantee that the stated result actually holds, but Wallis apparently found no difficulty in asserting his theorems after an enumeration of the initial cases. As might be suspected, Berkeley found such a procedure to violate the standards of rigor appropriate to mathematical demonstration.

Leibniz and the Differential Calculus

The powerful *calculus differentialis* of Leibniz and his followers extended the basic ideas behind the method of indivisibles and resulted in a powerful algorithmic technique for the solution of problems in analysis. The key concept in the differential calculus is that of the *difference* of a variable. The difference of the variable x (expressed dx) is defined as the difference between two values of the variable which are infinitely close to one another. Two differences, although themselves infinitesimal quantities, can be compared with one another to form finite ratios, so that the expression dy/dx is a finite quantity expressing the relationship between the infinitesimal differences dy and dx.

The best way to understand the Leibnizian differential calculus is to begin by observing how finite differences can be used to give approximate solutions to problems of tangency and quadrature. If we take the curve $\alpha\beta$ with ordinate y and abscissa x (fig. 4.5) we can approximate the tangent at any point and the area under the curve between any two points by dividing the abscissa into a finite collection of equal subintervals, say $x_1, x_2, \ldots, x_7$. The tangent at point p can be

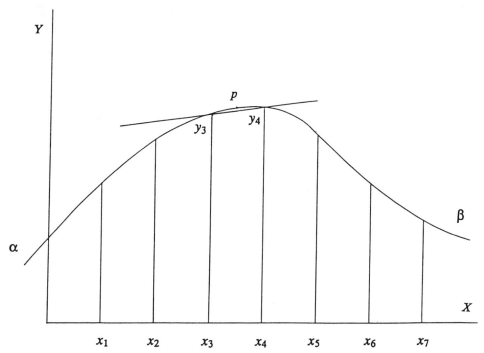

Fig. 4.5

approximated by taking the line through points y_3 and y_4, corresponding to the points x_3 and x_4. Similarly, the quadrature of the curve can be approximated by summing up the rectangles whose bases are the subintervals of the abscissa. Leibniz proposes to transform such approximations into exact results by taking infinitesimal differences between the variables x and y. In the infinitesimal case, we will have a "differential triangle" at point p, whose legs dx and dy determine the slope of the tangent as dy/dx. Alternatively, we can treat the tangent construction as being obtained by considering the differential increments of y and x. At any point on the curve, the tangent can be constructed by taking $((y + dy) - y)/dx$ and then canceling out the terms y to obtain the result dy/dx, as before.

The Marquis de l'Hôpital's 1696 *Analyse des infiniment petits pour l'intelligence des lignes courbes* is the most forthright statement of the foundations of the differential calculus. He opens with definitions of the two key concepts, making no apology for his use of infinitesimals:

> DEFINITION I: *Variable* quantities are those which increase or diminish continually; and *constant* quantities are those which

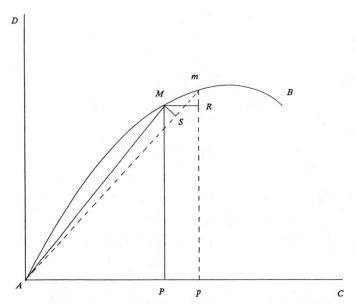

Fig. 4.6 Based on L'Hôpital, *Analyse des infiniment petits* (1696), postulate 1.

remain the same while others change. Thus in a parabola the ordinate and abscissa are variable quantities, while the parameter is a constant quantity.

DEFINITION II: The infinitely small portion by which a variable quantity continually increases or diminishes is called the *difference*. Let there be, for example, any curved line *AMB*, which has the line *AC* as its axis or diameter, and the right line *PM* as one of its ordinates, and let *pm* be another ordinate infinitely near to the former [fig. 4.6]. This being granted, if *MR* is drawn parallel to *AC*, and the chords *AM*, *Am* are drawn; and about the center *A* with distance *AM* the small circular arc *MS* is described: then *Pp* will be the difference of *AP*, *Rm* that of *PM*, *Sm* that of *AM*, and *Mm* that of the arc *AM*. And similarly, the small triangle *MAm* which has as its base the arc *Mm* will be the difference of the segment *AM;* and the small space *MPpm* will be the difference of the space contained by the right lines *AP*, *PM*, and the arc *AM*. (L'Hôpital 1696, 1–2)

These definitions are then supplemented with two postulates:

POSTULATE OR SUPPOSITION: It is postulated that one can take indifferently for one another two quantities which differ from one another by an infinitely small quantity: or (which is the same thing) that a quantity which is augmented or diminished by another quantity infinitely less than it, can be considered as if it remained the same. It is postulated, for example, that one can take *Ap* for *AP*, *pm* for *PM*, the space *Apm* for the space *APM*, the small space *MPpm* for the rectangle *MPpR*, the small sector *AMm* for the small triangle *AMS*, the angle *pAm* for the angle *PAM*, etc.

POSTULATE OR SUPPOSITION: It is postulated that a curved line can be considered as an infinite collection of right lines, each infinitely small: or (which is the same thing) as a polygon of an infinite number of sides, each infinitely small, which determine the curvature of the line by the angles they make with one another. It is postulated, for example, that the portion of the curve *Mm* and the arc of the circle *MS* can be considered as right lines because they are infinitely small, so that the small triangle *mSM* can be supposed to be rectilinear. (L'Hôpital 1696, 2–3)

The great strength of the calculus lies in its generality, and the differential calculus reduces the problem of tangency to a simple algorithm (known as differentiation) for curves which can be represented analytically by an equation. For example, suppose we take the equation

$$y = x^3 + 5x^2 - 4x + 1. \tag{4.1}$$

Using the differences dy and dx, we treat the tangent as a straight line coincident with one of the infinitely small right lines which compose the curve. We can substitute infinitesimal increments into the equation for the curve in order to find the relationship between dy and dx, and their ratio will give us the equation for the tangent. Substituting $x + dx$ and $y + dy$ for x and y in equation (4.1) we obtain

$$(y + dy) = (x + dx)^3 + 5(x + dx)^2 - 4(x + dx) + 1.$$

Expanding yields

$$(y + dy) = x^3 + 3x^2dx + 3xdx^2 + dx^3 + 5x^2 \\ + 10xdx + 5dx^2 - 4x - 4dx + 1. \tag{4.2}$$

Now, equation (4.2) represents (4.1) augmented by the increments dy and dx, but the increment itself can be obtained by subtracting (4.1) from (4.2). This yields

$$dy = 3x^2dx + 3xdx^2 + dx^3 + 10xdx + 5dx^2 - 4dx. \tag{4.3}$$

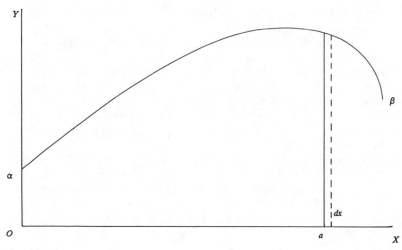

Fig. 4.7

Simplifying (4.3) by dividing through by dx will give the ratio of the two increments dy and dx, which is expressed

$$\frac{dy}{dx} = 3x^2 + 3xdx + dx^2 + 10x + 5dx - 4. \qquad (4.4)$$

But because dx is infinitely small when compared to x, we can drop the terms containing it from the right side of equation (4.4) and retain only terms in x, which results in

$$\frac{dy}{dx} = 3x^2 + 10x - 4$$

and this equation gives the tangent at any point on the curve.

Even more strikingly, the calculus can relate the problems of quadrature and tangency by showing that computing a quadrature and finding a tangent are inverse operations. This inverse relationship can be brought out by treating the area under the curve $\alpha\beta$ on the interval $[0, a]$ as the sum of the ordinates y over the interval (fig. 4.7). Using the familiar notation for integration, this area can be written $\int_0^a y dx$. Alternatively, the integral can be treated as the sum of the infinitely narrow rectangles formed by the ordinate y and the differential dx. To show that quadrature and tangency are inverse operations, we introduce a new curve defined as the integral with a variable upper bound x. Using the modern notation of functions, this curve will be given by the equation $G(x)$, where

$$G(x) = \int_0^x y\,dx.$$

Then the tangent to this curve $G(x)$ at an arbitrary point with abscissa x_0 can be written

$$\frac{G(x_0 + dx) - G(x_0)}{dx}.$$

But $G(x_0 + dx)$ is the area between 0 and $x_0 + dx$, and $G(x_0)$ represents the area between 0 and x_0. Similarly, $G(x_0 + dx) - G(x_0)$ gives the area between x_0 and $x_0 + dx$. Then

$$G(x_0 + dx) - G(x_0) = y\,dx. \tag{4.5}$$

But dividing through equation (4.5) by dx gives

$$\frac{G(x_0 + dx) - G(x_0)}{dx} = \frac{y\,dx}{dx} = y.$$

Thus, the instantaneous rate of change of the function which represents the area under the curve $\alpha\beta$ (or the tangent to the curve which represents the area under the original curve $\alpha\beta$) is the same as the ordinate of the curve $\alpha\beta$. Using the analytic representations of the curves, we can express this inverse relationship by saying that the equation which expresses the area under the curve $\alpha\beta$ on the interval [0, x] is one that, when differentiated, yields the ordinate of the curve $\alpha\beta$ at x.

The Newtonian Method of Fluxions

I come now to a treatment of one of Berkeley's principal targets in his critique of the calculus—the Newtonian method of fluxions. The most important element in Newton's presentation of the calculus is the conception of geometric magnitudes as produced by continuous motion. This kinematic conception is not a Newtonian innovation; it dominates Barrow's *Lectiones Geometricæ*[12] and can be found in other

---

12. See Barrow (1670, lectures 1 and 2), where he considers the "generation of magnitudes."

writings as well. Indeed, the view that lines are produced by the continuous motion of a point was common in antiquity. The classical definitions of the "mechanical" curves such as the quadratrix are almost always given in terms of the continuous motion of a point. Barrow declares in the second of his *Lectiones Geometricæ* that geometric magnitudes such as lines, surfaces, and solids are properly conceived as being produced by continuous local motion. Thus, the motion of a point traces a line, the motion of a line sweeps out a surface, and the motion of a surface produces a solid.

Newton (who helped prepare Barrow's *Lectiones Geometricæ* for publication) was obviously familiar with this treatment of geometric magnitudes and came to adopt it in his presentation of the calculus. In the introductory paragraph to his important treatise *On the Quadrature of Curves,* he explicitly opposes this theory to one which treats magnitudes as composed of infinitesimal parts:

> I don't here consider Mathematical Quantities as composed of Parts *extreamly small,* but as *generated by a continual motion.* Lines are described, and by describing are generated, not by any apposition of Parts, but by a continual motion of Points. Surfaces are generated by the motion of Lines, Solids by the motion of Surfaces, Angles by the Rotation of their Legs, Time by a continual flux, and so in the rest. These *Geneses* are founded upon Nature, and are every Day seen in the motion of Bodies. (Newton 1964–67, 1:141)

An example will clarify the role of this kinematic conception in the fluxional calculus. Suppose we have the curve $\alpha\beta$, generated by the continuous motion of a point (fig. 4.8). As the point traces out the curve, its velocity can be resolved into two components $\dot{x}$ and $\dot{y}$, parallel to the axes OX and OY. Taken together, the components $\dot{x}$ and $\dot{y}$ give the instantaneous velocity of the point at any stage in the generation of the curve $\alpha\beta$ or, in Newton's terminology, the *fluxion* of the flowing quantity (or "fluent") $\alpha\beta$. If we take a line of length κ, we can form the products $\dot{x}\kappa$ and $\dot{y}\kappa$ which will stand in the same proportion as the components $\dot{x}$ and $\dot{y}$. Now consider the instantaneous velocity at the point (x_1, y_1) on the curve. The tangent to the curve at (x_1, y_1) can be found by completing the parallelogram formed by the lines $\dot{x}_1\kappa$ and $\dot{y}_1\kappa$ at (x_1, y_1) and taking its diagonal. The slope of the tangent at (x_1, y_1) is thus $\dot{y}_1/\dot{x}_1$. Given this conception of curves, the next step in the Newtonian development of the calculus is to devise a general method for computing tangents and then to extend this method to solve a number of important traditional problems in analysis. Tangency problems thus become problems of finding the fluxions $\dot{x}$ and

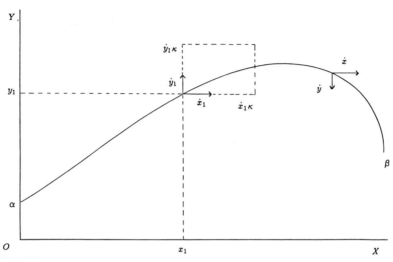

Fig. 4.8

$\dot{y}$ when given an equation which describes the relationship between the fluents x and y. A quadrature will be an inverse problem, that of determining the fluents when the fluxions are given.

In solving these problems, Newton developed two devices—the doctrine of moments and the theory of prime and ultimate ratios. The moment of a fluent is defined as its "momentaneous synchronal increment," or the amount by which a fluent is increased in an "indefinitely small" period of time. These periods of time are represented by the symbol "o," and the moment of the fluent x by $o\dot{x}$. Thus, the fluents x and y will be augmented by their moments in an indefinitely small period of time to become $x + o\dot{x}$ and $y + o\dot{y}$.

The theory of prime and ultimate ratios is closely tied to the kinematic conception of magnitudes and involves the consideration of ratios between magnitudes as they are generated by motion. The prime ratios of nascent magnitudes are those which hold as the magnitudes are just beginning to be generated, while the ultimate ratios of evanescent magnitudes are ratios holding between magnitudes which diminish to nothing and vanish. Newton gave several expositions of this theory, with one fairly straightforward account, illustrated by figure 4.9, in the Introduction to the *Quadrature of Curves:*

> Fluxions are very nearly as the Augments of the Fluents, generated in equal, but infinitely small parts of Time; and to speak exactly, are in the *Prime Ratio* of the nascent Augments: but they may be expounded by any Lines that are propor-

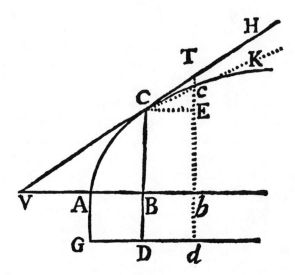

Fig. 4.9 From Newton, *Treatise on the Quadrature of Curves,* Introduction (Newton 1964–67, 1:141).

tional to 'em. As if the *Areas ABC, ABDG* be described by the Ordinates *BC, BD*, moving with an uniform motion along the Base *AB*, the Fluxions of these *Areas* will be to one another as the describent Ordinates *BC* and *BD*, and may be expounded by those Ordinates; for those Ordinates are in the same Proportion as the Nascent Augments of the Areas. Let the Ordinate *BC* move out of its Place *BC* into any new one *bc:* Compleat the Parallelogram *BCEb*, and let the Right Line *VTH* be drawn which may touch the Curve [at] *C* and meet *bc* and *BA* produced in *T* and *V;* and then the just now generated Augments of the Abscissa *AB*, the Ordinate *BC*, and the Curve Line *ACc*, will be *Bb, Ec* and *Cc;* and the Sides of the Triangle *CET*, are in the *Prime Ratio* of these Nascent Augments, and therefore the Fluxions of *AB, BC* and *AC* are as the Sides *CE, ET* and *CT* of the Triangle *CET*, and may be expounded by those Sides, or which is much at one, by the Sides of the Triangle *VBC* similar to it. (Newton 1964–67, 1:141)

What we find here is a procedure that appears to avoid the use of infinitesimal quantities, since it apparently gives us a way of reasoning about fluxions that allows the fluxions to be "expounded by lines proportional to 'em." Whether, in fact, the method relies upon infinitesimal considerations is a question that I will leave open for the present,

Berkeley and the Calculus: The Background

except to note that Berkeley's comments in the *Analyst* show that he viewed the theory of prime and ultimate ratios as a clever bit of mathematical trickery designed to disguise the use of infinitesimals in the Newtonian calculus. The chief problem with the doctrine of ultimate ratios is that it simply takes for granted that there is such a thing as an ultimate ratio between two quantities which vanish—an assumption that is certainly less than obvious.[13]

By exploiting the kinematic conception of curves, the doctrine of moments, and the theory of prime and ultimate ratios, Newton developed an algorithmic procedure for solving problems in analysis and showed that quadratures are the inverse of such tangency problems. In his *Treatise of the Method of Fluxions and Infinite Series* (1736), these two fundamental problems of the calculus are stated as follows:

> Now it remains, that for an illustration of the Analytic Art, I should give some specimens of Problems, especially such as the nature of Curves will supply. Now in order to this, I shall observe that all the difficulties hereof may be reduced to these two Problems only, which I shall pose, concerning a Space describ'd by local Motion, any how accelerated or retarded.
>
> I. The length of the space describ'd being continually (that is, at all times) given; to find the velocity of the motion at any time propos'd.
>
> II. The velocity of the motion being continually given; to find the length of the Space describ'd at any time propos'd. (Newton 1964–67, 1:48–49)

---

13. Newton recognized that his doctrine might meet with skepticism. In the *Principia* he writes: "Perhaps it may be objected that there is not ultimate proportion of evanescent quantities; because the proportion, before the quantities have vanished, is not the ultimate, and when they are vanished, is none. But by the same argument it may be alleged that a body arriving at a certain place, and there stopping, has no ultimate velocity; because the velocity, before the body comes to the place, is not its ultimate velocity; when it has arrived, there is none. But the answer is easy; for by the ultimate velocity is meant that with which the body is moved, neither before it arrives at its last place and the motion ceases, nor after, but at the very instant it arrives; that is, that velocity with which the body arrives at its last place, and with which the motion ceases. And in like manner, by the ultimate ratio of evanescent quantities is to be understood the ratio of the quantities not before they vanish, nor afterwards, but with which they vanish. In like manner the first ratio of nascent quantities is that with which they begin to be. And the first or last sum is that with which they begin and cease to be (or to be augmented or diminished). There is a limit which the velocity at the end of the motion may attain, but not exceed. This is the ultimate velocity. And there is the like limit in all quantities and proportions that begin and cease to be. And since such limits are certain and definite, to determine the same is a problem strictly geometrical. But whatever is geometrical we may use in determining and demonstrating any other thing that is also geometrical" (*Principia* I, 1, Scholium).

Chapter Four

The first of these problems concerns the construction of tangents and can be reduced to a simple algorithm. The second deals with the problem of quadratures; it does not have a simple algorithmic solution, but Newton shows how a large number of cases can be solved algorithmically.

Newton's *Quadrature of Curves* contains a discussion of both problems, and it is important to consider his approach to some basic results in the fluxional calculus in some detail before proceeding. The presentation of the calculus in the *Quadrature of Curves* begins with the problem, *"Having given an Equation involving any number of fluent or flowing Quantities, to find their Fluxions,"* with the solution stated in the following algorithm:

> Multiply every Term of the Equation by the Index of the Power of each flowing Quantity contained in that Term, and in each Multiplication change the Root of the Power into its Fluxion; and then the Aggregate of all the Products under their proper Signs will be the new Equation. (Newton 1964–67, 1:143)

Newton's thesis here is that for any analytic representation of a curve in the form of a polynomial

$$a_0 x^n + a_1 x^{(n-1)} y + \ldots + a_{(n-1)} x y^{(n-1)} + a_n y^n = 0 \qquad (4.6)$$

with flowing quantities x and y and constants a_k, we first replace each occurrence of x^m in equation (4.6) with $m x^{(m-1)} \dot{x}$ to obtain the new equation

$$a_0 n x^{(n-1)} \dot{x} + a_1 (n-1) x^{(n-2)} \dot{x} y + \ldots + a_{(n-1)} \dot{x} y^{(n-1)} + 0 = 0 \qquad (4.7)$$

and then replace each occurrence of y^m with $m y^{(m-1)} \dot{y}$, obtaining

$$0 + a_1 x^{(n-1)} \dot{y} + \ldots + a_{(n-1)} (n-1) x y^{(n-2)} \dot{y} + n a_n y^{(n-1)} \dot{y} = 0. \qquad (4.8)$$

Combining equations (4.7) and (4.8), we have

$$a_0 n x^{(n-1)} \dot{x} + a_1 (n-1) x^{(n-2)} \dot{x} y + \ldots + a_{(n-1)} \dot{x} y^{(n-1)}$$
$$+ a_1 x^{(n-1)} \dot{y} + \ldots + a_{(n-1)} (n-1) x y^{(n-2)} \dot{y} + n a_n y^{(n-1)} \dot{y} = 0. \qquad (4.9)$$

Equation (4.9) determines the relation of the fluxions $\dot{x}$ and $\dot{y}$ to the flowing quantities x and y, or the total differential of x and y.

· 149 ·
Berkeley and the Calculus: The Background

Thus far, the whole procedure for finding fluxions is little more than an algorithm to be taken on faith. Newton was clearly aware that some justification for the algorithm is necessary, and he provides as an example the equation

$$x^3 - xy^2 + a^2z - b^3 = 0 \qquad (4.10)$$

with fluents x, y, and z and constants a and b. He begins by adding the moments $o\dot{x}$, $o\dot{y}$, and $o\dot{z}$ of the flowing quantities as increments into equation (4.10). This yields

$$(x + o\dot{x})^3 - (x + o\dot{x})(y + o\dot{y})^2 + a^2(z + o\dot{z}) - b^3 = 0$$

which expands to produce

$$x^3 + 3x^2o\dot{x} + 3xo^2\dot{x}^2 + o^3\dot{x}^3 - xy^2 - o\dot{x}y^2 - 2xo\dot{y}y$$
$$- 2\dot{x}o^2\dot{y}y - xo^2\dot{y}^2 - \dot{x}o^3\dot{y}^2 + a^2z + a^2o\dot{z} - b^3 = 0. \qquad (4.11)$$

Equation (4.11) represents equation (4.10) plus the increments $\dot{x}o$, $\dot{y}o$, and $\dot{z}o$. The difference between (4.11) and (4.10) is thus the increment of the original equation. Subtracting (4.10) from (4.11) yields

$$3x^2o\dot{x} + 3xo^2\dot{x}^2 + o^3\dot{x}^3 - o\dot{x}y^2 - 2xo\dot{y}y$$
$$- 2\dot{x}o^2\dot{y}y - xo^2\dot{y}^2 - \dot{x}o^3\dot{y}^2 + a^2o\dot{z} = 0.$$

Dividing by o, we obtain:

$$3x^2\dot{x} + 3xo\dot{x}^2 + o^2\dot{x}^3 - \dot{x}y^2 - 2x\dot{y}y - 2\dot{x}o\dot{y}y$$
$$- xo\dot{y}^2 - \dot{x}o^2\dot{y}^2 + a^2\dot{z} = 0.$$

If we now "let the quantity o be lessened infinitely" and neglect the "evanescent terms" which contain o as a factor, we obtain

$$3x^2\dot{x} - \dot{x}y^2 - 2x\dot{y}y + a^2\dot{z} = 0$$

as the equation which determines the fluxion of the original equation. This procedure can then be extended to an algorithm exactly analogous to the procedure of differentiation in the Leibnizian calculus.

The second problem—determining a fluent, given the fluxion—proceeds inversely and can be used to determine quadratures in essentially the same way that the Leibnizian calculus takes integration as the inverse of differentiation. The quadrature of the curve $\alpha\beta$

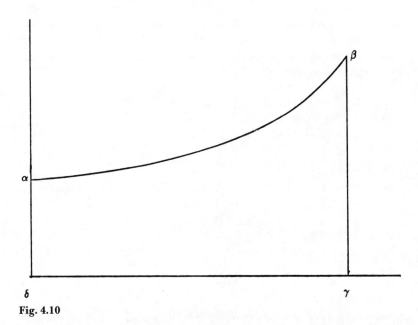

Fig. 4.10

(fig. 4.10) will yield the area $\alpha\beta\gamma\delta$ bounded by the curve, the abscissa and the two ordinates. According to the kinematic conception of magnitudes, the area is swept out by the ordinate moving from $\delta\alpha$ to $\gamma\beta$. The equation for the curve gives us the value of the ordinate at any point in its traversal of the abscissa, and the fluxion of the area will be the ordinate since this is the rate of the increase of the area. Thus, if we are given the equation for the curve we are thereby given the equation for the fluxion of the area, and the quadrature of the curve can be obtained by finding an equation whose fluxion is the equation of the curve $\alpha\beta$.

This is the fundamental theorem of the calculus, which shows the problems of quadrature and tangency to be inversely related. Given an algorithm for determining fluxions, the fundamental theorem allows an extraordinarily wide range of classical problems in analysis to be solved and opens up a whole class of new problems. The mathematical significance of this new method can hardly be overestimated, and Berkeley's attack on the foundations of the calculus amounts to an assault on the most important body of mathematical work in the seventeenth and eighteenth centuries.

My investigation is most concerned with the apparent reliance of Newton's fluxional calculus on infinitesimal considerations. All of the demonstrations in which Newton presents the algorithm for finding

Berkeley and the Calculus: The Background

fluxions include a maneuver which attracted Berkeley's attention. As described, finding a fluxion begins with taking a "very small" increment o and constructing from it the moments $o\dot{x}$, $o\dot{y}$, and $o\dot{z}$. Later the equation is divided by o and the terms containing it "vanish" or "become evanescent" by decreasing o to nothing. What troubles Berkeley is that we have introduced a quantity o which we treat as greater than zero (if very small) when we divide but as equal to zero when we multiply and cancel terms which contain it. The Newtonian moments look suspiciously like Leibnizian infinitesimals, since they are effectively treated as being greater than zero but less than any finite quantity. Indeed, one major thrust of Berkeley's critique of Newton is that the Newtonian procedures are in fact inadmissible, rendering Newton's pretense to a more rigorous treatment of the calculus a sham.

This completes my overview of the background to Berkeley's critique of the calculus. As we will see, Berkeley was familiar with all of the principal issues in the calculus, and his attack on the new methods of "the modern analysis" proceeds by contrasting them to the rigor of such ancient techniques as the method of exhaustion.

CHAPTER FIVE

Berkeley and the Calculus: Writings before the *Analyst*

Berkeley's critique of the calculus did not originate in 1734 with the publication of the *Analyst*. He made a series of observations on the calculus in the *Philosophical Commentaries*, devoted §§130–32 of the *Principles* to a discussion of its foundations, and attacked the doctrine of infinitesimals in a 1709 paper, "Of Infinities." The object of the present chapter is to review these writings and to draw together some of Berkeley's sources. Berkeley's rejection of the infinitesimal calculus in the *Commentaries* and the *Principles* should come as no surprise. The epistemological constraints which led him to deny the thesis of infinite divisibility would obviously rule out the much stronger doctrine of infinitesimal magnitudes or the theory of fluxions. Nevertheless, it is worthwhile to explore the issues raised by the pre-*Analyst* texts, precisely because we can find both continuity and change in his treatment of the calculus from 1709 to 1734. I discuss the works chronologically in this chapter, beginning with the *Philosophical Commentaries*, then considering "Of Infinities," the *Principles*, and related writings.

Berkeley was a philosopher with a broad understanding of the mathematical issues surrounding the calculus, and his reservations about the new methods were longstanding. This is not to say that all of the main points in the *Analyst* are simply a restatement of earlier concerns; there is much in the later work which is not foreshadowed in Berkeley's pre–1734 writings on the calculus. But these earlier writings are of interest in tracing the development of Berkeley's views on the calculus, and they show that the *Analyst* was not the work of a mathematical incompetent who critiqued theories he had neither studied nor understood.

Berkeley and the Calculus: Writings before the *Analyst*

The Calculus in the *Philosophical Commentaries*

Berkeley first raised his objections to the calculus in an interesting series of entries in the *Philosophical Commentaries* in which he attacked the doctrines of fluxions, differentials, and infinitesimals. His treatment of the calculus in these notebooks has two major parts: he rejects the doctrine of infinitesimals and fluxions as inconsistent, and he explores the possibility of replacing the calculus with his own geometry of the minimum sensible. Thus, there is a clear analogy between his account of geometry in the *Commentaries* and his early views on the calculus. This is natural enough, since Berkeley (like many of his time) treated the calculus as a fundamentally geometric method and therefore handled both subjects in essentially the same manner. Nevertheless, a separate consideration of Berkeley's early views on the calculus can help us in understanding his philosophy of mathematics and its sources.

The first mention of the calculus in the *Commentaries* is Berkeley's declaration: "Mem: to enquire into the Angle of Contact. & into fluxions &c." (*Commentaries*, 168). Much later, he writes, "Hays's Keil's &c. method of proving the infinitesimals of ye 3^d order absurd, & perfectly contradictious," which is followed immediately by a cryptic and apparently unrelated remark: "Angles of Contact, & vēly all angles comprehended by a right line & a curve, cannot be measur'd, the arches intercepted not being similar" (*Commentaries*, 308–9). These last two entries constitute an interesting objection to the doctrine of infinitesimals of all orders, and all three relate the calculus to the ancient problem of the angle of contact between a right line and a curve.

The place to begin unraveling this objection is with the mention of "Hays" and "Keil." The entry refers to Charles Hayes's 1704 *Treatise of Fluxions* and to lecture 4 of John Keill's *Introductio ad Veram Physicam*. Both of these works undertake to prove the existence of infinitesimals with an argument based on Newton's discussion of the angle of contact between a curve and a right line (*Principia* I, 1, 11). The relevant texts reveal that the argument Berkeley opposes originated with Newton and was modified by Keill, with Keill's presentation taken over almost verbatim by Hayes.[1]

Although Hayes's account is not original, I will concentrate on his treatment of infinitesimals because it makes an interesting connection between infinite divisibility and the foundational concepts of the cal-

---

1. Newton's account uses his doctrine of evanescent magnitudes to investigate ratios between the tangent and subtangent formed at the angle of contact. Keill extends Newton's comments to argue for the existence of infinitesimals of higher orders, and Hayes simply repeats Keill's account of the matter.

culus. Hayes's *Treatise* is chiefly remarkable for its terminological confusions and unhesitating use of infinitesimals.[2] He opens his account of fluxions with a declaration, under the heading "The Nature of Fluxions," which must surely have attracted Berkeley's notice:

> Magnitude is divisible in *infinitum,* and the Parts after this infinite Division, being infinitely little, are what Analysts call *Moments* or *Differences;* And if we consider Magnitude as Indeterminate and perpetually Increasing or Decreasing, then the infinitely little Increment or Decrement is call'd the *Fluxion* of that Magnitude or Quantity: and whether they be call'd *Moments, Differences,* or *Fluxions,* they are still suppos'd to have the same Proportion to their Wholes as a Finite Number has to an Infinite; or as a finite Space has to an infinite Space. Now those infinitely little Parts being extended, are again infinitely Divisible; and these infinitely little Parts of an infinitely little Part of a given Quantity, are by Geometers call'd *Infinitesimæ Infinitesimarum* or *Fluxions of Fluxions.* Again, one of those infinitely little Parts may be conceiv'd to be Divided into an infinite Number of Parts which are call'd *Third Fluxions,* &c. (Hayes 1704, 1)

This passage clearly invokes the thesis of infinite divisibility and goes on to develop an account of fluxions, differentials, and moments as a natural consequence of infinite divisibility.[3] Although it is unlikely that Hayes's explicit connection of infinite divisibility and the foundational concepts of the calculus was the sole motivation for Berkeley's attack, it is certainly noteworthy that Berkeley's critique of the calculus begins by dismissing as "absurd and perfectly contradictious" a text which links infinite divisibility and the calculus in such an obvious way.[4] Another point which deserves comment here is Hayes's use

---

2. Cajori (1919, 56) cites Hayes's *Treatise* as an example of the kind of confused presentation of the calculus which provided an ideal target for Berkeley's attack in the *Analyst.* After discussing Hayes's work and that of others, Cajori declares, "What an opportunity did this medley of untenable philosophical doctrine present to a close reasoner and skillful debater like Berkeley!" Guicciardini (1989, chap. 1) also contains an account of Hayes's role in the development of the calculus in Britain.

3. However, Hayes departs from the classical tradition by assuming that the doctrine of infinite divisibility is equivalent to the claim that there are infinitely small parts which result from an infinite division. Classically, every magnitude is divisible and there are no infinitely small magnitudes.

4. In many ways, Hayes's *Treatise* is the natural place for Berkeley to have begun an investigation of the calculus. It was, in 1704, one of the first presentations of the calculus in English and a standard secondary work on the calculus in Berkeley's day. Despite his criticisms, Berkeley cannot have held the work in very low esteem, since it is one of the books he donated to Yale University in 1734. See Keough (1934), where

Berkeley and the Calculus: Writings before the *Analyst*

of the term fluxion for the "infinitely little increment or decrement" of a flowing quantity and his subsequent pronouncement that moments, differences, and fluxions are all infinitesimal quantities. Even by the comparatively lax standards of 1704, this is a serious confusion of terms, because Newton continually insisted that fluxions are to be regarded not as infinitely small quantities but as finite ratios of quantities.[5] In the dispute which followed the publication of the *Analyst*, some defenders of Newton charged that Berkeley had misunderstood the Newtonian doctrine and confused the allegedly rigorous Newtonian theory with the infinitesimal methods of the continental mathematicians. I will investigate that charge later, noting now that Hayes (as well as other writers of the period) drew no distinction between the Newtonian infinitesimals, Leibnizian differentials, and Newton's "moments of flowing quantities." Recall that Berkeley dismissed Hayes's (and Keill's) demonstration of infinitesimals of the second order as "absurd and perfectly contradictious." Hayes acknowledges that the doctrine of higher-order infinitesimals "may seem hard to most readers at first" and promises to "prove that there are Quantities infinitely less than any given Quantity, which are also infinitely greater than another Quantity." Hayes and Keill present two demonstrations of this doctrine, the second of which Berkeley attacks more explicitly. As Hayes states it:

> [F]or a clearer Illustration of this Doctrine, take the following example from the Incomparable Mr. *Newton*, which I find Demonstrated by a late ingenious author thus: Let *AC* [see Fig. 5.1] be a common Parabola, *AB* its Axis, and *AF* a line touching the same in the principle Vertex *A*. Then it is evident from the nature of the Curve, that the Angle of Contact *FAC* is less than any rectilineal Angle. To the same Axis *AB* and Vertex *A*, describe a Parabola of another kind, *v.g.* a cubical Parabola *AD*, whose Ordinates encrease in a subtriplicate Proportion of the intercepted Diameters; I say the Angle of Contact *FAD* will be infinitely less than the Angle of Contact *FAC*. Or which is the same thing, it is impossible to diminish the Angle of Contact (of the Apollonian Parabola *AC*,) *FAC*,

---

Hayes's work is listed among the Berkeley donations. Presumably, Berkeley regarded the *Treatise of Fluxions* as an adequate presentation of the fundamental results of the calculus and no more flawed than any other work of the period.

5. Note, for example, that Newton's treatise *On the Quadrature of Curves*, first published in 1704 (the same year as Hayes's *Treatise*), sought to establish the calculus of fluxions as a rigorous alternative to the Leibnizian calculus. See Cajori (1919, chap. 1) for a survey of Newton's pronouncements on fluxions and on the failure of authors such as Hayes to distinguish between fluxions and infinitesimals.

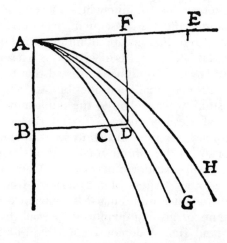

Fig. 5.1 From Hayes, *Treatise of Fluxions* (1704, 2).

that it shall be equal to or less than the Angle of Contact *FAD*, let the Parameter of *AC* be never so great. (Hayes 1704, 2)

Berkeley responds to this argument with the claim that the angle of contact cannot be measured, since the "arches intercepted" are not similar (*Commentaries*, 309).

The point at issue here is important, and I think Berkeley's response reflects well on his competence. The root of the difficulty is the notion of an "angle of contact" between a curve and a straight line—a problem that had a long history in geometry and was much discussed in Berkeley's day. According to Euclid, "A *plane angle* is the inclination to one another of two lines in a plane which meet one another and do not lie in a straight line" (*Elements* I, def. 8). This definition covers cases beyond the intersection of two straight lines. In particular, it allows for angles formed by the intersection of straight and curved lines. It is with the question of the so-called "horn angle" (i.e., an angle formed by a straight line and a curve) that paradoxical results threaten. The first theorem relating to the angle of contact (*Elements* III, 16) proves that the angle between circle and tangent is less than any given rectilinear angle. The details of the proof need not concern us,[6] but the basic idea can be indicated easily enough. We begin with a circular arc *AB* and tangent *AC* as in figure 5.2. It can be shown that any rectilinear angle *CAE* must be greater

---

6. They can be found in Euclid ([1925] 1956, 2:37–43). Another discussion of these problems is Thomason (1982).

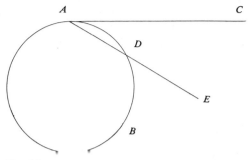

Fig. 5.2

than the angle of contact CAB by showing that an arbitrarily chosen rectilinear AE cuts the circle at some point D. But because $\angle CAE = \angle CAD$, the angle of contact must lie within the rectilinear angle CAE, and thus the angle of contact CAB is less than any rectilinear angle CAE.

The paradoxical aspect of the angle of contact is that it violates the "axiom of Archimedes," discussed in chapter 4, which in one of its many equivalent formulations reads: "Given two magnitudes A and B such that $A < B$, if B is reduced by half and the remainder is again reduced by half, and if this process be continued, then there will eventually remain a magnitude that is less than A" (*Elements* X, 1). It should be clear that the angle of contact cannot satisfy this condition. If it did we could produce a rectilinear angle smaller than it by continual bisection of a rectilinear angle greater than it. But we have already shown that the angle of contact is less than any rectilinear angle. To maintain a consistent geometry we must either deny that the angle of contact is really an angle at all or admit non-Archimedean (infinitesimal) magnitudes.

Berkeley's course is to deny that there can be such a magnitude as the angle of contact (*Commentaries*, 309). In a manner that accords well with modern conceptions of rigor, he asserts that there can be only a measurable angle (i.e., a magnitude) where "the arches intercepted are similar." Reading "similar" as "capable of being brought into coincidence," his argument amounts to a declaration that we can speak of the measure of an angle only for angles which are formed by (curved or straight) lines which can be brought into coincidence, and all such angles can be shown to satisfy the Archimedean axiom.[7]

---

7. To show that any such congruent lines can satisfy the Archimedean principle, it suffices to observe that all rectilinear angles are Archimedean magnitudes. Similarly, angles formed by congruent curves will reduce to the rectilinear case, because the angle

Thus, the basis for Berkeley's dismissal of Hayes's demonstration of the existence of higher order infinitesimals is essentially a demand that only finite magnitudes be admitted in analysis.

At this stage, Berkeley has simply rejected one attempt to use the angle of contact as a means to underwrite a theory of infinitesimals, without providing an independent reason for disallowing infinitesimal magnitudes as an object of mathematical investigation. To round out his case against the calculus, then, Berkeley should provide some grounds for his view that only finite quantities can be used in the solution of problems in analysis. Given Berkeley's insistence in the *Commentaries,* that only immediately perceived extension can be taken as the object of geometry, we should expect him to reject infinitesimals, because they cannot be objects of our experience, as metaphysically and epistemologically problematic.

In essence, this is Berkeley's procedure although in other entries from the *Commentaries* he rejects infinitesimals on the grounds that they are simply incomprehensible:

> Axiom. No reasoning about things whereof we have no idea. Therefore no reasoning about Infinitesimals.[8]
> Nor can it be objected that we reason about Numbers wch are only words & not ideas, for these Infinitesimals are words of no use if not supposed to stand for Ideas.
> Much less infinitesimals of infinitesimals &c.(*Commentaries,* 354–55)

Berkeley argues here that we can frame no idea of infinitesimals and, thus, that there is no legitimate purpose served by introducing signs such as dx or $o\dot{x}$ into mathematical discourse. His criticisms clearly depend upon the "axiom" that no word is to be used without an idea.

---

formed by their intersection will be the same as the angle formed by their tangents at the point of intersection. Berkeley apparently felt that the mathematicians of his era had failed to understand the nature of angles: "Mathematicians have no right idea of angles. hence angles of Contact wrongly apply'd to prove extension divisible ad infinitum" (*Commentaries,* 381). Similarly, he complains, "Angle not well Defin'd see Pardies Geometry by Harris etc: this one ground of Trifling" (*Commentaries,* 432). The reference is to the translation of Ignatius Pardies' *Eléments de Géometrie* by John Harris, which reads: "When two lines meet in a Point, the *Aperture, Distance,* or *Inclination* between them, is call'd an *Angle.* Which, when the Lines forming it are right or strait ones, is called a *Rectilineal Angle.* . . . But if they are crooked, 'tis called a *Curvilineal* One. . . . And when one is strait and the other crooked, 'tis called a *Mix'd Angle*" (Pardies 1746, 8). This clearly borrows from Euclid's definition, but it is not fully clear what Berkeley found objectionable.

8. This entry is repeated at *Commentaries,* 421.

Berkeley and the Calculus: Writings before the *Analyst*

When Berkeley later repudiates this axiom, it might be thought that he is no longer entitled to this kind of critique of infinitesimals.[9] I think, however, that this conclusion is unduly hasty.

The grounds for a distinction must surely lie in Berkeley's conception of the fundamental difference between arithmetic and geometry. Berkeley declares that numbers are only words, consistent with his formalistic philosophy of arithmetic. But he quickly adds that infinitesimals (by which he must mean terms such as dx and $o\dot{x}$ which are supposed to denote infinitesimals) are of no use if not supposed to stand for ideas. His clear implication is that the use of signs such as "237^{93}" which do not correspond to any collection of ideas is acceptable in arithmetic but unacceptable in analysis. On Berkeley's view, arithmetic is a "purely nominal" science dealing with computational procedures and rules for the combination of symbols; geometry is a science of perceived extension, and propositions in geometry or analysis must ultimately describe perceivable extension. Because we are incapable of discerning infinitely small parts of extension, the introduction of terms which supposedly denote these infinitesimals must be ruled out in geometry and analyis. To put it another way, Berkeley's requirements for rigorous geometric demonstration demand that the terms of the calculus be interpreted in terms of our experiences, but terms such as dx lack such an interpretation and are therefore deemed illegitimate.

Berkeley raises another objection to the methods of the calculus when he attacks as incomprehensible L'Hôpital's postulate that a curve can be treated as a polygon with an infinite number of sides:[10]

> W^t do the Mathematicians mean by Considering Curves as Polygons? either they are Polygons or they are not. If they are why do they give them the Name of Curves? why do not they constantly call them Polygons & treat them as such. If they are not polygons I think it absurd to use polygons in their Stead. Wt. is this but to pervert language to adapt an idea to a name that belongs not to it but to a different idea? (*Commentaries*, 527)

---

9. See Belfrage (1986) on Berkeley's account of the meaning of terms, and particularly reasons for his abandonment of the doctrine that every word must correspond to an idea.

10. Although this postulate can be found in other sources, it plays a very prominent role in L'Hôpital's *Analyse des infiniment petits*, where it is listed as the second postulate. Hayes adopts the same approach in his *Treatise of Fluxions*, where he claims that "[I]n this Method we consider all Curve-lines, as Compos'd of an infinite Number of infinitely little Straight-lines, or as Polygons of an infinite Number of Sides" (Hayes 1704, 4).

It is clear in this entry that Berkeley is unwilling to grant L'Hôpital's postulate based on the view that it is simply inconceivable that a curve could be a polygon. Now assuming that curves and polygons are fundamentally different things, the postulate seems downright contradictory. Thus, Berkeley charges that L'Hôpital has essentially "perverted" language in the attempt to save an incoherent theory.

Berkeley's negative appraisal of the calculus in the *Commentaries* is accompanied by suggestions for accomplishing its tasks legitimately by making the minimum sensible play the role of differentials, fluxions, and moments. In many respects, this resembles his program for a geometry of minima. Berkeley's notes on this reworking of the calculus read as follows:

> Qu. why may not the Mathematicians reject all the extensions below the M. as well as the dds &c wch. are allow'd to be Something & consequently may be magnify'd by glasses into inches, feet &c as well as the quantitys next below the m? (*Commentaries*, 324)
>
> Newton's fluxions needless. Anything below a M. might serve for Leibnitz's Differential Calculus. (*Commentaries*, 333)
>
> We need not force our Imagination to conceive such very small lines for infinitesimals. They may every whit as well be imagin'd big as little since that the integer must be infinite. (*Commentaries*, 415)

Some discussion of terms is required before my interpretation of these entries. In his note to entry 324, Luce writes "M is apparently 'moment,' Newton's term for differential; . . . dd would then be differential of the second order." He makes a similar interpretation in his note to entry 333 (*Works* 1:121). On this interpretation, entries 324 and 333 would not be suggestions for reformulating the calculus; thus, it is important to explain why I disagree with Luce's reading of these entries.

I consider that taking "M" as an abbreviation for "moment" renders the entries nonsensical.[11] On Luce's interpretation, entry 324 has Berkeley asking why mathematicians may not reject extensions below a moment as well as second-order differentials. But as we have seen, mathematicians working in the Newtonian tradition routinely *did* reject moments (as well, obviously, as quantities less than moments) in

---

11. The error does not originate with Luce. Johnston's edition of the *Commentaries* (which he calls *The Commonplace Book*) makes the same interpretation of this entry (Berkeley 1930, 129). Moked (1988, 134) continues this misinterpretation in part, claiming that "'m' is apparently *minimum*, 'M.' means *moment*."

Berkeley and the Calculus: Writings before the *Analyst*

the derivation of theorems in the calculus, and Berkeley was clearly aware of this. Thus, it appears that Luce's reading of entry 324 uncharitably portrays Berkeley as uninformed about the basic procedures of the fluxional calculus. Similarly, Luce's reading of entry 333 comes out with Berkeley first dismissing fluxions as needless and then declaring that anything below a moment (which is the *product* of a fluxion and a very small quantity) "might serve for Leibnitz's differential calculus." But this too is such an absurd reading that we are obliged to find another way of interpreting the "M" in these entries.

I suggest that we read the "M" here as an abbreviation for "minimum," a reading which renders both entries readily intelligible.[12] Berkeley's suggestion here is that the rejection of such infinitesimal quantities as differentials from solutions in the calculus can be mimicked in a new calculus of indivisibles by simply neglecting quantities below the minimum sensible in the solution of problems in analysis. His remarks are admittedly brief and programmatic, offering no detailed consideration of the difficulties involved in such a project. Nevertheless, it is clear that Berkeley feels that infinitesimal methods can be eliminated in favor of a method requiring only finite minima. The other entries cited above support this scheme, as does Berkeley's claim that "Sir Isaac owns his book could have been demonstrated on the supposition of indivisibles" (*Commentaries*, 374). This is clearly a comment on Newton's claim that his method of prime and ultimate ratios provides a rigorous geometric equivalent to the method of indivisibles. Newton writes:

> These Lemmas are premised to avoid the tediousness of deducing involved demonstrations *ad absurdum,* according to the method of the ancient geometers. For demonstrations are shorter by the method of indivisibles; but because the hypothesis of indivisibles seems somewhat harsh, and therefore that method is reckoned less geometrical, I chose rather to reduce the demonstrations of the following Propositions to the first and last sums and ratios of nascent and evanescent quantities, that is to the limits of those sums and ratios, and so to premise, as short as I could, the demonstrations of those limits. For hereby the same thing is performed as by the method of indivisibles; and now those principles being demonstrated, we may use them with greater safety. Therefore if hereafter I should happen to consider quantities as made up of particles, or should use little curved lines for right ones, I would not be understood to mean indivisibles, but evanescent

---

12. Breidert (1969) suggests the same reading as I present here.

divisible quantities; not the sums and ratios of determinate parts, but always the limits of sums and ratios; and that the force of such demonstrations always depends on the method laid down in the foregoing Lemmas. (*Principia* I, 1, 11)

Berkeley's entry 374 clearly refers to this Newtonian passage and seems to claim that it is indeed possible to present the calculus in a form of indivisibles. Given his claim in entry 346 that "all might be demonstrated by a new method of indivisibles," it is clear that Berkeley is here again declaring his interest in developing an alternative to the calculus in which the minimum sensible will play the role of differences or moments. Because Berkeley eventually gave up his program for a geometry of the minimum sensible, these brief remarks on a possible alternative to the calculus are not worked out in detail. In considering the *Analyst*, I will demonstrate that Berkeley retained his interest in the possibility of gaining the results of the calculus by noninfinitesimal means, but by 1734 he was prepared to accept all of classical geometry and took classical methods to be the only alternative to the calculus.

Two other entries in the *Commentaries* show Berkeley to have been concerned with explaining the success of the calculus in delivering reliable results, despite its reliance upon what he takes to be unintelligible assumptions about the existence of infinitesimal magnitudes: "How can they hang together so well since there are in them (I mean the mathematiques) so many Contradictoriae argutiae," he asks (*Commentaries*, 334). Soon after, he writes, "The not Leading men into mistakes no argument for the truth of the infinitesimals. they being nothings may, perhaps, do neither good nor harm. Except w^n they are taken for somthing. & then the contradiction begets a Contradiction" (*Commentaries*, 337). Eventually, Berkeley concludes that the success of the calculus is due to a "compensation of errors" in which two mutually cancelling mistakes are made whenever the calculus is applied to problems in analysis. This thesis is first advanced in the *Analyst*, and I will deal with it in detail later, but it is interesting to see that even while writing the *Commentaries* Berkeley felt the need to explain the success of the calculus.

The Essay "Of Infinities"

Another early source we have for Berkeley's views on the calculus is his short essay "Of Infinities," delivered before the Dublin Philosophical Society on 19 November 1707. "Of Infinities" thus dates from the period when Berkeley was writing the *Philosophical Commentaries,* and while it develops some of the themes from the notebooks, it also contains material inconsistent with their doctrines. A proper reading of "Of Infinities" is essential to understanding the Berkeleyan philosophy of mathematics, but the mathematical content of the essay has generally been ignored by commentators reading it against the background of Berkeley's case for immaterialism or the development of his account of linguistic meaning. Thus, Luce declares, "Here is no abstract mathematical problem, accidentally connected with Berkeley's studies. It touches the heart of his philosophy, and is vitally connected with the massive argument for immaterialism, which Berkeley is known to have been shaping in the years 1707–8" (*Works* 4:233). In a similar vein, Belfrage insists that "It has not hitherto been observed that this paper could have generated a great controversy. To a modern reader it probably appears a purely academical discourse concerning some mathematical subtleties. But . . . we may be confident that it was received as an extremely controversial paper, relevant to issues far outside the narrow field of mathematics" (Belfrage 1986, 118). Certainly the positions Berkeley takes in "Of Infinities" have some interesting connections to the other parts of his philosophy, but there is no excuse for ignoring the essay's primary and obvious intention, to critique the use of infinitesimal methods in analysis. It therefore behooves us not to dismiss the mathematical content of "Of Infinities" but rather to pay close attention, in particular, to the grounds Berkeley advances for his rejection of the infinitesimal.

The argument in "Of Infinities" has three principal points: First, Berkeley insists, based on a distinction (borrowed from Locke) between "infinity" and "infinite" and a semantic thesis that no word should be used without an idea corresponding to it, that talk of infinitely small magnitudes is, quite literally, unintelligible. Second, he establishes that the introduction of infinitesimal magnitudes has produced dispute and confusion among mathematicians by reviewing the controversy between Leibniz and Bernard Nieuwentijt over the foundations of the differential calculus. And third, he argues that there is no need to use infinitesimal magnitudes in the solution of problems in analysis, a point he believes to have been conceded by both Leibniz and Newton in some of their pronouncements on the calculus.

Berkeley's case against infinitesimals in "Of Infinities" begins with the observation that the techniques of the calculus have enabled "prodigious advances" in the solution of important problems but that the new methods have also given rise to disputes. Because the notion of an infinitely small magnitude is not clearly conceivable, the introduction of infinitesimals violates a criterion of rigor which demands that the objects of mathematical investigation be clearly conceived. Berkeley suggests that the disputes over the infinitesimal would cease if Locke's distinction between "the idea of the infinity of space" and "the idea of space infinite" were applied to infinitely small quantities as well as infinitely great ones. Locke's distinction is essentially as follows: we can frame an idea of the infinity of space by recognizing that, for any space we can imagine, we can extend that idea and imagine a greater space. Thus, infinity is an "endless growing idea" which cannot be delimited. But to frame the "idea of space infinite" would be to imagine a space that is infinite in extent, and this is beyond our power (*Essay* II, xvii, 7).

Applying this distinction to infinitely small magnitudes, Berkeley concludes that we should distinguish between the claim "given any magnitude, we can imagine a smaller one" and the claim "we can imagine a magnitude smaller than any given magnitude." The former is unproblematically true, while the latter is simply incoherent:

> For he that, with Mr. Locke, shall duly weigh the distinction there is betwixt infinity of space & space infinitely great or small, & consider that we have an idea of the former, but none at all of the later, will hardly go beyond his notions to talk of parts infinitely small or *partes infinitesimæ* of finite quantitys, & much less of *infinitesimæ infinitesimarum,* and so on. (*Works* 4:235)

Berkeley acknowledges that proponents of the infinitesimal calculus have contrived a notation for infinitely small magnitudes but insists that this notation is meaningless in that there are no ideas corresponding to the symbols. Invoking the axiom from the *Commentaries* that no sign should be used without an idea, he concludes that we must reject infinitesimals altogether:

> 'Tis plain to me we ought to use no sign without an idea answering to it; & 'tis as plain that we have no idea of a line infinitely small, nay, 'tis evidently impossible there should be any such thing, for every line, how minute soever, is still divisible into parts less than itself; therefore there can be no such thing as a line *quavis data minor* or infinitely small. (*Works* 4:235–36)

There are two important features of this passage: it appears to endorse the thesis of infinite divisibility and it relies upon the doctrine that no sign should be used without an idea. Although the first feature was touched on in chapter 2, both merit some discussion here. We know that Berkeley in the *Commentaries* rejects the doctrine of infinite divisibility in favor of the minimum sensible, so his public pronouncement in "Of Infinities" is at variance with his private thoughts on the matter. Furthermore, although he initially accepted the claim that all terms must denote ideas, Berkeley seems to have abandoned this tenet by the time he finished the *Commentaries*.[13]

We could take Berkeley's endorsement here of the thesis of infinite divisibility as simply disingenuous: after all, it would not have served his purpose to present the radical view of the *Commentaries* in public. This hypothesis is strengthened by recalling Berkeley's comments in the final section of the first editions of the *New Theory of Vision*, where he expressed reluctance to make public a conception of geometry which was "far out of the common road." We might imagine that he intended "Of Infinities" as a critique of infinitesimal theories based on classical geometry, even though he was prepared to reject classical geometry *in toto* when he wrote it. Continuing this line of thought, it is no great stretch of the imagination to think that Berkeley might well have wanted to avoid offending the mathematical sentiments of George Ashe, Bishop of Clogher, a member of the Dublin Philosophical Society with an interest in geometry and later a patron of Berkeley. Ashe presented two papers on geometry to the Society some months before Berkeley read his "Of Infinities."[14] He ordained Berkeley in February 1709 and later enlisted him as a tutor to his son on a tour of the continent (1716–20), so the two must have known each other well, and Berkeley may (rightly) have viewed him as a potential source of income and influence and therefore kept his radical geometry of minima to himself.

Alternatively, we could take the passage at face value and see it as symptomatic of misgivings about the radical project Berkeley proposed for the overthrow of traditional geometry. Knowing that he eventually abandoned this enterprise, one could read "Of Infinities" as a step toward the more sane doctrine we find in the *Principles*. However we choose to resolve this difficulty, it should be clear that Berkeley's case against infinitesimals does not hinge upon the doctrine of the minimum sensible. Infinitesimals are rejected, not simply because

---

13. This is the "clash on semantics" which interests Belfrage, and it is clear that Berkeley did eventually acknowledge the usefulness of terms which do not denote distinct ideas. Clearly, by the time of the *Principles*, Berkeley was prepared to accept the use of terms which have the function of the "raising of some passion, the exciting to, or deterring from an action, the putting the mind in some particular disposition" (Introduction, §20). Moreover, he grants that we can speak of God or the self, without framing distinct ideas to correspond to the relevant words.

14. British Library Additional MS. 4811, 53–57.

we cannot see them, but because we cannot imagine them—the very description of an infinitely small magnitude contains a contradiction, and there is no more reason to construct a theory of infinitesimals than to develop a geometry of round squares.

Furthermore, although Berkeley later abandons the principle that all terms must correspond to ideas, his abandonment of the general thesis seems not to have extended to the calculus. He complains in the *Analyst* that proponents of "the modern analysis" have devised a notation for fluxions and infinitesimals to which no ideas correspond and takes this to imply that there is no genuine science of fluxions or infinitesimals. I will discuss this issue in more detail when I consider Berkeley's rejection of a possible formalistic account of the calculus.

The dismissal of infinitesimal magnitudes in "Of Infinities" goes beyond that of the *Commentaries* in one important respect. Rather than content himself with arguing that we can frame no idea of infinitely small magnitudes, Berkeley advances an independent argument to show "that an infinitesimal even of the first degree is meerly *nothing*." The argument is ingenious and proceeds by an analysis of Wallis's results on the quadrature of the hyperbola in his *Arithmetica Infinitorum*.

Wallis considers asymptotic curves and their quadrature in propositions 91–105 of the *Arithmetica Infinitorum,* proceeding in accordance with his account of indivisibles. Take the curve $\beta\beta\beta$, asymptotic to the lines AD and $A\delta$ (fig. 5.3). The area enclosed by the curve and the asymptotes is considered as a sum of lines, each term of the sum being a reciprocal of the ordinate distance. Wallis represents the series with a parameter a in the form

$$\frac{1}{0a} + \frac{1}{1a} + \frac{1}{2a} + \frac{1}{3a} + \ldots$$

and calls it a series of right lines of the first order:[15]

> The plane figure consisting of a series of right lines of the first order in reciprocal proportion is infinite. And the same holds for all other reciprocal series.
>
> For since the first term in the series of lines of the first order is 0, the first term in the reciprocal series will be ∞ or infinite (just as, in division, if the divisor is 0, the quotient will

---

15. He considers series of other orders as well, such as the series $\frac{1}{0a^2} + \frac{1}{1a^2} + \frac{1}{4a^2} + \ldots$ which would be a series of the second order.

Berkeley and the Calculus: Writings before the *Analyst*

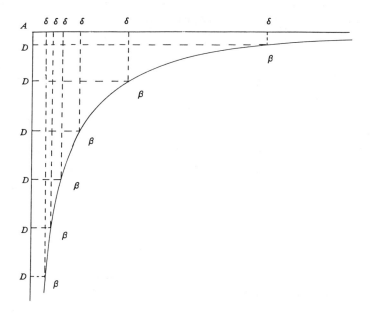

Fig. 5.3 Based on Wallis, *Arithmetica Infinitorum*, Proposition 95 (1693–99, 1:404)

be infinite). And furthermore, the right line $A\delta$ and the curve $\beta\beta\beta$ will only meet after an infinite distance (that is to say, never).

For the same reason neither will the curve $\beta\beta\beta$ meet the line AD (however closely they are brought together) until the distance $D\beta$ vanishes, which will make the right line $A\delta$ infinite. (Wallis 1693–99, 1:405).

Wallis concludes that the area enclosed by the hyperbola is infinite, since it is represented by the series of lines of the first order. Berkeley challenges this method, claiming that it shows the infinitesimal part of a line to be a mere nothing, or a line of zero length. The argument proceeds as follows: the first term in Wallis' series for the quadrature of the hyperbola represents the asymptote—a line of infinite length. Interpreting the terms in the series as representing the ratios between the lengths of lines, Berkeley observes that the first term gives the result of dividing a line of unit length by a line of length zero. Now when Wallis takes the term $1/0$ to be infinite, he is similarly committed to holding that $1/\infty = 0$, a result which Berkeley claims follows "from the nature of division." But the infinitesimal part of a line is supposed to have the same ratio to a finite line that a finite line has to an infinite line. Thus, if we take α to be the length of the infinitesimal line, we have $\alpha/1 = 1/\infty = 0$, and the infinitesimal part of a finite line must be a line with zero length. Berkeley adds, "Now a man speaking of lines infinitely small will hardly be suppos'd to mean nothing by them, and if he understands real finite quantitys he runs into inextricable difficulties." (*Works* 4:236)

This argument is perceptive, for it shows Berkeley to have been aware of a major problem surrounding the doctrine of infinitesimals. If infinitesimal magnitudes are introduced into analysis, the question arises whether they obey the ordinary laws of addition, subtraction, multiplication, and division. The thrust of Berkeley's argument here is that if the infinitesimals are assumed to obey the laws of arithmetic, it can be shown that an infinitesimal quantity is in fact just zero.[16] But if this were case, then it seems that the standard procedures in the calculus could be rewritten with zeros in place of the infinitesimals. This, of course, would wreak mathematical havoc. However, if the infinitesimal does not obey the ordinary laws of arithmetic, one must ask whether it is a genuine quantity at all. Even worse, the procedures of the calculus which rely upon the division and addition of infinitesimal magnitudes would then appear to be entirely without foundation. In selecting Wallis as the target for his attack, Berkeley is using the seventeenth century's most reckless practitioner of infinitesimal methods to make his point but, nevertheless, he presents a problem that remained unresolved in his day—that of finding a plausible account of magnitudes which makes room for the infinitesimal without making it mysteriously both greater than and equal to zero.

Berkeley's second claim of "Of Infinities" is that acceptance of the infinitesimal has produced disputes "to the great scandal of the so much celebrated evidence of Geometry." He cites the controversy between Nieuwentijt and Leibniz over the foundations of the calculus, and proposes that both parties have been led into error by assuming that there really are infinitesimal magnitudes. It is, I think, significant that Berkeley should focus upon this dispute. His remarks show him to have been familiar with the mathematical literature of the period and reveal that he understood the major issues. Before I proceed to Berkeley's account, an overview of the dispute is necessary.

Bernard Nieuwentijt was a Dutch mathematician of some note who published a critique of the Leibnizian calculus in 1694. His first fun-

---

16. The same point could be made by taking L'Hôpital's postulate that infinitesimal differences can be neglected to imply that infinitesimals are zero. If an infinitesimal ε and a real number ρ obey the law $\rho + \varepsilon = \rho$, then the ordinary laws of subtraction imply that $\varepsilon = \rho - \rho = 0$.

Berkeley and the Calculus: Writings before the *Analyst*

damental claim against the differential calculus was that infinitely small magnitudes were treated as zero or nothing when they were rejected. Moreover, he argued, even if the notion of an infinitely small magnitude were admitted, all higher-order differences must be regarded as equal to zero. Nieuwentijt sought to develop a version of the infinitesimal calculus that would contain only differences of the first order and carry the infinitely small differences through in a calculation without rejecting them. Thus, for example, he would take the derivative of the equation $y = 2x^2 + 3x$ as $dy/dx = 4x + 2dx + 3$, retaining the term $2dx$ in the right side of the equation.

Nieuwentijt substantiated his claim that the calculus treats first-order infinitesimals as zero by taking an example from Barrow's *Lectiones Geometricæ* where an infinitesimal is rejected in the calculation of a tangent. He insists that "The celebrated author sets forth this thesis in his reasoning: *If a determinate quantity has a ratio greater than any assignable to any other quantity, the latter will be equal to zero*" (Nieuwentijt 1694, 6). He further insists upon the principle that two quantities can only be called equal if their difference is absolutely nothing, so that two magnitudes which differ by an infinitely small amount are not, strictly speaking, equal. His principle is set forth as follows: "I declare that this proposition is indubitable, and carries with it most evidently the certain signs of truth: *Only those quantities are equal, whose difference is zero, or is equal to nothing*" (Nieuwentijt 1694, 10). The intent of this principle is to bar the practice of rejecting infinitely small quantities in analysis, which was of course a common practice of the time.

But Nieuwentijt was not opposed to the use of infinitesimals, provided that only those of the first order are admitted. He seems to have thought that the infinite divisibility of quantity made talk of infinitesimal magnitudes legitimate, but he introduced a principle, enunciated in his treatise *Analysis infinitorum,* which would make any supposed higher-order differentials equal to zero:

> *Whatever is taken so often as there are numbers, but cannot be multiplied so as to be equal to any given quantity, however small its magnitude, is not a quantity and is simply nothing in geometrical matters.*
> (Nieuwentijt 1695, 2)

The idea here is that a first-order difference such as dx can be multiplied infinitely so as to be equal to some given quantity. But a second-order difference such as dx^2 would remain less than any given quantity even if multiplied infinitely so it must be genuinely nothing.

Leibniz's response to Nieuwentijt is revealing. He first takes issue with the claim that two quantities are equal if their difference is zero. On the Leibnizian scheme, quantities which differ by an infinitesimal amount are also equal:

> I think that those things are equal not only whose difference is absolutely nothing, but also whose difference is incomparably small; and although this difference need not be called absolutely nothing, neither is it a quantity comparable with those whose difference it is. Just as when you add a point of one line to another line, or a line to a surface, you do not increase the magnitude: it is the same thing if you add to a line a certain line, but one incomparably smaller. Nor can any increase be shown by any such construction. (Leibniz [1848–53] 1962, 322)

This is the weak link in Leibniz's response to criticisms of his calculus, since he seems to be confessing that an infinitesimal (or incomparable) difference is a difference which is not really a difference at all.

But whatever the weakness in his own position, Leibniz notes that Nieuwentijt is on no stronger ground. By admitting only infinitesimals of the first order, Nieuwentijt must explain how comparisons of ratios can be made between the infinitesimal differences. If the ratio dy/dx is legitimate (as Nieuwentijt allows), then the quantities dy and dx are capable of being multiplied, divided, added, and compared with other quantities, finite or infinite. But then we can construct a proportion using the finite quantity x, the infinitesimal dx, and an unknown quantity ξ to get $x : dx :: dx : \xi$. Clearly, ξ is infinitely smaller than dx, and in Leibniz's treatment of the calculus can be treated as the product of two infinitesimals of the first order.

Berkeley's response to this controversy is direct: both Nieuwentijt and Leibniz hold incoherent positions, and the root of their errors is the shared assumption that infinitesimal magnitudes are legitimate in analysis. In admitting only first-order infinitesimals, Nieuwentijt must claim that "the square, cube, or other power of a positive real quantity is equal to nothing; wch is manifestly absurd" (*Works* 4:236). Leibniz, on the other hand, both denies the obvious truth that equal quantities have no difference and adopts a mysterious ontology of higher-order infinitesimals:

> *Quemadmodem* (says he) *si lineae punctum alterius lineae addas quantitatem non auges*. But if lines are infinitely divisible, I ask how there can be any such thing as a point? Or Granting there are points, how can it be thought the same thing to add an indivisible point as to add, for instance, the *differentia* of an ordinate, in a parabola, wch is so far from being a point that

it is itself divisible into an infinite number of real quantitys, whereof each can be subdivided *in infinitum*, and so on, according to Mr Leibniz. (*Works* 4:237)

Even worse, the Leibnizian formulation of the calculus entirely abandons the plainest criterion of mathematical rigor—the demand that the principles of demonstration must be clear and beyond doubt. Berkeley accuses both Leibniz and L'Hôpital of offending against this requirement and being more eager to apply the new methods than scrupulous in examining their foundations. Leibniz had suggested that an excess of scruples in foundational matters might hinder the progress of analysis, while L'Hôpital simply takes the Leibnizian pronouncements at face value and declines to undertake a rigorous examination of his two fundamental postulates.[17]

Thus far, Berkeley's strategy in "Of Infinities" has been only to argue that the doctrine of infinitesimals is incomprehensible and productive of disputes. But he then takes up the question of whether the abandonment of infinitesimal methods will not result in the loss of important results in analysis. He notes George Cheyne's argument for the acceptance of infinitesimals in *Philosophical Principles of Natural Religion*:

> Besides, the whole *abstract Geometry* depends upon the possibility of *infinitely* great, and small Quantities, and the Truth's discover'd by *Methods,* which depend upon these Suppositions, are confirm'd by other *Methods,* which have other Foun-

17. Berkeley quotes Leibniz's dictum "an excess of scruple may impede the art of discovery," responding sarcastically, "as if a man could be too scrupulous in Mathematics, or as if the principles of Geometry ought not to be as incontestable as the consequences drawn from them" (*Works* 4:237). He also makes an oblique reference to L'Hôpital's comments at the end of the preface to the *Analyse des infiniment petits* with the remark that "if we may believe the French author . . . , Mr. Leibnitz has sufficiently established & vindicated his principles. Tho' 'tis plain he cares not for having 'em called in question" (*Works* 4:237). L'Hôpital had declared: "When I was preparing to print the final sheet of this treatise, the book of M. Nieuwentijt came into my hands. Its title, *Analysis infinitorum,* gave me the curiosity to look through it, but I found that it was far from that contained here. [This author] absolutely rejects the second, third, etc. differences. As I have based the better part of this work on that foundation, I would feel obliged to respond to his objections and see that these foundations are more solid, if M. Leibniz had not already done this quite satisfactorily in the *Acta* of Leipzig. Concerning the other two postulates or suppositions which I have placed at the beginning of this treatise, and upon which it is founded, they seem so evident that I do not think they can leave any doubt in the mind of the attentive reader. I would have demonstrated them easily in the manner of the ancients, if I had not proposed to be brief about those things which are already known and attend principally to those which are new," (L'Hôpital 1696, xiv–xv).

dations; and they are too well supported, to allow of any Hesitation, in those that are thoroughly acquainted with this Science. Instances are needless, since whoever does understand these things, need not be told this. (Cheyne 1705, 10)

Berkeley's response to this argument is to declare:

[T]he supposition of quantitys infinitely small is not essential to the great improvements of the Modern Analysis. For Mr. Leibnitz acknowledges that his *Calculis differentialis* might be demonstrated *reductione ad absurdum* after the manner of the ancients; & Sir Isaac Newton in a late treatise informs us his method can be made out *a priori* without the supposition of quantitys infinitely small. (*Works* 4:237)

Leibniz's claim that his methods can be demonstrated by an Archimedean procedure appear in his reply to Nieuwentijt.[18] Newton's "late treatise" is the *Quadrature of Curves*, first published in 1704 as an appendix to the Latin translation of the *Optics*. There Newton declares:

In Finite Quantities so to frame a Calculus, and thus to investigate the Prime and Ultimate Ratio's of Nascent or Evanescent Finite Quantities, is agreeable to the Geometry of the Ancients; and I was willing to shew, that in the Method of Fluxions there's no need of introducing Figures infinitely small into Geometry. (Newton 1964–67, 1:143)

This point is of great importance in understanding Berkeley's views on infinitesimals. It is natural to read his declaration as an endorsement of the Archimedean method of exhaustion as a correct and rigorous approach to problems in analysis. By referring to Leibniz's and Newton's claims that their methods can be justified without resort

---

18. He writes: "Of course I hold with Euclid (*Elements* V, def. 5) that those homogeneous quantities are comparable which cannot be made to exceed one another by finite multiplication. And things which do not differ by such comparable quantities I take to be equal, which even Archimedes assumes, and all after him. And this is what is said to be a difference less than any given difference. In an Archimedean procedure the matter can still be confirmed by a *reductio ad absurdum*. But because the direct method is more quickly understood and more useful in discovering new results, it suffices if the direct method is once and for all reduced to the indirect method and later applied, in which incomparably small quantities are neglected. This procedure is sound and carries its demonstration with it according to a lemma communicated by me in February of 1689," (Leibniz [1848–1853] 1962, 6:322). The reference to a lemma from February of 1689 is apparently to his "Tentamen de motuum coelestium causis," in Leibniz ([1848–53] 1962, 6:161–75, esp. §5).

Berkeley and the Calculus: Writings before the *Analyst*

to infinitesimals, Berkeley suggests that the calculus be rewritten so as to conform to the Archimedean standard. He naturally makes no attempt to carry out such a monumental task in this essay, but he sounds a theme which will reappear in the *Analyst*—namely, that the results achieved by infinitesimal methods are not in dispute but that the introduction of infinitesimals has created such intolerable confusion in the foundations of mathematics that an alternative method of demonstrating the results must be found.

The *Principles* and Other Works

Apart from what we find in the *Commentaries* and "Of Infinities," Berkeley has little to say on the calculus before the *Analyst*. He makes some programmatic comments on analysis in the *Principles* and briefly mentions some aspects of infinitesimal mathematics in *De Motu* and *Alciphron* but otherwise is largely silent about the calculus between 1710 and 1734. This should not be taken to mean that he abandons his criticisms of "the modern analysis"; the same fundamental attitude toward infinitesimal mathematics remains intact throughout these other pre–1734 writings but receives its fullest articulation in the *Analyst*.

Berkeley's account of the calculus in the *Principles* is a simple repetition of the key claims in "Of Infinities."[19] He reviews the Leibniz-Nieuwentijt dispute, observing that "Of late speculations about infinities have run so high, and grown to such strange notions, as have occasioned no small scruples among the geometers of the present age," unambiguously outlining their positions but without mentioning their names. Some mathematicians (notably Leibniz and L'Hôpital) hold that there are infinitesimal quantities of all orders and "assert that there are infinitesimals of infinitesimals of infinitesimals, without ever coming to an end." Others (for example, Nieuwentijt) oppose this view and "hold all orders of infinitesimals below the first to be nothing at all" (*Principles*, §130).

Following the line of argument from "Of Infinities," Berkeley rejects both points of view as absurd (*Principles*, §131). But in answering the objection that the denial of infinitesimals is tantamount to a rejection of the whole science of geometry, he does not exactly follow the

---

19. This has not, to my knowledge, been remarked upon in the literature. Luce and Belfrage are concerned to place "Of Infinities" within the context of the *Commentaries* but take no notice of the fact that the entire argument in *Principles* (§§130–32) is taken almost verbatim from "Of Infinities."

line of thought he pursued in "Of Infinities." There, he held out hope for a reduction of the calculus to classical methods, but in the *Principles* he writes:

> [I]t may be replied, that whatever is useful in geometry and promotes the benefit of human life, doth still remain firm and unshaken on our principles. That science considered as practical, will rather receive advantage than any prejudice from what hath been said. But to set this in a due light, may be the subject of a distinct inquiry. For the rest, though it should follow that some of the more intricate and subtle parts of *speculative mathematics* may be pared off without any prejudice to truth; yet I do not see what damage will thence be derived to mankind. (*Principles*, §131)

It is worth observing that in the manuscript version of the *Principles* the second sentence of the passage differs from this formulation, and the difference suggests that Berkeley had essentially the same attitude as when he wrote "Of Infinities." It reads,

> But to set this in a due light and show how Lines and Figures may be measured and their properties investigated without supposing Finite Extension to be infinitely divisible will be the proper business of another place. (British Library Additional MS. 39304, fol. 94ʳ)

This manuscript statement can plausibly be read as indicating that Berkeley believes the results of the calculus (and, indeed, of geometry generally) can be obtained by "considering lines and figures" in accordance with his theory of representative generalization, that is, treating lines and figures in geometric proofs as representatives of all other lines and figures of any size, using careful comparison of figures and their ratios to underwrite such classical techniques of proof as the method of exhaustion.

Berkeley's intimation that the results of the calculus can be obtained by other methods is not confined to obscure manuscript jottings. In §132 of the *Principles,* a section apparently added after the manuscript version, he takes up the question of whether theorems of the calculus can be demonstrated by other methods:

> If it be said that several theorems undoubtedly true, are discovered by methods in which infinitesimals are made use of, which could never have been, if their existence included a contradiction in it. I answer, that upon a thorough examination it will not be found, that in any instance is it necessary to make use of or conceive infinitesimal parts of finite lines, or

> even quantities less than the *minimum sensibile:* nay, it will be evident this is never done, it being impossible. And whatever mathematicians may think of fluxions or the differential calculus and the like, a little reflexion will shew them, that in working by those methods, they do not conceive or imagine lines or surfaces less than what are perceivable to sense. They may, indeed, call those little and almost insensible quantities infinitesimals or infinitesimals of infinitesimals, if they please: but at bottom this is all, they being in truth finite, nor does the solution of problems require the supposing any other. But this will be more clearly made out hereafter. (*Principles*, §132)[20]

Again, Berkeley takes up the line of thought from "Of Infinities." But rather than cite Newton and Leibniz in support of his claim that the main results of the calculus can be obtained without the supposition of quantities infinitely small, he proposes that his case "will be more clearly made out hereafter," presumably in the proposed third part of the *Principles*. Although this third part was never written, Berkeley does attempt to show how to eliminate infinitesimal magnitudes from analysis in the *Analyst*.

Two interesting passages in Berkeley's 1721 treatise *De Motu* complicate the question of his attitude toward the calculus. As we saw in chapter 2, Berkeley's account of geometry acquires an element of instrumentalism, rooted in his claim that we must treat lines and figures in geometric demonstrations "as if they contained parts which really they do not," by the time of the *Principles*. Nevertheless, his instrumentalism does not extend to allowing the introduction of infinitesimal methods in the *Principles*. But in *De Motu* his instrumentalistic leanings seem to permit the introduction of infinitesimal magnitudes. Most notably, he declares:

> Just as a curve can be considered as consisting of an infinity of right lines, even if in truth it does not consist of them but because this hypothesis is useful in geometry, in the same way circular motion can be regarded as traced and arising from an infinity of rectilinear directions, which supposition is useful in the mechanical philosophy. (*De Motu*, §61)

Earlier in the same work, he insists that

> [J]ust as geometers for the sake of their discipline contrive many things which they themselves can neither describe, nor

---

20. The final three sentences of this passage were deleted from the second edition (1734). Since this edition appeared after the *Analyst*, it seems reasonable to treat the first edition as the definitive statement of Berkeley's views before 1734.

find in the nature of things, for just the same reason the mechanician employs certain abstract and general words, and imagines in bodies force, action, attraction, solicitation, &c. which are exceedingly useful in theories and propositions, as also in computations concerning motion, even if in the very truth of things and in bodies actually existing they are sought in vain, no less than those things geometers frame by abstraction. (*De Motu*, §39)

These passages suggest a vastly more tolerant attitude toward infinitesimal mathematics. L'Hôpital's postulate that a curve can be treated as an infinitary polygon, which he had once denounced, he now takes to be a hypothesis useful to geometry. Similarly, §39 of *De Motu* seems to acknowledge that infinitesimal methods such as the method of indivisibles or the differential calculus are "devices" or "fictions" useful in the solution of problems even though not to be "found in the nature of things."

These comments seem to indicate a mathematical instrumentalism much more thoroughgoing than that of the *Principles*. If Berkeley were truly willing to accept the methods he had earlier critiqued this would mark a substantial change in his views on the nature of mathematical demonstration and the criteria of geometric rigor. It remains possible, however, that Berkeley is not endorsing infinitesimal methods, but simply pointing out that there are false assumptions which can lead to useful results without implying that these assumptions are or should be acceptable in a well-developed mathematical theory. When I consider the main argument in the *Analyst*, it will become clear that Berkeley did, in fact, accept the key results of the calculus but thought that its principles were obscure and therefore unacceptable. Thus, I take the two pronouncements in *De Motu* as Berkeley's acknowledgement that infinitesimal mathematics can deliver accurate results but not as declaring that the methods of "the modern analysis" should be accepted simply on the grounds of their utility.

In this I differ from Warnock and Urmson. Both take Berkeley's statements in *De Motu* as indicating a substantial (and apparently permanent) change in view. Warnock claims that "even the use of the concept of infinity, formerly so much condemned, is taken to be legitimate" (Warnock 1953, 221). Urmson thinks that "By the time he wrote *De Motu* Berkeley had come to hold a view of geometry similar to his view of arithmetic" (Urmson 1982, 89). But both of these interpretations ignore the polemics in the *Analyst* against infinitary methods, as well as Berkeley's explicit rejection of the possibility of an instrumentalistic or formalistic account of the calculus. These issues will be taken up in greater detail in the next chapter.

Berkeley and the Calculus: Writings before the *Analyst*

This concludes my survey of Berkeley's writings on the calculus before the publication of the *Analyst* and puts me at last in a position to undertake an interpretation of his major work in the philosophy of mathematics. Some of the themes in these earlier writings reemerge in the later work but with differences in emphasis and a wealth of new material. I will now proceed to an account of the *Analyst*.

CHAPTER SIX

Berkeley and the Calculus: The *Analyst*

The publication of the *Analyst* marked a continuation of two Berkeleyan projects, one theological and the other mathematical. Theologically, the *Analyst* was part of Berkeley's battle against freethinking, while the mathematical intent was to sharpen his critique of the calculus and to show how the results of "the modern analysis" can be obtained rigorously within the context of classical geometry. For my purposes, the mathematical aspect of the *Analyst* is central and I will be primarily concerned with interpreting and evaluating its mathematical claims. Nevertheless, it will be helpful to summarize Berkeley's main theological concerns before we proceed.

Taken as a theological work, the *Analyst* attempts to show that certain criticisms of revealed religion are unjust. In particular, Berkeley is concerned to show that freethinkers who deride religion for its mysteries cannot consistently accept the calculus, since it is at least as incomprehensible as any point in revealed religion. Berkeley's opposition to freethinking was longstanding,[1] but became much more active during his stay in Newport (1729–31), where he apparently encountered numerous heterodox opinions among the colonial population. In *Alciphron* he attempts to defend Anglican Christianity by arguing that the freethinkers have no monopoly on reason but are victims of the same prejudice, irrationality, and unsound reasoning which they pur-

---

1. In a 1713 series of essays (including one entitled "Minute Philosophers") in Sir Richard Steele's *Guardian,* Berkeley attacked freethinkers and made a case for revealed religion which is similar to that found in *Alciphron.* See *Works* 7:181–228.

port to find in Christianity. This strategy is also central to the *Analyst*, as is indicated by the full title of the work attributed to "The Author of *The Minute Philospoher*": *The Analyst; or A Discourse addressed to an Infidel Mathematician; Wherein It is examined whether the Object, Principles, and Inferences of the Modern Analysis are more distinctly conceived or more evidently deduced, than Religious Mysteries and Points of Faith.*

The identity of the "infidel mathematician" is somewhat uncertain, although it seems clear that Berkeley has a specific person in mind; he declares as much in his *Defence of Free-Thinking in Mathematics:*

> Whether there are such infidels, I submit to the judgement of the reader. For my own part I make no doubt of it, having seen some shrewd signs thereof by others. . . . [The late celebrated Mr. Addison] assured me that the infidelity of a certain noted mathematician, still living, was one principal reason assigned by a witty man of those times for his being an infidel. Not that I imagine geometry disposeth men to infidelity; but that, from other causes, such as presumption, ignorance, or vanity, like other men geometricians also become infidels, and that the supposed light and evidence of their science gains credit to their infidelity. (*Defence*, §7)

Joseph Stock, an early but unreliable biographer of Berkeley, identifies Edmund Halley (1656–1742) as the "infidel mathematician" whose conduct was reported by Addison:

> The occasion was this: Mr. Addison had given the Bishop an account of their common friend Dr. Garth's behaviour in his last illness, which was equally unpleasing to both those excellent advocates for revealed religion. For when Mr. Addison went to see the Doctor, and began to discourse with him seriously about preparing for his approaching dissolution, the other made answer "Surely, Addison, I have good reason not to believe those trifles, since my friend Dr. Halley who has dealt so much in demonstration has assured me that the doctrines of Christianity are incomprehensible and the religion itself an imposture." The Bishop therefore took arms against this redoubtable dealer in demonstration, and addressed the *Analyst* to him, with a view of shewing, that Mysteries in Faith were unjustly objected to by mathematicians, who admitted much greater Mysteries, and even falsehoods in Science, of which he endeavoured to prove that the doctrine of fluxions furnished an eminent example. (Stock [1776] 1989, 29–30)

The difficulty with this account is that Samuel Garth died in January and Addison in June of 1719, fifteen years before the publication of the *Analyst* and during a period when Berkeley was in Italy. Addison could only have informed Berkeley of the incident by letter, but no such letter survives in the (admittedly scanty) collection of Berkeley's correspondence from the period.

Ultimately, the identity of the "infidel mathematician" is of no great consequence in interpreting the *Analyst*. The significant point is that Berkeley saw the enemies of revealed religion as taking mathematics for a paradigm of sound reasoning and then rejecting religion because it fails to live up to the mathematical standard. The core of Berkeley's theological strategy in the *Analyst* is thus to show that the calculus is no less mysterious than Christianity.[2] Berkeley sees mystery as admissible (indeed, essential) in religion, while his criteria for mathematical rigor demand that the object of a genuine science be clearly conceived. He thus makes a fundamental distinction—theology considers mysteries beyond, but not contrary to, human reason; science, only things evident to reason—stated clearly in query 62 at the end of the *Analyst:* "Whether mysteries may not with better right be allowed of in Divine Faith than in Human Science?"

In fact, Berkeley consistently rejects demands for mathematical proofs of theological results and, in particular, the idea that mysteries could be excised from religion or rendered more intelligible through mathematical demonstration. In the *Philosophical Commentaries,* he declares:

> There may be Demonstrations used even in Divinity. I mean in the reveal'd Theology, as contradistinguish'd from natural. for tho' the Principles may be founded in Faith yet this hinders not but that legitimate Demonstrations might be built thereon. Provided still that we define the words we use & never go beyond our Ideas. Hence 'twere no very hard matter for those who hold Episcopacy or Monarchy to be establish'd jure Divino, to demonstrate their Doctrines if they are true. But to pretend to demonstrate or reason any thing about the Trinity is absurd here an implicit Faith becomes us. (*Commentaries,* 584)

In other words, certain doctrines may be deduced within theology from premises founded only on faith, but there can be no room for quasi-mathematical demonstrations of the mysterious premises them-

---

2. Cantor (1984) calls this the "Matthew strategy," as it is drawn from the Biblical verse Matthew 7.5: "Thou hypocrite, first cast out the beam out of thine own eye; and then shalt thou see clearly to cast out the mote out of thy brother's eye." The verse appears on the title page of the *Analyst*.

Berkeley and the Calculus: The *Analyst*

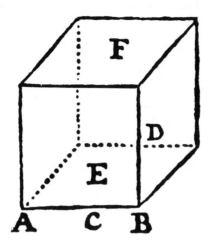

Fig. 6.1 From Wallis, *The Doctrine of the Trinity Briefly Explained, in a Letter to a Friend* (1690, 11).

selves, which in this respect function rather like the unquestionable axioms or postulates of a geometric system.

This placed Berkeley in opposition to writers who sought to explain or demonstrate religious mysteries through mathematical analogies, such as the analogy between the infinity of God and the mathematical infinite. For example, John Wallis, who in addition to holding the Savilian chair in geometry at Oxford was a prominent Doctor of Divinity, thought that the doctrine of the trinity could be explained by analogy with the three-dimensionality of space. In *The Doctrine of the Blessed Trinity Briefly Explained*, he asks the reader to consider a cube (fig. 6.1) and declares:

> If in this (supposed) Cube, (we suppose in Order, not in Time) its first Dimension, that of Length, as *A.B.*, and to this Length be given an equal Breadth (which is the true generation of a Square) as *C.D.*, which compleats the square Basis of this Cube; and to this Basis (of Length and Breadth) be given (as by a further Procession from Both) an equal Heighth *E.F.*, which compleats the Cube; and all this eternally, (for such is the Cube supposed to be,) here is a fair Resemblance (if we may *parvis componere magna*) of the *Father*, (as the Fountain or Original;) the *Son*, (as generated of him from all Eternity;) and of the *Holy-Ghost*, (as eternally Proceeding from Both:) And all this without any Inconsistence. This *longum, latum, profundum* (Long, Broad, and Tall,) is but *One* Cube; of *Three*

> *Dimensions,* and yet but *One Body:* And this *Father, Son* and *Holy-Ghost; Three Persons,* and yet but *One God.* And as, there, the Dimensions are not (in the Abstract) predicated or affirmed each of other, or the Cube of either, (the Length is not the Breadth or the Heighth, nor either of these a Cube;) but (in the Concrete) Cube is affirmed of all; this *longum, latum, profundum* is a *Cube,* and the *same Cube:* So here, (in the Abstract) the Personality of the *Father* is not that of the *Son,* nor either of these that of the *Holy-Ghost,* nor the *Deity* or *Godhead* any of these; but (in the Concrete) though the Personalities are not, yet the Persons are, each of them *God* and the *same God.* (Wallis 1690, 13–14)

Berkeley rejects such attempts to elucidate theological mysteries. Indeed, he comments on "The danger of Expounding the H: Trinity by extension" (*Commentaries,* 310).[3] Berkeley felt that once mystery was rationalized out of religion, the path away from orthodoxy had been opened, and there was no logical stopping place along the road to atheism.[4]

Despite an undeniably theological element, *The Analyst* is by no means solely an exercise in Christian apologetics[5] Berkeley pursues a number of important philosophical points in the work, outlining his standard for rigorous demonstration and raising epistemological and metaphysical issues relevant to mathematics. Given Berkeley's claim that mathematicians have been led into error by the doctrine of ab-

---

3. Luce (*Works* 1:120) comments on this entry that it concerns "*E.g.* St. Patrick's exposition of Trinitarianism (according to the legend) by aid of the shamrock leaf." I think, however, that it must be directed at Wallis and the attempt to rationalize mystery through mathematics.

4. This path is outlined in *Alciphron,* where the freethinking minute philosopher Alciphron remarks: "[H]aving observed several sects and subdivisions of sects espousing very different and contrary opinions, and yet all professing Christianity, I rejected those points wherein they differed, retaining only that which was agreed to by all, and so became a Latitudinarian. Having afterwards, upon a more enlarged view of things, perceived that Christians, Jews, and Mahometans had each their different systems of faith, agreeing only in the belief of one God, I became a Deist. Lastly, extending my view to all the other various nations which inhabit this globe, and finding they agreed in no one point of faith, but differed one from another, as well as from the forementioned sects, even in the notion of a God, in which there is as great diversity as in the methods of worship, I thereupon became an atheist; it being my opinion that a man of courage and sense should follow his argument wherever it leads him, and that nothing is more ridiculous than to be a free-thinker by halves" (*Alciphron,* dialogue 1, §8).

5. Cantor (1984) concentrates exclusively on these theological elements and places Berkeley in the context of the eighteenth-century debate over deism. I find this treatment rather one-sided, however, as the most interesting theses in the *Analyst* are completely independent of theological considerations.

stract ideas and the supposition of objects "without the mind," one would expect the critique of abstract ideas to play a prominent role in the *Analyst*. In fact, however, the significance of Berkeley's epistemological tenets emerges most clearly only after he has presented a case against the calculus which does not explicitly hinge upon his own epistemology.

Beyond these theological and philosophical concerns, the *Analyst* advances a number of important mathematical claims: First, that, contrary to the pronouncements of Newton and some of his followers, the Newtonian calculus of fluxions does indeed require the postulation of infinitesimal quantities. Second, that the proofs of even the most basic results in the calculus are fallacious and sophistical. Third, that the success of the calculus can be explained by a "compensation of errors" which arises whenever infinitesimal differences or evanescent increments are used in a calculation but dismissed from the result.

All of these theological, philosophical, and mathematical issues are intertwined in the *Analyst,* but the argumentation quite clearly follows the rhetorical conventions of the eighteenth-century polemical piece.[6] After a brief introduction (§§1–2), Berkeley launches into his case against the calculus, arguing that its object is obscure and its principles and demonstrations unscientific (§§3–20) and that a "compensation of errors" in which two false suppositions cancel one another explains the success of the calculus (§§21–29). He anticipates possible responses to his critique, considering various interpretations of the calculus which might avoid his arguments but concluding that each is unsatisfactory (§§30–47). He concludes with a blanket indictment of "these modern analytics" as obscure and unscientific (§§48–50) and adds a series of sixty-seven "Queries" (obviously modeled on the list of queries at the end of Newton's *Opticks*) which range widely over topics in the philosophy of mathematics and related fields. Given this structure, it makes sense to organize my investigation along the lines Berkeley proposed, beginning with his treatment of the "object" of the calculus and proceeding from there.

The Object of the Calculus

Berkeley's fundamental charge in the *Analyst* is that the calculus is unrigorous. Mathematical rigor is a notoriously difficult concept to

---

6. For a study of Berkeley's relationship to the rhetorical conventions of the eighteenth century, see Walmsley (1990). Unfortunately, he does not consider the *Analyst*, despite its considerable rhetorical virtues.

articulate, but I propose that it must include both metaphysical and logical criteria. On one hand, a demonstration might be rejected for invoking objects which are thought to be conceptually or metaphysically problematic, as when a constructivist argues that we have no concept of an infinite totality and thus cannot use infinitary proofs. Berkeley intends such an attack when he contrasts the "object of the modern analysis" with religious mysteries, as I will discuss in this section. On the other hand, a mathematical procedure might be deemed logically unrigorous for relying upon invalid or mistaken inferences. Berkeley attempts such a critique of the calculus when he asks whether the "principles and inferences of the modern analysis" are as evidently demonstrated as points of faith. I will take this up later. Berkeley's standard for rigorous demonstration is not particularly novel in the context of seventeenth- and eighteenth-century mathematics. Similar declarations can be found throughout the literature of the period. Barrow, for example, summarizes his lecture on the nature of demonstration in this way:

> But I fear that you may become annoyed with this long and prolix comparison. From which nevertheless it can appear what method of demonstration is used by mathematicians; namely such that they take up only those things for consideration of which they have clear and distinct ideas in their minds; and these they designate with proper, adequate, and invariable names. For investigating their affections and drawing true conclusions about them they set forth very few axioms which are most known and certain; and similarly they lay down a very few hypotheses, which are most highly consonant with reason, nor are these to be refused by a sane mind. They also assign causes or generations, easily understood and readily admitted by all. And they preserve an exquisite order in their demonstrations, so that every proposition follows readily from things previously supposed or proved; and lastly they reject everything, however probable or plausible, which they cannot infer and deduce in this manner. (*Lectiones*, 4:76)

These criteria, derived in part from Aristotle's discussion of demonstration in the *Posterior Analytics*, were widely recognized in Berkeley's day.

Berkeley acknowledges that some parts of traditional mathematics live up to this "official" standard of rigor, but suggests that the modern methods of analysis fall far short:

> It hath been an old remark that Geometry is an excellent Logic. And it must be owned, that when the Definitions are clear; when the Postulata cannot be refused, nor the Axioms

Berkeley and the Calculus: The *Analyst*

> denied; when from the distinct Contemplation and Comparison of Figures, their Properties are derived, by a perpetual well-connected chain of Consequences, the Objects being still kept in view, and the attention ever fixed upon them; there is acquired an habit of Reasoning, close and exact and methodical: which habit strengthens and sharpens the Mind, and being transferred to other Subjects, is of general use in the inquiry after Truth. But how far this is the case of our Geometrical Analysts, it may be worth while to consider. (*Analyst*, §2)

I think we can read this as endorsing the method of exhaustion over the modern methods of analysis. Berkeley's specific criticisms of the calculus and his suggestions for a rigorous alternative strengthen this interpretation. Among other things, Berkeley claims that the practitioners of the calculus have neglected the true object of the calculus (finite geometric figures and their relations) while reasoning about empty symbols. His proposal for reform of the calculus involves taking sequences of approximations, very much in the style of classical methods.

With this metaphysical standard of rigor as a background, Berkeley outlines some basic concepts of the Newtonian calculus of fluxions, particularly Newton's doctrine of moments and definition of fluxions as the velocities with which geometric magnitudes are produced.[7] Moments are not finite particles but "the nascent Principles of finite Quantities." They are not to be treated as having magnitude, yet ratios between moments are considered. Fluxions are not average velocities taken over a given time, but rather *instantaneous* velocities, defined as the prime ratio of nascent increments of time and distance: "These Fluxions are said to be nearly as the Increments of the flowing Quantities, generated in the least equal Particles of time; and to be accurately in the first Proportion of the nascent, or in the last of the evanescent, Increments" (*Analyst*, §3). And what is more, higher-order fluxions may be introduced by taking a fluxion as itself a flowing quantity.

As might be expected, Berkeley finds these objects wholly mysterious, inconceivable in terms of the epistemological principles he apparently takes to be unproblematic:

> Now as our Sense is strained and puzzled with the perception of Objects extremely minute, even so the Imagination,

---

7. He cites the Introduction to the *Quadrature of Curves* in describing how lines are supposed to be generated by the motion of points. The account of the doctrine of moments is a paraphrase of the *Principia* (II, 2, 2).

which Faculty derives from Sense, is very much strained and puzzled to frame clear Ideas of the least Particles of time, or the least Increments generated therein: and much more so to comprehend the Moments, or those Increments of the flowing Quantities in *statu nascenti,* in their very first origin or beginning to exist, before they become finite Particles. And it seems still more difficult, to conceive the abstracted Velocities of such nascent imperfect Entities. But the Velocities of the Velocities, the second, third, fourth and fifth Velocities, &c. exceed, if I mistake not, all Humane Understanding. The further the Mind analyseth and pursueth these fugitive Ideas, the more it is lost and bewildered; the Objects, at first fleeting and minute, soon vanishing out of sight. (*Analyst,* §4)

In claiming that moments and fluxions are incomprehensible, Berkeley assumes that "extremely minute" objects cannot be clearly apprehended by sense and that the faculty of imagination "derives from" sensation. I grant that Berkeley is right about the first claim and might even grant that what can be imagined depends in a crucial way upon what can be perceived. The epistemological model here is familiar: What we perceive by sense is the raw material from which imaginative ideas are constructed. The mental faculties consist of sense and imagination, with no faculty of "pure intellect" able to frame ideas independently.[8] Berkeley shares this picture of human cognition with many of his contemporaries but applies it restrictively to mathematical reasoning, counter to the mathematical epistemologies standard at the time. To convince mathematicians that fluxions are inconceivable and the calculus unrigorous Berkeley must therefore carefully construct an argument based on acknowledged principles of rigorous demonstration.

Continuing his critique of analysis, Berkeley next argues that Leibniz's *calculus differentialis* also offends against metaphysical criteria of rigor, paraphrasing L'Hôpital and then observing:

Now to conceive a Quantity infinitely small, that is, infinitely less than any sensible or imaginable Quantity, or than any the least finite Magnitude, is, I confess, above my Capacity. But to conceive a Part of such infinitely small Quantity, that shall be

---

8. The famous opening sentence of the *Principles* gives the best gloss of this theory: "It is evident to any one who takes a survey of the objects of human knowledge, that they are either ideas actually imprinted on the senses, or else such as are perceived by attending to the passions and operations of the mind, or lastly ideas formed by help of memory and imagination, etiher compounding, dividing, or barely representing those originally perceived in the aforesaid ways" (*Principles,* §1). Obviously, the role of memory here can be neglected and we can take the mental faculties to consist of sensation and imagination.

still infinitely less than it, and consequently though multiply'd infinitely shall never equal the minutest finite Quantity, is, I suspect, an infinite Difficulty to any Man whatsoever; and will be allowed such by those who candidly say what they think; provided they really think and reflect, and do not take things upon trust. (*Analyst*, §5)

Here, Berkeley seems to be on stronger ground than in his attack on the Newtonian calculus of fluxions. Berkeley is hardly voicing a novel opinion when he rejects the infinitesimal as incomprehensible. I take it as obvious that infinitesimal magnitudes are difficult to conceive, and their exclusion from mathematics is very much a part of the classical conception of rigor.

Of course, presentations of the infinitesimal calculus generally ignore these reservations. Some authors, including L'Hôpital, attempt to illustrate the fundamental tenets of the infinitesimal calculus with diagrams in which finite lines and figures are labeled dx or dy. These are clearly conceivable, but that is precisely because the diagrams misrepresent the infinitesimal calculus by attempting to illustrate it. Strictly speaking, differences such as dx must be infinitely less than the finite quantities in the diagram, and there is no way that the relationship between an ordinate x and its difference dx could adequately be depicted in a diagram. Whatever else they might be, infinitesimal differences are not the kind of thing we can perceive.

Although Berkeley's claims against the calculus are frequently at the level of first-person reports of his inability to conceive fluxions, moments, or differences, he is clearly aware that more argument is needed. To bolster his case, he first insists that his own inability to conceive these objects is *prima facie* evidence that they offend against metaphysical criteria of rigor:

> There is a natural Presumption that Mens Faculties are made alike. It is on this Supposition that they attempt to argue and convince one another. What, therefore, shall appear evidently impossible and repugnant to one, may be presumed the same to another. (*Analyst*, §7)

There is certainly something to this argument, but one would be hesitant to place too much weight on it. A natural reply would be to grant that all men's faculties are similar, but to claim that in his attempts to comprehend fluxions Berkeley simply is not applying himself with the requisite effort. His failure to conceive the object of the calculus could thus be taken as a sign of his own mathematical incompetence rather than any fundamental lack of rigor in the calculus.

Berkeley extends his argument to acknowledge that the points he finds incomprehensible are readily admitted by mathematicians as clear and incontestable. He argues, however, that their confidence is misplaced and that the proponents of the calculus have merely devised a handy notation to which no objects can be conceived to correspond:

> But, notwithstanding all these Assertions and Pretensions, it may be justly questioned whether, as other Men in other Inquiries are often deceived by Words or Terms, so they likewise are not wonderfully deceived and deluded by their own peculiar Signs, Symbols, or Species. Nothing is easier than to devise Expressions or Notations, for Fluxions and Infinitesimals of the first, second, third, fourth and subsequent Orders, proceeding in the same regular form without end or limit $\dot{x}, \ddot{x}, \dddot{x}, \ddddot{x}, \&c.$ or $dx, ddx, dddx, ddddx, \&c.$ These Expressions indeed are clear and distinct, and the Mind finds no difficulty in conceiving them to be continued beyond any assignable Bounds. But if we remove the Veil and look underneath, if laying aside the Expressions we set ourselves attentively to consider the things themselves, which are supposed to be expressed or marked thereby, we shall discover much Emptiness, Darkness, and Confusion; nay, if I mistake not, direct Impossibilities and Contradictions. (*Analyst*, §8)

The strength of this argument is that it ties the metaphysical objections against the calculus to the logical objections. If confusion, impossibilities, and contradictions follow logically from the basic claims of the calculus, then there is good reason to suspect that the objects which it supposes do not (indeed, cannot) exist. Imagine, for example, someone who claims to be able to conceive a round square, introduces notation and a theory of round squares, and views my complaints that round squares are inconceivable as indicative of atrophied or underdeveloped mental powers. The only obvious way to convince him of his error is to show that his alleged geometry of round squares is inconsistent, from which it would seem to follow that he had confused a symbol for the round square with the existence of its alleged object.[9]

Proving inconsistency in the calculus would clearly be much more effective for Berkeley than simply reporting that he cannot conceive fluxions or differences, especially considering the history of mathematics. Many supposedly inconceivable objects—irrational, negative,

---

9. Of course, a truly recalcitrant geometer of the round square may refuse to recognize my proof of inconsistency or may dispute the logic by which it was derived. In such a case, reason fails and the dispute must be resolved by other means.

and complex numbers are obvious examples from well before Berkeley's day—have gained respectability as their usefulness became manifest.[10] Thus, a staunch eighteenth-century defender of the calculus might retort that the metaphysical criterion of rigor is simply irrelevant. But if these metaphysical objections are joined with cogent logical objections, then Berkeley's case is significant.

Still, Berkeley need not abandon his metaphysical critique of the calculus. His metaphysical objections to the Leibnizian calculus hold a significant degree of cogency, since the postulation of infinitesimal magnitudes is difficult to reconcile with prevailing eighteenth-century criteria of mathematical rigor. In fact, partisans of Newton who attacked Leïbniz frequently claimed that his formulation of the calculus was vastly less rigorous than Newton's, because it depended upon the supposition of infinitesimal magnitudes.[11] Thus if Berkeley can show that there is no meaningful distinction between fluxions and differences, the rigor of the Newtonian method will also be impugned. Newton's defenders could not then consistently claim that the method of fluxions is superior to the *calculus differentialis*, and their complaints about the obscurity of the Leibnizian method would apply equally to the foundations of the fluxional calculus.

Berkeley's claims for the inconceivability of the object of the calculus reappear when he considers alternative accounts of the "modern analysis" which might overcome his objections. However, I will first consider Berkeley's logical objections, since it is here that his case against the calculus comes most fully to light.

The Principles and Demonstrations of the Calculus

When Berkeley shifts his attention from the object of the calculus to its principles and demonstrations, he finds "Emptiness, Darkness, and Confusion . . . direct Impossibilities and Contradictions" (*Analyst*, §8). He evaluates two Newtonian proofs of elementary theorems in the

---

10. See Nagel (1935) for an account of the ways in which changes in mathematical theories have made previously "inconceivable" objects mathematically legitimate. Sherry (1991) is a discussion and revision of Nagel's claims.

11. Raphson (1715) contains a typical "history" which credits Newton with the invention of the calculus and charges Leibniz with both stealing the method and making it unrigorous by introducing infinitesimals. Taylor's attitude is typical, when he contrasts the Newtonian method with the continental approaches: "Cavalieri and more recent authors consider these parts as diminished in *infinitum*. But all of them, in considering the generation of quantities by the addition of parts, do not sufficiently observe that ἀκρίβεια of the geometers" (Taylor 1715, Preface). A study of the Newton-Leibniz dispute can be found in Hall (1980).

calculus, the first of which is a method for finding the fluxion of a product of two flowing quantities, Newton's version of the "product rule" for differentiation of a product (*Principia* II, 2, 2). In modern notation the rule asserts that, given functions $f(x)$ and $g(x)$, the derivative of the product $f(x)g(x)$ is $f'(x)g(x) + f(x)g'(x)$. Newton treats the product as a rectangle whose sides are the flowing quantities A and B, with moments a and b. The proof considers the case where each side lacks one-half of its moment and the resulting rectangle has an area of

$$\left(A - \frac{1}{2}a\right) \times \left(B - \frac{1}{2}b\right).$$

Multiplying through, this becomes

$$AB - \frac{1}{2}aB - \frac{1}{2}bA + \frac{1}{4}ab. \tag{6.1}$$

Newton then takes the rectangle formed after the flowing quantities have been increased by the remaining halves of their moments, viz:

$$\left(A + \frac{1}{2}a\right) \times \left(B + \frac{1}{2}b\right).$$

When expanded, this becomes

$$AB + \frac{1}{2}aB + \frac{1}{2}bA + \frac{1}{4}ab. \tag{6.2}$$

Newton claims that the moment of the product will be the difference between equations 6.2 and 6.1 or $aB + bA$.

Berkeley dismisses this proof as a sham. He rightly points out that the "direct and true" method of finding the increment of the area is to compare the product AB to the product $(A + a) \times (B + b)$. His augmented rectangle has an area of

$$AB + aB + bA + ab,$$

thus the increase in area is $aB + bA + ab$, which differs from Newton's result by the additional term ab.

Berkeley astutely reveals a fundamental flaw in Newton's procedure. Newton begins his discussion with the declaration that:

> These quantities I here consider as variable and indetermined, and increasing or decreasing, as it were, by a continual motion of flux; and I understand their momentary increments or decrements by the name of moments; so that the increments may be esteemed as added or affirmative moments; and the decrements as subtracted or negative ones. (*Principia*, II, 2, 2)

Newton straightforwardly declares that the moment of a flowing quantity is its momentary increment; thus, the moment of a product AB must be its momentary increment. Now if the moments of the quantities A and B are a and b, the moment of the product is the difference between AB and $(A + a) \times (B + b)$ as Berkeley contends. Newton's procedure here is utterly mysterious, since he actually takes the increment of the rectangle $(A - \frac{1}{2}a) \times (B - \frac{1}{2}b)$. Not only does Newton take the increment of the wrong product, but his procedure depends upon the confusing supposition that we can divide momentary increments of negligible magnitude into parts.

Berkeley insists that no matter how we interpret the doctrine of moments, Newton's procedure requires the use of infinitely small quantities and his denial of using infinitesimals is simply sophistical:

> The Points or mere Limits of nascent Lines are undoubtedly equal, as having no more Magnitude one than another, a Limit as such being no Quantity. If by a Momentum you mean more than the very initial Limit, it must be either a finite Quantity or an Infinitesimal. But all finite Quantities are expressly excluded from the Notion of a Momentum. Therefore the Momentum must be an Infinitesimal. And indeed, though much Artifice hath been employ'd to escape or avoid the admission of Quantities infinitely small, yet it seems ineffectual. For ought I see, you can admit no Quantity as a Medium between a finite Quantity and nothing, without admitting Infinitesimals. An Increment generated in a finite Particle of Time, is it self a finite Particle; and cannot therefore be a Momentum. You must therefore take an Infinitesimal Part of Time wherein to generate your Momentum. It is said, the Magnitude of Moments is not considered: And yet these same Moments are supposed to be divided into Parts. This is not easy to conceive, no more than it is why we should take Quantities less than A and B in order to obtain the Increment of AB, of which proceeding it must be owned the final Cause or Motive is very obvious; but it is not so obvious or easy to explain a just and legitimate Reason for it, or shew it to be Geometrical. (*Analyst*, §11)

There is nothing to contest in this passage, and with it Berkeley has gone a long way toward establishing his central claim for the absence of rigor in the calculus.[12] Certainly, Newton's mysterious procedure is motivated by a desire to avoid embarrassing questions about infinitesimal magnitudes, but in setting out a proof of this sort Newton has instead shown how unrigorous the calculus really is. We can thus grant that Berkeley is right on two counts: the procedures of the calculus are not properly demonstrated and the Newtonian apparatus of fluxions and moments is indistinguishable from the infinitesimal calculus of Leibniz.

It is worth noting that Berkeley was not alone in finding this Newtonian demonstration indirect and unconvincing. After setting out the same proof, Hayes declares:

> There is yet another way to find the Fluxion of any Rectangle XZ; which is thus, the Fluxions of the sides X and Z are $\dot{x}$ and $\dot{z}$, and therefore the Sides of the Rectangle Become $X + \dot{x}$ and $Z + \dot{z}$, and the Rectangle itself is $XZ + \dot{x}Z + \dot{z}X + \dot{x}\dot{z}$, from which subtracting the given Rectangle XZ, the remainder $\dot{x}Z + \dot{z}X$ (the term $\dot{x}\dot{z}$ being infinitely little in comparison of either of these) is the Fluxion of the Rectangle XZ. Q.E.I. (Hayes 1704, 6)[13]

Hayes follows Berkeley's procedure for finding the increment of the product, except that he dismisses the term ab (or $\dot{x}\dot{z}$ in his notation) as "infinitely little" relative to the other terms, in this following the practice of L'Hôpital and other continental analysts who routinely dismissed the products of two infinitesimals. Nieuwentijt objected to the Newtonian proof for substantially the same reasons as Berkeley, so this part of the *Analyst* is not entirely original.[14]

---

12. Commentators have noted the strength of Berkeley's case on this point. Blay observes: "But Newton gives no justification for his procedure, if it is not on the one hand that $\frac{1}{2}a - (-\frac{1}{2}a) = a$, and on the other hand that it permits him to avoid terms of the second order without neglecting them, since they have disappeared by themselves in the Newtonian calculus" (Blay 1986, 245–46). Sherry (1987), although otherwise unsympathetic to Berkeley's case against Newton, grants that Berkeley's shows this Newtonian proof to be merely "window dressing." It is important to note that the term "limit" in this passage is not used in the modern sense but reflects Euclid's definition of a point as the limit or terminus of a line.

13. As I noted earlier, Hayes uses the term "fluxion" for the "infinitely little increment or decrement" of a flowing quantity, a sense that differs significantly from Newton's. However, if we read "moment" where Hayes writes "fluxion," the passage makes reasonable sense.

14. See Nieuwentijt (1694, 24–27). It is not clear whether Berkeley read any of Nieuwentijt's critique of the calculus, since he mentions it only in citing such secondary sources as Leibniz's response in the *Acta Eruditorum*. Berkeley refers to Nieuwentijt's

Berkeley and the Calculus: The *Analyst*

Having dealt with the *Principia* proof of the product rule, Berkeley goes on to consider Newton's rule for finding the fluxion of any power, as demonstrated in the Introduction to the *Quadrature of Curves*. Berkeley's interest is understandable: the *Quadrature of Curves* contains a much more complete statement of the calculus than does the *Principia* and employs a significantly different method of proof. Berkeley insists that the obscurity of the proof in the *Principia* is intended to mask the use of infinitesimals, but suggests that Newton must have suffered "some inward Scruple or Consciousness of defect in the foregoing Demonstration" and, in view of the fundamental importance of the result for the whole calculus, resolved "to demonstrate the same in a manner independent of the foregoing Demonstration" (*Analyst*, §12).

Berkeley prefaces his objection to the second proof with a lemma which he regards as "so plain as to need no Proof." The lemma reads:

> If with a View to demonstrate any Proposition, a certain Point is supposed, by virtue of which certain other Points are attained; and such supposed Point be it self afterwards destroyed or rejected by a contrary Supposition; in that case, all the other Points, attained thereby and consequent thereupon, must also be destroyed and rejected, so as from thence forward to be no more supposed or applied in the Demonstration. (*Analyst*, §12)

In essence this lemma asserts the unexceptionable principle that contradictory premises are not to be admitted in a demonstration; to claim otherwise would permit the use of obviously fallacious patterns of argumentation. Berkeley argues that Newton violates this principle by employing contradictory assumptions in the following proof:

> *Let the Quantity x flow uniformly, and let the Fluxion of x^n to be found.* In the same time that the Quantity x by flowing becomes $x + o$, the Quantity x^n will become $(x + o)^n$, that is, by the method of infinite series
>
> $$x^n + nox^{n-1} + \frac{nn - n}{2} oox^{n-2} + \&c.,$$

Religious Philosopher in sections 190 and 198 of *Siris* but makes no direct reference to any of his mathematical works. It is clear that Berkeley was familiar with a wide range of late seventeenth-century mathematical literature, but it is quite possible that he did not have access to a copy of Nieuwentijt's work. Thus, Berkeley's critique of Newton's proof may well be independent of Nieuwentijt's. Vermeulen (1985) suggests that Berkeley's critique may derive from a reading of Nieuwentijt but admits that there is no firm evidence to link them.

And the Augments

$$o \text{ and } nox^{n-1} + \frac{nn-n}{2}oox^{n-2} + \&c.,$$

are to one another as

$$1 \text{ to } nx^{n-1} + \frac{nn-n}{2}ox^{n-2} + \&c.$$

Now let those Augments vanish, and their ultimate Ratio will be the Ratio of 1 to $nx^{(n-1)}$ and therefore the Fluxion of the Quantity x is to the Fluxion of the Quantity x^n as 1 to $nx^{(n-1)}$. (Newton 1964–67, 1:142)

Berkeley contends that Newton violates his foregoing lemma by making contradictory assumptions concerning o. Initially Newton treats o as a positive quantity, computing increments and comparing their ratios on the supposition that o is greater than zero. But after simplifying the ratios of the increments by dividing out the common term o, he makes a contradictory assumption: that o is equal to zero. According to Berkeley's lemma, when this new assumption, contrary to the original, is introduced, all consequences drawn from the original assumption must be rejected. But, in fact, important consequences are retained—consequences that cannot be derived from the new assumption. Berkeley declares:

> Hitherto I have supposed that x flows, that x hath a real Increment, that o is something. And I have proceeded all along on that Supposition, without which I should not have been able to have made so much as one single Step. From that Supposition it is that I get at the Increment of x^n, that I am able to compare it with the Increment of x, and that I find the Proportion between the two Increments. I now beg leave to make a new Supposition contrary to the first, *i.e.* I will suppose that there is no Increment of x, or that o is nothing; which second Supposition destroys my first, and is inconsistent with it, and therefore with every thing that supposeth it. I do nevertheless beg leave to retain nx^{n-1}, which is an Expression obtained in virtue of my first Supposition, which necessarily presupposeth such Supposition, and which could not be obtained without it: All which seems a most inconsistent way of arguing, and such as would not be allowed of in Divinity. (*Analyst*, §14)

Berkeley goes on at great length in sections 15 and 16, insisting that this Newtonian method of proof is an entirely sophistical exercise in the "shifting of hypotheses." He allows that this procedure involves a certain degree of finesse—if the shift is made too early, either all terms will reduce to zero or a division by zero will be required—but this skill is no substitute for sound and convincing demonstrations. Berkeley roundly condemns proponents of the Newtonian calculus for repeating this kind of argument, concluding that only mathematicians (and not theologians) could accept such blatant fallacies as demonstrations.

Berkeley concludes his attack by declaring that Newton's proof from the *Quadrature of Curves* is essentially the same method employed by the proponents of the differential calculus. He insists that when Newton supposes the increment o to be infinitely diminished and then rejects it, he is effectively rejecting an infinitesimal. As Berkeley rightly observes, it requires a "marvellous sharpness of Discernment" to distinguish between an evanescent increment and an infinitesimal difference. Indeed, the role of the mysterious quantity o can only be interpreted as that of an infinitesimal quantity, since it is treated as being both positive and less than any assignable magnitude.

As noted above, this attack appeals to arguments familiar to British mathematicians from the Newton-Leibniz priority dispute. Newton himself declared his approach more rigorous than the Leibnizian calculus because it was devoid of infinitesimal magnitudes.[15] This is the import of Newton's dictum, "Errours, tho' never so small, are not to be neglected in Mathematicks," a slogan which Berkeley picks up and uses to great rhetorical effect. Recall that Newton explicitly declares his intention to banish infinitesimals from the calculus at the end of the Introduction to the *Quadrature of Curves,* where he writes:

> In Finite Quantities so to frame a Calculus, and thus to investigate the Prime and Ultimate Ratio's of Nascent or Evanescent Finite Quantities, is agreeable to the Geometry of the Ancients; and I was willing to shew, that in the Methods of Fluxions there's no need of introducing Figures infinitely small into Geometry. (Newton 1964–67, 1:143)

Berkeley is not stating an entirely new view in claiming that there is no meaningful distinction between Newton's evanescent increments and Leibniz's infinitesimal differences. As we have seen, Hayes drew no distinction between differences and fluxions,[16] and several other

---

15. Thus, in his "Account of the *Commercium Epistolicum*" he argues that indivisibles and differences "have no Being either in Geometry or Nature" and cannot provide an adequate basis for analysis (Newton 1722, 205). See Hall (1980), where the "Account" is reprinted as an appendix.

16. Indeed, Hayes writes that "By the Doctrine of *Fluxions,* we understand the *Arithmetick* of infinitely small Increments or Decrements of Indeterminate or variable Quan-

works on the calculus from the early part of the eighteenth century also treat fluxions, moments, evanescent increments, and infinitesimal differences as essentially the same thing.[17] But in the context of remarks on the rigor of the Leibnizian methods, Berkeley's refusal to distinguish Newton's fluxions from the differential calculus has the effect of denying the admissibility of Newton's methods.

Having pointed out the compelling similarities between Newtonian evanescent increments and Leibnizian infinitesimal differences, Berkeley asserts that such infinitesimal methods do not provide the kind of accuracy required in mathematical demonstration: "It may perhaps be said that the Quantity being infinitely diminished becomes nothing, and so nothing is rejected. But according to the received Principles it is evident, that no Geometrical Quantity, can by any division or subdivision whatsoever be exhausted, or reduced to nothing" (*Analyst*, §17). This remark shows Berkeley to be unwilling to depart from the classical standards of rigor. The "received Principles" are evidently those of Greek geometry, and the claim that no geometric quantity can be exhausted by division can only be read as saying that no *finite* number of divisions can reduce a finite magnitude to nothing. Of course, allowing an infinite process of division would permit a finite quantity to be exhausted, but Berkeley is clearly not prepared to admit such infinite subdivision. I hardly need mention that Berkeley's declaration of the authority of received principles is inconsistent with his position in the *Philosophical Commentaries*, which posited a geometric magnitude, the minimum sensible, that cannot be further reduced. Presumably any "division" of the minimum would produce zero-magnitudes that can be rejected.

Berkeley suggests that Newton was himself aware of the shortcomings of his putative demonstrations. He interprets the different presentations employed by "the great Author of the Fluxionary Method" to mean that Newton was unsure of the cogency of his arguments and "not enough pleased with any one notion steadily to adhere to it." In support of this thesis, he cites a letter from Newton to Collins in which Newton claims to have a remarkable method for solving problems in analysis, but declines to demonstrate his basic results.[18] Berkeley

---

tities, and by such Quantities we understand those which in the Generations of (v.g.) a curve by local Motion perpetually increase or decrease" (Hayes 1704, 3).

17. See Cajori (1919, chapter 2) for a brief overview of these works.

18. The letter was published as part of the *Commercium Epistolicum* (Newton, 1712), a collection of letters published by Newton in support of his claim of priority in the invention of the calculus. The relevant passage is "I say there is no such curve line, but I can, in less than half a quarter of an hour, tell whether it may be squared, or what are

entertains the hypothesis that Newton settled for shaky "inductions" analogous to Wallis's procedures in the *Arithmetica Infinitorum*. Needless to say, Newton and his followers do not profit by this comparison:

> Whether this Satisfaction arose from tentative Methods or Inductions; which have often been admitted by Mathematicians (for instance by Dr. *Wallis* in his Arithmetic of Infinities) is what I shall not pretend to determine. But, whatever the Case might have been with respect to the Author, it appears that his Followers have shewn themselves more eager in applying his Method, than accurate in examining his Principles. (*Analyst*, §17)

Having critiqued the Newtonian principles of demonstration with considerable care, Berkeley quickly dismisses the efforts of the continental school of analysis. He finds, however, an amusing irony in contrasting the attitudes of Newton and Leibniz toward infinitesimal magnitudes. Newton denies relying upon infinitesimals and makes valiant but futile efforts to avoid them, while the infinitely small is "admitted and embraced by others without the least repugnance." Berkeley claims that the Leibnizian calculus employs the same inconsistent premises as its Newtonian counterpart, treating infinitesimals as sometimes positive and sometimes zero. He observes:

> *Leibnitz* and his Followers in their *calculus differentialis* making no manner of Scruple, first to suppose, and secondly to reject, Quantities infinitely small; with what clearness in the Apprehension and justness in the Reasoning, any thinking Man, who is not prejudiced in favour of those things, may easily discern. (*Analyst*, §18)

He characterizes this procedure as an abandonment of the stringent rigor of traditional geometrical methods, which allowed nothing to be "neglected" in the course of a demonstration.

What are we to make of Berkeley's critique of the calculus? Some authors have argued that Berkeley fails to make much of a case, at

---

the simplest figures it may be compared with, be those figures conic sections or others.... This may seem a bold assertion, because it is hard to say a figure may or may not be squared or compared with another, but it is clear to me by the fountain I draw it from though I will not undertake to prove it to others" (Rigaud [1841] 1965, 1:403–4). Clearly, Berkeley is more than a bit uncharitable when he reads this as Newton's acknowledgement that he was "satisfied concerning certain Points, which nevertheless he could not undertake to demonstrate to others." It could more plausibly be read as an indication of Newton's notorious reluctance to make his results public than as an expression of his inability to prove them.

least in his attack on Newton. Philip Kitcher, for example, has written that Berkeley's reading of Newton is "competent but uncharitable," and he finds the logical criticism to depend upon a deliberate misunderstanding of Newton's views. Moreover, he takes the doctrine of prime and ultimate ratios to be a reasonably rigorous articulation of a limiting process which can plausibly be read as a precursor to the Cauchy-Weierstrass presentation of the calculus (Kitcher 1984, 236–41). David Sherry similarly argues that Newton's talk of evanescent increments is "plainly . . . indicative of a limiting process" and his seemingly inconsistent principles can be seen to "function differently in different but related contexts" drawn from the science of kinematics (Sherry 1987). On Sherry's view, Newton's apparently contradictory assumptions about evanescent increments can be interpreted as a harmless equivocation which nevertheless promotes understanding.

I think that such responses to Berkeley miss an important point, namely that the effectiveness of his criticisms should be judged within the context of the mathematical work of the 1730s. Once the Berkeleyan critique is considered in its historical context it is really quite impressive. Nobody today doubts that the calculus has been made rigorous; the theory of limits and Robinson's nonstandard analysis have vindicated the procedures of eighteenth century calculus, and these versions of the calculus can overcome Berkeley's logical objections. But observing that the calculus can be made more rigorous based on the modern theory of limits does not blunt Berkeley's criticism that it lacked rigor in 1734: the apparatus of sequences and the relevant definitions of convergence were only developed a century after the publication of the *Analyst*. Similarly, it is of no use to show that charitable interpretations of Newton can evade Berkeley's critique if such interpretations were not part of the mathematical landscape of the 1730s.

In fact, the aptness of Berkeley's case can be gathered by noting that no figure in eighteenth-century British mathematics could give a widely-accepted exposition of the Newtonian theory of prime and ultimate ratios. Far from it, the efforts of Newton's successors to elucidate the mysteries of evanescent magnitudes were so notably unsuccessful that they led to a bitter controversy among British mathematicians in the mid-eighteenth century. In this dispute avowed Newtonians exchanged vituperative essays intended to set forth the true meaning of the Newtonian doctrine.[19] The confusion engen-

---

19. See Cajori (1919) for an account of the controversies. Much of the dispute was carried on in volumes 16 through 18 of the *Present State of the Republick of Letters* (1735–

Berkeley and the Calculus: The *Analyst*

dered by Newton's pronouncements on the nature of evanescent magnitudes was so great that John Wright, in his 1833 commentary on Newton's *Principia* was moved to declare:

> After all, however, neither [Newton] himself nor any of his Commentators, though much has been advanced upon the subject, has obviated this objection. Bishop Berkeley's ingenious criticisms in the Analyst remain to this day unanswered. He therein facetiously denominates the results, obtained from the supposition that the quantities, before considered finite and real, have vanished, the "Ghosts of Departed Quantities"; and it must be admitted there is reason as well as wit in the appelation. The fact is, Newton himself ... had no knowledge of the *true nature* of this Method of Prime and Ultimate Ratios. (Wright [1833] 1972, 2–3)[20]

Such an assessment of the century after the *Analyst* is hardly what we could expect if the Newtonian procedure was as rigorous and open to favorable interpretation as Kitcher and Sherry maintain. If anything, the history of the calculus after the *Analyst* suggests that (at least in Britain) the theory was considered unrigorous and was the object of conflicting interpretations, all attempting to find a rigorous presentation of the calculus which would overcome Berkeley's objections. The lack of agreement on how best to proceed lends considerable weight to Berkeley's characterization of the calculus as a mathematical theory whose principles were obscure and mysterious.

The Compensation of Errors Thesis

Berkeley's attack on the methods of reasoning employed in the calculus poses an interesting problem: If the procedures of the calculus are indeed founded upon sophistical arguments how can they deliver reliable results? Berkeley announces that he is not challenging the

---

36) and volumes 20 and 21 of the *History of the Works of the Learned* (1737). I will be concerned with some of these controversies in chapter 7 when I consider the various responses to the *Analyst*. Documentation of the debate includes Jurin (1735b, 1736a–c, 1737a–d), Robins (1736a–b), and Pemberton (1737a–f).

20. It should be noted that Wright misapplied Berkeley's famous reference to "ghosts of departed quantities" to the results obtained from the supposition that finite magnitudes have vanished, whereas Berkeley is characterizing evanescent magnitudes: "And what are these Fluxions? The Velocities of evanescent Increments? And what are these same evanescent Increments? They are neither finite Quantities, nor Quantities infinitely small, nor yet nothing. May we not call them the Ghosts of departed Quantities?" (*Analyst*, §35).

truth of any theorems in the calculus but only the manner of their demonstrations:

> I have no Controversy about your Conclusions, but only about your Logic and Method. How you demonstrate? What Objects you are conversant with, and whether you conceive them clearly? What Principles you proceed upon; how sound they may be; and how you apply them? It must be remembered that I am not concerned about the truth of your Theorems, but only about the way of coming at them; whether it be legitimate or illegitimate, clear or obscure, scientific or tentative. (*Analyst*, §20)

But accepting the results of the calculus while objecting to the reasoning which leads to these results requires Berkeley to explain how correct results can be obtained by flawed reasoning. Berkeley's explanation requires a sometimes confusing interweaving of algebraic and geometric argumentation. He claims that the calculus yields *inexact* results when its algorithms are applied to analytic expressions for curves but that this analytic error is balanced by a compensating geometric error when the resulting equation is used in the solution of a problem. Much of my work here will be devoted to reconstructing Berkeley's argumentation. The compensation of errors thesis is intended as a completely general explanation of the success of the calculus, but Berkeley's presentation of it is confined to four examples which he takes to illustrate the general rule. I propose to work through each example, trying to clarify the Berkeleyan thesis and evaluating its explanatory success.

Berkeley first considers the problem of finding a subtangent to a parabola as solved by the Leibnizian method of infinitesimal differences (*Analyst*, §21–23).[21] He begins with the parabolic curve AB (fig. 6.2) whose analytical expression is $y^2 = px$, where p is a constant and x and y are variables. For a point B on the curve, the subtangent will be the line PT. The problem is to obtain an expression for PT in terms of x and y. Denote AP by x, PB by y, the difference PM by dx, and the difference RN by dy.

First, treat the curve as a polygon, BN as a straight line coincident with the tangent, and the "differential triangle" BRN as similar to the triangle TPB. From the similarity of the triangles, take the proportion $RN : RB :: PB : PT$. Expressed in the language of differences this

---

21. Berkeley actually characterizes the problem as that of drawing the tangent to a parabola, but this problem quickly reduces to that of finding an appropriate expression for the subtangent.

Berkeley and the Calculus: The *Analyst*

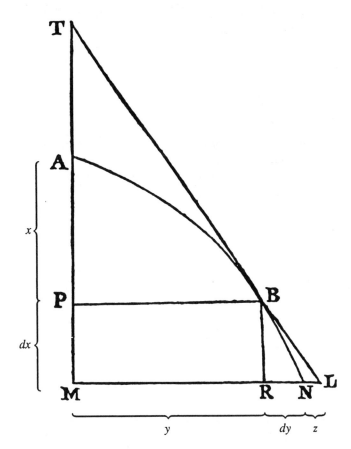

Fig. 6.2 From Berkeley, *The Analyst* (1734). Lowercase labels mine.

provides $PT = y\,dx/dy$ as the value of the subtangent. But this is only a partial solution, since the subtangent PT must be expressed solely in terms of x and y.

Next, find a convenient expression for dy which can be used to eliminate dx and dy from our subtangent equation by taking the derivative of the equation $y^2 = px$, or $2y\,dy = p\,dx$. Thus $dy = p\,dx/2y$. Replacing dy with $p\,dx/2y$ in the earlier subtangent equation, we find

$$PT = \frac{y\,dx}{p\,dx/2y} = \frac{2y^2}{p} \tag{6.3}$$

and our problem is solved.

Berkeley claims that two compensating errors have been made in this procedure, one in each of the first two steps. First, we falsely assumed that *BRN* was similar to *TPB*. In fact, *BRL* is similar to *TPB*. Thus, if we denote the line *NL* by z, the true expression for the subtangent is

$$PT = \frac{ydx}{dy + z} \tag{6.4}$$

and the value for the subtangent at the end of the first step appears to have been too large. The second error occurs in the differentiation of $y^2 = px$, where a second order infinitesimal was discarded. Taking the increments dx and dy of x and y, we should get

$$(y + dy)^2 = p(x + dx),$$

or

$$y^2 + 2ydy + dy^2 = px + pdx. \tag{6.5}$$

The increment of the equation is thus $2ydy + dy^2 = pdx$. Rearranging terms and simplifying, we get

$$dy = \frac{pdx}{2y} - \frac{dy^2}{2y}.$$

Here, it seems that the value obtained for dy is also too large, since it differs by the term $dy^2/2y$. However, if we can show that $z = dy^2/2y$ then the two errors will cancel and we will get the same result as before, since substituting $(pdx/2y - z)$ for dy in equation (6.4) yields

$$PT = \frac{ydx}{(pdx/2y - z) + z} = \frac{2y^2}{p}.$$

Berkeley proves that $z = dy^2/2y$ by invoking Apollonius' *Conics* (book 1, proposition 33). The Apollonian theorem states that the subtangent to a parabola is bisected at the vertex.[22] Thus, the Apollonian

---

22. The Apollonian statement reads: "If in a parabola some point is taken, and from it an ordinate is dropped to the diameter, and, to the straight line cut off by it on the diameter from the vertex, a straight line in the same straight line from its extremity is made equal, then the straight line joined from the point thus resulting to the point taken will touch the section" (Apollonius [1939] 1952, 640).

Berkeley and the Calculus: The *Analyst*

theorem implies that $TP = 2AP = 2x$. Now let BR or dx be denoted by m, and RN or dy be denoted by n. Using the tangent-axis theorem together with the similarity of the triangles TPB and BRL, we have

$$\frac{2x}{y} = \frac{m}{(n+z)}$$

and so

$$(n+z) = \frac{my}{2x}. \tag{6.6}$$

Returning to the expanded expression for the parabola in equation (6.5) but substituting n for dy and m for dx in accordance with our notational stipulation above, we get

$$y^2 + 2yn + n^2 = px + mp.$$

Because $y^2 = px$ we can subtract and get

$$2yn + n^2 = mp.$$

Dividing by p yields

$$m = \frac{2yn + n^2}{p}.$$

From the equation for the parabola ($y^2 = px$) we have $x = y^2/p$. Substituting these values for m and x into equation (6.6) we get

$$n + z = \frac{my}{2x} = \frac{2y^2np + yn^2p}{2y^2p}.$$

Cancelling out the common term py on the right half of the equation yields

$$n + z = \frac{2yn + n^2}{2y}$$

which can be reduced to an expression for z, namely

$$z = \frac{n^2}{2y} = \frac{dy^2}{2y}.$$

Q.E.D.

Berkeley concludes that there is, indeed, a compensation of errors in the solution of the subtangency problem:

> Now I observe in the first place, that the Conclusion comes out right, not because the rejected Square of *dy* was infinitely small; but because this error was compensated by another contrary and equal error. I observe in the second place, that whatever is rejected, be it ever so small, if it be real and consequently makes a real error in the Premises, it will produce a proportional real error in the Conclusion. Your Theorems therefore cannot be accurately true, nor your Problems accurately solved, in virtue of Premises, which themselves are not accurate. . . . I observe in the last place, that in case the Differences are supposed finite Quantities ever so great, the Conclusion will nevertheless come out the same: inasmuch as the rejected Quantities are legitimately thrown out, not for their smallness, but for another reason, to wit, because of contrary errors, which destroying each other do upon the whole cause that nothing is really, though something is apparently thrown out. (*Analyst*, §23)

I want to discuss this first example in some detail before proceeding, because Berkeley worked it out most carefully and I think that it ultimately provides the best case for his general claim.

The first thing to note is that Berkeley has shown that the problem of finding the subtangent can be solved without the use of infinitesimals. If we interpret dx and dy as denoting finite differences (as Berkeley's own diagram strongly suggests they be interpreted), we see that the tangent-axis theorem of Apollonius implies that the value of the subtangent is correctly given by equation (6.3). This result suggests the possibility of developing the calculus without relying upon infinitesimals. If Berkeley's procedure in this case is an instance of a more general technique of solving problems in analysis, then it should be possible to eliminate infinitesimal methods from the calculus both in principle and in practice. Berkeley, of course, was convinced that the results of the calculus could (in principle) be derived without resort to infinitesimals, although he himself never undertook the task of recasting its basic methods. It should be clear, however, that he considers this a necessary project, since he regards the theorems as true but the methods as flawed.[23]

---

23. In fact, the compensation of errors thesis can be extended to a more general case. Grattan-Guinness has shown that it can be made to work for any curve representable by a Taylor expansion. See Grattan-Guinness (1969).

Berkeley and the Calculus: The *Analyst*

Berkeley has shown that the subtangent problem can be solved with classical methods, but if we are to accept his argument that the calculus-based solution to the problem depends upon a compensation of errors we must first acknowledge that two errors are made when the differential calculus is used to solve the subtangent problem. It is not immediately obvious that this is the case. If we take the modern viewpoint and see the calculus as based upon the theory of limits, we can deny that an error is made in either of the two steps, since both cases involve taking the limit of a quantity $k + z$ as z approaches zero. Similarly, if we rely upon a theory of infinitesimals, where z is an infinitesimal which obeys the law $p + z = p$ for any real number p, the supposed errors turn out to be no errors at all.

Of course, invoking the modern theory of limits or of infinitesimals to undermine the compensation of errors thesis is an exercise in anachronism. But we must also ask whether the eighteenth-century theory of infinitesimals can by itself be invoked to deny that any errors appear in the course of the solution. Berkeley simply takes for granted that the balancing quantities in his analysis are finite.[24] However, a proponent of infinitesimal analysis could claim that the balancing errors, being infinitely small, are not real errors at all: they are equal and opposite, but either could have been discarded without requiring compensation.

Thus, this first illustration of the compensation of errors thesis makes an interesting point concerning the eliminability of infinitesimals from a specific problem in analysis but does not establish the more general thesis that the calculus involves errors *simpliciter*. A more modest (but still quite interesting) conclusion would seem to follow: if we grant that there are no infinitesimal magnitudes, the procedures of the differential calculus can be vindicated (at least for a certain class of problems) by interpreting them within the framework of classical geometry, with compensating finite quantities doing the work of discarded infinitesimals.

Berkeley's second example of the compensation of errors considers a problem similar to the first but follows a slightly different procedure in that the second appears to rely upon the rejection of only a single infinitesimal (*Analyst*, §§24–25). Berkeley takes this to be a harder case but attempts to show that two compensating errors have indeed been made. We begin by considering the parabola *ARS* (fig. 6.3), with

---

24. This is reflected in his remark: "whatever is rejected, be it ever so small, if it be real and consequently makes a real error in the Premises, it will produce a proportional real error in the Conclusion" (*Analyst*, §23).

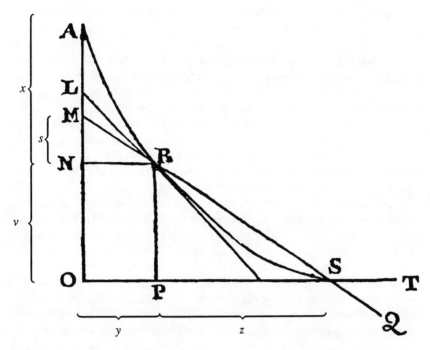

Fig. 6.3 From Berkeley, *The Analyst* (1734). Lowercase labels mine.

ordinates NR and OS, tangent LR at point R, and secant MQ cutting AR at points R and S. Denote AN by x, NR by y, NO by v, PS by z, and the subsecant MN by s. We take the equation $y = x^2$ as the analytic expression of the curve and proceed to find the subtangent NL. In distinction to the previous case, we begin by considering finite increments z and v of the quantities x and y, but then diminish v and discard a single infinitesimal. Including the increments z and v in the equation $y = x^2$, we get $y + z = x^2 + 2xv + v^2$, so $z = 2xv + v^2$. Because the triangles PRS and NMR are similar, we have $PS : PR :: NR : NM$. Expressed in the language of differences, this becomes $z/v = y/s$, so we have $s = vy/z$. But because $y = x^2$ and $z = 2xv + v^2$, we can substitute and get a value for the subtangent in terms of x and v, since

$$s = \frac{vx^2}{2xv + v^2} = \frac{x^2}{2x + v}.$$

Now we suppose the quantity v to be infinitely diminished and discard it, getting

Berkeley and the Calculus: The *Analyst*

$$s = \frac{x^2}{2x} = \frac{x}{2}.$$

At the moment v is infinitely diminished, secant MQ cuts AR only at R; thus MQ coincides with tangent LR and subsecant NM with subtangent NL. Hence the value for the subsecant, $x/2$, is also the value for the subtangent.

Berkeley finds this case more problematic because it seems to show that a single infinitesimal difference can be ignored without error. He acknowledges that $x/2$ is the true value of the subtangent, and observes that

> [S]ince this was obtained by one only error, *i.e.* by once rejecting one only Infinitesimal, it should seem, contrary to what hath been said, that an infinitesimal Quantity or Difference may be neglected or thrown away, and the Conclusion nevertheless be accurately true, although there was no double mistake or rectifying of one error by another, as in the first Case. (*Analyst*, §24)

He attempts to show that a less apparent second error has been made in supposing that the subsecant NM ultimately becomes equal to the subtangent NL. He declares:

> For in the first place, it was supposed, that when NO is infinitely diminished or becomes an Infinitesimal then the Subsecant NM becomes equal to the Subtangent NL. But this is a plain mistake, for it is evident, that as a Secant cannot become a Tangent, so a Subsecant cannot be a Subtangent. Be the Difference ever so small, yet still there is a Difference. And if NO be infinitely small, there will even then be an infinitely small Difference between NM and NL. Therefore NM or s was too little for your supposition, (when you supposed it equal to NL) and this error was compensated by a second error in throwing out v, which last error made s bigger than its true value, and in lieu thereof gave the value of the Subtangent. This is the true State of the Case, however it may be disguised. (*Analyst*, §24)

A number of points call for comment here. First, when Berkeley says that a secant cannot become a tangent, he is objecting to the supposition that gradually diminishing the arc cut off by the secant (by rotating it about one of the points in which it cuts the curve) will eventually cause the two points of intersection to coincide, so that the line has only one point in common with the curve and is therefore a tangent. Just why this procedure should be found objectionable is not at all clear. The real object of Berkeley's concern here is the related claim that ratios determined by the properties of the secant will continue to hold when the secant is brought into coincidence with the tangent. Such ratios are nothing other than Newton's ultimate ratios of evanescent quantities, and the substance of Berkeley's criticism here is that we cannot speak of ratios of evanescent quantities, because there can be no ratios when the quantities have vanished or become nothing.[25]

Second, Berkeley does not show in this case that the errors supposedly involved in this case are equal and opposite. In the previous case, he could show that the errors introduced (when the differences are taken as finite quantities) mutually cancel, and as a consequence of this result the first subtangency problem can be solved without resort to infinitesimals. Here, however, he merely asserts that the errors balance, although it is clear that this will not happen if we use only finite differences. Berkeley claims that $NL - NM = NO$ in the infinitesimal case, but this does not hold for arbitrary finite values of NO or finite differences $NL - NM$.[26] Berkeley seems simply to assume here that all infinitesimal quantities are equal—an assumption obviously not made by those who are prepared to introduce infinitesimal quantities and compare their ratios. Thus, there is a rather major defect in Berkeley's analysis of the second case, where he conveniently assumes that the two errors here are equal without giving any reason to think that they are, which is not present in the previous case. This shortcoming in Berkeley's argument does not imply that the problem under consideration is unsolvable without the supposition of infinitesimal magnitudes. By exploiting the classical theory of conic sections in more or less the same manner as in the first problem, the subtangent can be found without resort to infinitesimal considerations. However, Berkeley fails to establish his stronger thesis that this application of the calculus involves compensating errors.

In his third example of the compensation of errors thesis Berkeley moves from a consideration of the subtangent to a discussion of the Newtonian method of fluxions as applied to the computation of quadratures. The problem is to determine the ordinate from an equation

---

25. It is clear that Berkeley's presentation of the case is intended to apply to the Newtonian method of fluxions, whereas the first case was intended as applying to the Leibnizian differential calculus. Indeed, the case Berkeley discusses is practically drawn from book I of the *Principia* (lemmas V–VIII), where Newton presents that part of his doctrine of ultimate ratios which is necessary for the derivation of the inverse square law.

26. This is easily seen by observing that NO can be made as large as desired, NL is constant, and NM is never greater than NL. The two quantities will become equal in the limiting case, but Berkeley does nothing to show this.

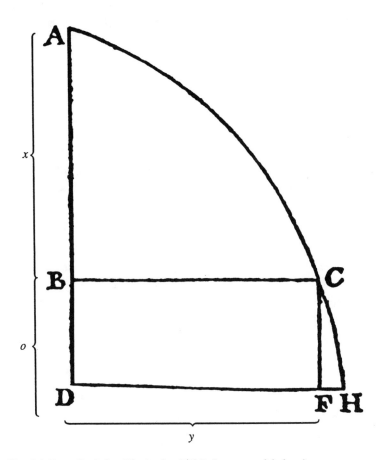

Fig. 6.4 From Berkeley, *The Analyst* (1734). Lowercase labels mine.

for the area under a curve—the familiar fundamental theorem of the calculus. In the case at hand, we want to show that the ordinate is the fluxion of the area. As we will see, Berkeley's argument here depends upon an interesting assumption and fails to establish anything like a generalized compensation of errors thesis.

Berkeley begins by promising that "the Doctrine premised may be further illustrated by the following simple and easy Case, wherein I shall proceed by evanescent Increments" (*Analyst*, §26). He takes the flowing quantity AC, letting $AB = x$, $BC = y$, and $BD = o$ as in fig. 6.4. Supposing that the area ABC is x^2, we want to show that $y = 2x$. Berkeley argues that when x flows to $x + o$, the area x^2 increases to $x^2 + 2xo + o^2$, and the area ABC becomes ADH, which is equal to $ABC + CBDH$. Thus, $2xo + o^2 = BCHD$, but $BCHD$ itself is in turn equal to $BCFD + CFH$. He next assumes that CFH has an area of qo^2, so that

$$2xo + o^2 = yo + qo^2.$$

Dividing by o, we get

$$2x + o = y + qo.$$

Following the method of fluxions, we then let the quantity o vanish and discard the evanescent terms containing it, getting $2x = y$ as the value of the ordinate, and the problem is solved.

As before, Berkeley insists that it is absurd and fallacious to assume that o vanishes while retaining the intermediate results which depend upon o being a positive quantity. Thus, he asks "how comes it to pass that the throwing out o is attended with no Error in the Conclusion?" His answer is that "because q being Unit, $qo = o$." This assumption allows Berkeley to conclude that

$$2x + o - qo = y = 2x.$$

As he puts it:

> Therefore o cannot be thrown out as an Infinitesimal, or upon the Principle that Infinitesimals may be safely neglected. But only because it is destroyed by an equal Quantity with a negative Sign, whence $o - qo$ is equal to nothing. And as it is illegitimate to reduce an Equation, by subducting from one Side a Quantity when it is not to be destroyed, or when an equal Quantity is not subducted from the other Side of the Equation: So it must be allowed a very logical and just Method of arguing, to conclude that if from Equals either nothing or equal Quantities are subducted, they shall still remain equal. And this is the true Reason why no Error is at last produced by the rejecting of o. (*Analyst*, §27)

There are two very obscure points here. The first is Berkeley's assumption that the area $CFH = qo^2$, the second his claim that $q = 1$. Neither of these assumptions holds generally, and it is hard to see why Berkeley introduces them, except for the obvious reason that they allow the problem to be solved in a special case.

John Wisdom has pointed out these shortcomings in Berkeley's argument, and it is worth quoting his analysis of the situation:

> Berkeley throws no light on the assumptions. They seem to be pure mistakes except where ACH is a straight line. It is very difficult, moreover, to understand why he introduced q at all when its value was supposed to be unity. Attention has never, so far as I am aware, been drawn to these points. A strange feature of the argument is that the area ABC, when its value is x^2, is a triangle; and the argument is obviously valid for a triangle. It would seem that Berkeley thought out the example in terms of the special case of the straight line, for which [both of the questionable assumptions] are true, then sought to make the illustration more general by replacing the straight line by a simple curve, but forgot to look for any alterations in his equations that such a change might necessitate. (Wisdom 1941, 60)

Wisdom may well have the correct account of these mysterious assumptions. There is certainly reason to think that Berkeley is here motivated more by a desire to produce compensating errors than any well-grounded mathematical reasoning.

Berkeley's final example of the compensation of errors is an attempt to generalize the foregoing case. He wants to show that the rule for finding the fluxion of x^n is derived by the same kind of compensating errors as before. He first assumes that the area ABC in the previous figure is equal to x^n and admits that the fluxion of the area is indeed $nx^{(n-1)}$, proposing "to inquire how it is arrived at." Of course, he cannot accept the Newtonian demonstration, but he argues that by considering the areas in the figure we can indeed see how the result should be properly demonstrated, since "if we fairly delineate the Area and its Increment, and divide the latter into two parts BCFD and CFH, and proceed regularly by Equations between the algebraical and geometrical Quantities, the reason of the thing will plainly appear" (*Analyst*, §28).

The strategy here is to consider the analytical and geometrical increments, and argue that the terms in the two expressions will balance. By the binomial theorem, the increment of $(x + o)^n$ is

$$nox^{(n-1)} + \frac{(n^2 - n)}{2} o^2 x^{(n-2)} + \dots$$

And the geometric increment is $BDFC + CFH$. Setting these equal to one another we have:

$$nox^{(n-1)} + \frac{(n^2 - n)}{2} o^2 x^{(n-2)} + \dots = BDFC + CFH$$

Chapter Six

We then retain only the first term from each side of the equation obtaining $nox^{(n-1)} = BDFC$. Dividing through by the increment of x, which is o in the case of the analytic expression and BD in the geometric case, we get $nx^{(n-1)} = BC$, as required. Berkeley notes that this procedure requires the assumption that the area CFH is equal to the algebraic expression $((n^2 - n)/2)o^2x^{(n-2)} + \ldots$. Hence, the compensating errors arise when the rejection of these terms in the analytic expression for the curve is balanced by similarly ignoring the geometrical quantity CFH.

Here again, Berkeley's case falls rather short of demonstration. There is no reason to assume that these two quantities are equal. He cannot justify his claim that $CFH = ((n^2 - n)/2o^2x^{(n-2)} + \ldots$, but resorts to the device of invoking the truth of the conclusion as a justification for a questionable procedure:

> If therefore the Conclusion be true, it is absolutely necessary that the finite space CFH be equal to the Remainder of the Increment expressed by
>
> $$\frac{n^2 - n}{2}o^2x^{n-2} + \&c.$$
>
> equal I say to the finite Remainder of a finite Increment. (*Analyst*, §28)

This is exactly the kind of argument Berkeley attacks in his critique of the Newtonian methods, but he seems to have forgotten his scruples in his haste to find compensating errors. Berkeley's argument is inadequate in that it moves from a premise of the form $\alpha + \beta = \gamma + \delta$ to a conclusion of the form $\alpha = \gamma$ and $\beta = \delta$. To avoid explicitly relying upon this invalid argument, Berkeley appeals to the fact that the value of the ordinate is $nx^{(n-1)}$ and the correct result can only be obtained if the terms in the geometric and analytic sums are allowed to be identified in this way. But it is exactly the correctness of the value for the ordinate that is in question, since it was supposedly derived by Newton's invalid methods.

Berkeley concludes his presentation of the compensation of errors thesis by claiming that the methods he has used to show the compensation can be extended so as to free the calculus from its reliance upon infinitesimal magnitudes. After he has discussed his fourth example of compensating errors, he writes:

> This hint may, perhaps, be further extended and applied to good purpose, by those who have leisure and curiosity for such Matters. The use I make of it is to shew, that the Analysis

cannot obtain in Augments or Differences, but it must also obtain in finite Quantities, be they ever so great, as was before observed. (*Analyst*, §29)

The clear intent of these remarks is to suggest revising the methods of the calculus to retain the standard results without the use of infinitesimals, and Berkeley refers back to these examples in query 37, when he proposes that the procedures of the calculus be replaced by methods which do not require "new *Postulata* or Assumptions."

Given the difficulties Berkeley encounters attempting to demonstrate his compensation of errors thesis, the prospects for this kind of rigorization program look rather dim. But this approach has some interesting consequences. First and foremost, although Berkeley's treatment of geometry in the *Principles* contains elements of instrumentalism, instrumentalism is hardly the dominant strain in his approach to the calculus at the time of the *Analyst*.

When Berkeley claims that "to the end any theorem may become universal in its use, it is necessary we speak of lines described on paper, as though they had parts which really they do not" (*Principles*, §128) he endorses a very weak instrumentalism, as I argued in chapter 2. To say that we must speak falsely in order to obtain a theorem which is universal in its use sounds very much like an admission that geometric theories, although literally false when taken as descriptions of what we actually perceive, are nevertheless true of idealizations toward which any approximations tend. But such a view is far removed from a radical instrumentalism that takes predictive success as the sole criterion for theory acceptance.

Berkeley's struggle to use compensating errors in accounting for the success of the calculus is also part of his rejection of a thoroughgoing mathematical instrumentalism. His effort to explain away the paradox of true conclusions drawn from false premises is exactly the kind of work an instrumentalist need not bother with. The committed mathematical instrumentalist would hold that the calculus is acceptable because it invariably delivers correct results, but the attempt to reveal the deeper truths which *really* account for these correct results runs counter to the fundamental tenets of thoroughgoing instrumentalism.

One reason for Berkeley's refusal to pursue an instrumentalistic reading of the calculus should be readily apparent. The instrumentalistic philosophy of science in *De Motu* and *Alciphron* is intended to make sense of natural sciences whose objects are not readily intelligible, and could at best overcome the metaphysical objections to the calculus. But the logical critique cannot be so easily avoided, unless we regard logical constraints on mathematical theories as wholly subordinate to practical usefulness. This simply makes a hash of the philosophy of mathematics (since it renders anything demonstrable) and takes all of the bite out of Berkeley's logical criticism. Small wonder, then, that Berkeley never considered such an option.

I suggested in chapter 2 that we read Berkeley's apparent endorsement of mathematical instrumentalism in the *Principles* as a very tame thesis, rather than a wholesale endorsement of the claim that mathematical theories must be judged purely on the basis of their utility. In essence, Berkeley's attitude toward geometry is much like that of a physicist who must ignore specific complicating features of an experimental setup. Nobody would pretend to apply Newtonian mechanics to the problem of determining a baseball's trajectory without ignoring the gravitational effect of Jupiter or the variable density of the air. Similarly, Berkeley's remark that we must treat perceived geometric figures "as if they had parts, which really they do not," can be interpreted as saying that geometric theorems are not (strictly speaking) true of perceivable objects, but only of idealizations to which perceivable lines and figures are approximations. Just as physics makes convenient (if false) assumptions about the characteristics and behavior of actual bodies, Berkeley's account of geometry requires that we make convenient assumptions about the characteristics of perceived magnitudes in order to construct a theory whose truth depends upon the limits toward which any actual sequence of approximations tends.

This picture of geometry as the science of approximations emerges more clearly in the queries at the end of the *Analyst*. Among other things, these queries state Berkeley's account of the "object and end" of geometry, as well as the methods by which the object is to be studied and its end pursued. The object of geometry is the proportions of finite extensions (query 1), its end or purpose is to measure such extensions (query 2). In measuring finite extension, we use diagrams to represent all other extensions of the same kind, and the diagrams therefore represent arbitrarily large magnitudes (queries 17–18). In finding proportions between magnitudes the end of geometry will be satisfied if we can generate "unlimited approximations" from the consideration of these arbitrarily large magnitudes and their mutual relations (query 53). It hardly needs to be stressed that such sequences of "unlimited approximations" are the fundamental tool in the method of exhaustion, and Berkeley's account of geometry is essentially that of the classical exhaustion method.

Berkeley sees the compensation of errors thesis as the key to a program for replacing the calculus with a method which can deliver accurate quadratures by use of unlimited approximations. He asks:

> Whether the greatest Genius wrestling with false Principles may not be foiled? And whether accurate Quadratures can be obtained without new *Postulata* or Assumptions? And if not, whether those which are intelligible and consistent ought not to be preferred to the contrary? *See Sect. 28 and 29.* (*Analyst,* query 37).

Sections 28 and 29, to which he refers here, are Berkeley's fourth illustration of the compensation of errors thesis and contain his general statement of the claim that demonstrations in the fluxional calculus can be accepted because the algebraic and geometric quantities in the demonstration balance each other out. Berkeley hints that others should attempt to carry out a program of interpreting the calculus in terms of his conception of geometry. Given what we have seen, the prospects for Berkeley's program of rigorizing the calculus are not promising. Nevertheless, we can now see what kind of theory would meet Berkeley's criteria of rigor.

The attempt to found the calculus on the theory of compensating errors did not end with Berkeley. The most famous proponent of error compensations was Lazare Carnot, who in *Réflexions sur la Métaphysique du Calcul Infinitésimal* (1797) attempted a project not unlike that which Berkeley suggested, although there is little direct evidence that Carnot was influenced by Berkeley.[27] Other interesting responses to Berkeley's demand that the calculus be rigorized came from various British mathematicians whose work I will discuss in chapter 7, but first I will consider the last twenty sections of the *Analyst,* where Berkeley rounds out his attack on the fluxional calculus.

Ghosts of Departed Quantities and Other Vain Abstractions

In the last sections of the *Analyst* Berkeley considers various possible responses to his case against the calculus and argues that all fail to satisfy appropriate standards of rigor. These sections connect Berkeley's metaphysical criticisms of the calculus to his antiabstractionist epistemology by arguing that certain formulations of the doctrine of

---

27. For an account of various eighteenth-century projects in the foundations of the calculus, including Carnot's attempted proofs by compensating errors, see Grabiner (1981, chapter 2). Carnot's scientific work is studied in Gillispie (1971). Ultimately, Carnot's arguments fall victim to the same problems we have seen in Berkeley's work on compensating errors.

fluxions are inadmissible because they require impossible abstractions. They also emphasize how different Berkeley's philosophy of geometry in the *Analyst* is from his formalistic treatment of algebra and his instrumentalistic approach to the physical sciences. It is worthwhile to consider Berkeley's strategy, since it draws together a number of important themes in his philsophy of mathematics.

The principal object of Berkeley's attack are presentations of the calculus which treat fluxions as velocities. He contends that:

> we have no Notion whereby to conceive and measure various Degrees of Velocity, beside Space and Time, or when the Times are given, beside Space alone. We have even no Notion of Velocity prescinded from Time and Space. When therefore a Point is supposed to move in given Times, we have no Notion of greater or lesser Velocities or of Proportions between Velocities, but only of longer or shorter Lines, and of Proportions between such Lines generated in equal Parts of Time. (*Analyst*, §30)

But defining velocity as an ultimate ratio of evanescent magnitudes offends against this principle, since it requires us to abstract the idea of velocity from the ideas of space and time. This is impossible on Berkeley's view of the matter, and he concludes that talk of instantaneous velocities or the proportions between such velocities is simply unintelligible:

> And if the Velocities of nascent and evanescent Quantities, *i.e.* abstracted from Time and Space, may not be comprehended, how can we comprehend and demonstrate their Proportions? Or consider their *rationes primæ* and *ultimæ*? For to consider the Proportion or *Ratio* of Things implies that such Things have Magnitude: That such their Magnitudes may be measured, and their Relations to each other known. But, as there is no measure of Velocity except Time and Space, the proportion of Velocities being only compounded of the direct Proportion of the Spaces, and the reciprocal Proportion of the Times; doth it not follow that to talk of investigating, obtaining, and considering the Proportions of Velocities, exclusively of Time and Space, is to talk unintelligibly? (*Analyst*, §31)

This style of criticism, essentially an argument that key concepts in the calculus of fluxions can only be introduced by impossible abstractions, depends upon Berkeley's antiabstractionist epistemology. Thus, Berkeley's metaphysical critique of the calculus ultimately depends upon his rejection of abstract ideas. But this should be no surprise:

Berkeley and the Calculus: The *Analyst*

after all, in objecting that a mathematical theory violates a metaphysical criterion of rigor one must eventually make explicit the epistemological or metaphysical doctrines against which it offends.

As Berkeley considers various presentations of the calculus of fluxions, he first sets out a gloss on each which might appear to render its key concepts more readily intelligible, then objects that the presentation nevertheless involves or presupposes an impossible abstraction. For instance, he investigates Newton's dictum in the Introduction to the *Quadrature of Curves* that fluxions may be "expounded" by finite lines which stand in the same proportion as the fluxions themselves, a strategy which, if successful, would seem to evade Berkeley's critique, since Berkeley is prepared to admit that finite lines and the proportion between such lines are easily conceived. Berkeley considers whether the Newtonian doctrine can produce these finite lines in an epistemologically acceptable fashion. I considered Newton's argument in chapter 4, but it bears repeating in this context.

We begin with the curve *AC* having abscissa *AB*, ordinate *BC*, and tangent *VCH* at the point *C* (fig. 6.5). Take *Bb* as a finite increment in the abscissa and construct the new ordinate *bc*, continuing it until it cuts *VCH* at *T*. Newton concludes that as the ordinate *bc* is brought closer to the ordinate *BC*, the line *CK* will approach the tangent; and when *BC* and *bc* meet the line *CK* will coincide with the tangent, the evanescent triangle *CEc* will become similar to the triangle *CET*, and the fluxions of the lines *AB*, *BC*, and *AC* will be in the same ratio as the finite lines *CE*, *ET*, and *CT*.

Berkeley objects to this procedure, arguing that it requires the incomprehensible thesis that a triangle is a point:

> It is particularly remarked and insisted on by the great Author, that the Points *C* and *c* must not be distant one from another, by any the least Interval whatsoever: But that, in order to find the ultimate Proportions of the Lines *CE*, *Ec*, and *Cc* (*i.e.* the Proportions of the Fluxions or Velocities) expressed by the finite Sides of the Triangle *VBC*, the Points *C* and *c* must be accurately coincident, *i.e.* one and the same. A Point therefore is considered as a Triangle, or a Triangle is supposed to be formed in a Point. Which to conceive seems quite impossible. Yet some there are, who, though they shrink at all other Mysteries, make no difficulty of their own, who strain at a Gnat and swallow a Camel. (*Analyst*, §34)

Here again, Berkeley uses Newton's pronouncements to great rhetorical effect. In noting that "the great Author" insists that the points *C* and *c* be brought into coincidence, he is referring to Newton's claim that

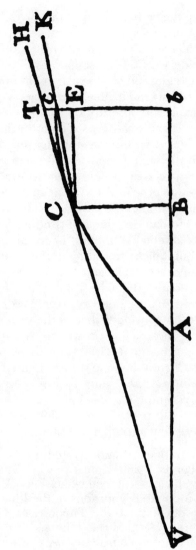

Fig. 6.5 From Berkeley, *The Analyst* (1734).

If the Points C and c be at any small distance from one another, then will CK be at a small distance from the Tangent CH. As soon as the Right Line CK coincides with the Tangent CH, and the ultimate Ratio's of the Lines CE, Ec and Cc be found, the Points C and c ought to come together and exactly to coincide. For Errours, tho' never so small, are not to be neglected in Mathematics. (Newton 1964–67, 1:141)

The underlying issue here again relates to abstraction. In order for Newton's presentation to work, it seems that we must admit ratios in abstraction from the quantities which form the ratios. To put it another way, Newton supposes that ratios determined by the sides of a triangle remain even after the triangle has vanished and become a point. Berkeley claims that although we can clearly conceive finite lines and understand what it is for them to stand in a given proportion, we cannot elucidate the doctrine of fluxions by simply declaring that fluxions are expressed by proportions between finite lines, particularly when the procedure for finding the requisite finite lines and their proportion requires us to assume that a figure can be formed in a point.

Berkeley considers and rejects other formulations of the calculus of fluxions, not all of which need concern us here. But one kind of argument is of interest because it shows that Berkeley was opposed to a purely formalistic foundation of the calculus in which algebraic argumentation would justify the procedures without geometric considerations. He observes that "some Men may hope to operate by Symbols and Suppositions, in such sort as to avoid the use of Fluxions, Momentums, and Infinitesimals" (*Analyst*, §35). He objects to the idea that we can simply replace geometric quantities with algebraic symbols and derive fluxions in a purely algebraic fashion. Thus, if we assume that x and z represent two abscissa of a curve, and take x^3 and z^3 for the corresponding areas, we can denote the increment of the abscissa as $z-x$ and the increment of the area as $z^3 - x^3$. Dividing the increment of the area by the increment of the abscissa, $(z^3 - x^3)/(z - x)$, yields $z^2 + zx + x^2$ as the quotient. Given this expression for the ratio of the two increments we can now assume that $z = x$ and substitute x for z in the expression for the ratio of the increments. This yields $3x^2$ as the ordinate of the curve, which we seem to have found without the use of fluxions, evanescent increments, or infinitesimals.

Berkeley responds with a logical criticism. The initial computation of increments and division by $z - x$ tacitly assumes that z and x are unequal; otherwise a division by zero would result. This assumption is later contradicted by the assumption that $z = x$ (or, equivalently, that $z - x = 0$), so this putative vindication of the calculus is as logically faulty as any of its predecessors. He adds a further criticism, to the effect that the "object and end" of geometry have been ignored by this purely algebraic maneuver:

> And there is indeed reason to apprehend, that all Attempts for setting the abstruse and fine Geometry on a right Foundation, and avoiding the Doctrine of Velocities, Momentums, &c. will be found impracticable, till such time as the Object and End of Geometry are better understood, than hitherto they seem to have been. (*Analyst*, §35)

Berkeley requires an understanding of the object and end of geometry explicitly opposed to formalism, and he contends in subsequent sections that various attempts to clarify the calculus have succeeded only in providing handy sets of signs or symbols to which no truly geometric ideas can correspond.

Berkeley's antiformalistic attitude toward the calculus is summed up nicely when he complains:

> Men too often impose on themselves and others, as if they conceived and understood things expressed by Signs, when in truth they have no Idea, save only of the very Signs themselves. And there are some grounds to apprehend that this may be the present Case. The Velocities of evanescent or nascent Quantities are supposed to be expressed, both by finite Lines of a determinate Magnitude, and by Algebraical Notes or Signs: but I suspect that many who, perhaps never having examined the matter, take it for granted, would upon a narrow scrutiny find it impossible, to frame any Idea or Notion whatsoever of those Velocities, exclusive of such finite Quantities and Signs. (*Analyst*, §36)

This passage prefaces Berkeley's consideration of several attempts to justify the calculus by introducing a finite right line and algebraic symbols intended to denote the velocities with which the line is produced by the motion of a uniformly accelerated point.

Thus, if we imagine the line *KP* (fig. 6.6) to be described by the

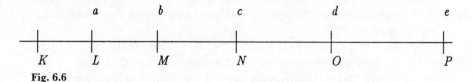

Fig. 6.6

Berkeley and the Calculus: The *Analyst*

motion of a continually accelerated point, we obtain the unequal intervals *KL, LM, MN, NO,* and *OP.* Furthermore, we can represent the velocity of the point at *L, M, N, O,* and *P* as *a, b, c, d,* and *e.* We can then treat the increasing velocity as a flowing quantity, and investigate the rate of increase in the velocity by comparing the differences of the respective velocities. This results in a sequence of expressions

$$a, (b - a), (c - 2b + a), (d - 3c + 3b - a),$$
$$(e - 4d + 6c - 4b + a), \ldots$$

These successive terms can then be called the fluxions of the velocity.

Such a sequence of algebraic expressions is quite intelligible, but Berkeley insists that it is not a proper explication of the doctrine of fluxions:

> Nothing is easier than to assign Names, Signs, or Expressions to these Fluxions, and it is not difficult to compute and operate by means of such Signs. But it will be found much more difficult, to omit the Signs and yet retain in our Minds the things, which we suppose to be signified by them. To consider the Exponents, whether Geometrical, or Algebraical, or Fluxionary, is no difficult Matter, But to form a precise Idea of a third Velocity for instance, in it self and by it self, *Hoc opus, hic labor.* Nor indeed is it an easy point, to form a clear and distinct Idea of any Velocity at all, exclusive of and prescinding from all length of time and space; as also from all Notes, Signs or Symbols whatsoever. This, if I may be allowed to judge of others by my self, is impossible. To me it seems evident, that Measures and Signs are absolutely necessary, in order to conceive or reason about Velocities; and that, consequently, when we think to conceive the Velocities simply and in themselves, we are deluded by vain Abstractions. (*Analyst,* §37)

The complaint here is a familiar one: the proponents of the calculus have neglected the true object of geometry, instead constructing a notational system which cannot be interpreted in terms of the properties of perceivable extension but only in terms of such "vain Abstractions" as instantaneous velocity.

I began by investigating Berkeley's case against abstract ideas, then considered how his antiabstractionist epistemology influenced his account of geometry and arithmetic. His metaphysical critique of the calculus at first seems to have no direct connection with his denial of abstract ideas and his logical critique none at all. But in elaborating his case against the calculus, Berkeley rejects the interpretation of fluxions as instantaneous velocities on grounds familiar from my discussion of his views on geometry and arithmetic.

Algebraic and arithmetical theories are legitimate, on Berkeley's account, precisely because they are "purely nominal" sciences which deal only with the manipulation of symbols. Their symbols denote no abstract ideas of number or quantity; their truths depend upon our arbitrary choice of notation and computational rules. But geometry has a distinct object—perceivable extension—and a geometric theory can be treated algebraically only to the extent that algebraic symbols correspond to a distinctly conceived geometric object. The doctrine of fluxions, on Berkeley's account, must be rejected because it introduces impossible abstractions as the object of geometry using notation to which no properly geometric ideas can correspond. This rejection of a formalistic treatment of the calculus is elaborated in the "Queries" at the end of the *Analyst*, where Berkeley hints that the use of algebraic methods in geometry must be limited by the particular properties of the geometric figures to which algebra is applied:

> *Qu.* 27. Whether because, in stating a general Case of pure Algebra, we are at full liberty to make a Character denote, either a positive or a negative Quantity, or nothing at all, we may therefore in a geometrical Case, limited by Hypotheses and Reasonings from particular Properties and Relations of Figures, claim the same License?
>
> *Qu.* 46. Whether, although Algebraical Reasonings are admitted to be ever so just, when confined to Signs or Species as general Representatives of Quantity, you may not nevertheless fall into Error, if, when you limit them to stand for particular things, you do not limit your self to reason consistently with the Nature of such particular things? And whether such Error ought to be imputed to pure Algebra? (*Analysis*, queries 27, 46)

These remarks, together with Berkeley's frequent complaint that proponents of the calculus have confounded the notation for fluxions with the fluxions themselves, emphasize that Berkeley is not a thoroughgoing formalist in the philosophy of mathematics. We can certainly take Berkeley's view of algebra as a forerunner of modern formalistic philosophies of mathematics, but to treat the *Analyst* as an exercise in formalism is to miss its point entirely.[28]

---

28. Thus, Warnock seems to be wide of the mark when he says of §20 of the *Analyst* that "[Berkeley] justified his criticism on the ground that a geometer's work must be assessed solely by the touchstone of logic" and concludes that "the interest of the dis-

Berkeley and the Calculus: The *Analyst*

While Berkeley's *Analyst* is not a formalist treatise, neither is it an endorsement of mathematical instrumentalism. I have indicated that Berkeley's compensation of errors thesis is not the kind of work one would expect from a mathematical instrumentalist, but there are other passages in the *Analyst* which more explicitly reject any form of mathematical instrumentalism. The most salient of these is in section 32, where Berkeley predicts that his critique will be rejected by those who hold that the use of the calculus does not in practice require one to frame ideas of such impossible objects as infinitesimals or evanescent increments, that, in so far as it is a practical tool, the calculus avoids the foundational problems raised in the *Analyst*. His reply is revealing, for he holds that without a clear conception of the fundamental principles of the calculus there can be no genuine science of fluxions. Berkeley admits that the calculus may be accepted and used by people unconcerned with its ultimate justification but insists they are no more mathematicians than the sailor who uses a navigational table is an astronomer. Again, this argument would hardly be expected from a dedicated instrumentalist.

Furthermore, in the *Analyst* Berkeley contradicts a key passage in *De Motu* which some have taken as evidence of mathematical instrumentalism. In *De Motu* Berkeley claims that

> Just as a curve can be considered as consisting of an infinity of right lines, even if in truth it does not consist of them but because this hypothesis is useful in geometry, in the same way circular motion can be regarded as traced and arising from an infinity of rectilinear directions, which supposition is useful in the mechanical philosophy. (*De Motu*, §61)

But in query 10 of the *Analyst,* he asks:

> Whether in Geometry it may not suffice to consider assignable finite Magnitude, without concerning our selves with Infinity? And whether it would not be righter to measure large Polygons having finite Sides, instead of Curves, than to suppose Curves are Polygons of infinitesimal Sides, a Supposition neither true nor conceivable? (*Analyst,* query 10)

As I argued in chapter 5, the passage from *De Motu* says only that a certain mathematical hypothesis is useful, not that a properly devel-

---

cussion is that it shows how far Berkeley had moved from his curious early views on geometry, and that in the end he became in this field also a pioneer of 'Formalism'" (Warnock 1953, 222).

Chapter Six

oped mathematical theory may contain such useful falsehoods. But even if we read it as espousing an instrumentalistic approach to mathematics, there is no such approach in the *Analyst*.

This might suggest that the *Analyst* is inconsistent with Berkeley's other views. He rejects the kind of instrumentalism he seems to champion in *Alciphron* and *De Motu* and his refusal to admit a formalistic interpretation of the calculus is a departure from the nominalistic account of arithmetic and algebra which he elsewhere maintains. Thus, one might suspect that Berkeley's views had changed significantly by the time he wrote the *Analyst* or that his criticisms of the calculus are somehow disingenuous. Such an interpretation is implausible: the title page of the *Analyst* attributes the work to "The Author of the *Minute Philospher*," and a footnote to the ninth query bids the reader "See a Latin treatise *De Motu,* published at London, in the year 1721." If Berkeley's views had changed importantly, it seems implausible that he would cite these earlier works so directly. In fact, I think that careful consideration will show the philosophy of science in *De Motu* and *Alciphron* to be consistent with the philosophy of mathematics implicit in the *Analyst*.

Berkeley's instrumentalistic approach to the natural sciences is based on his claim that these sciences do not have abstract ideas as their object. Whereas other philosophers had held that the basic terms of the sciences must refer to abstractions, Berkeley denies that any theory can be built on abstract ideas. On Berkeley's account, it is impossible to frame an idea of space abstracted from body, of time abstracted from events, or of velocity abstracted from space and time. In Berkeley's view, there is no need to link the terms of a scientific theory to abstractions; it suffices that the relevant terms are part of a useful theory which can be applied to experience. Furthermore, the attempt to find abstractions to serve as referents for the basic vocabulary of the sciences can only lead to confusion and dispute.

In *Alciphron*, Berkeley links this broadly instrumentalistic conception of science to a theory of demonstration in which signs are taken to be the immediate object of all demonstrable sciences. As his spokesman Euphranor puts it: "all sciences, so far as they are universal and demonstrable by human reason, will be found conversant about signs as their immediate object, though these in the application are referred to things" (*Alciphron,* dialogue 7, §13). The significant feature of this account is that the demonstrations in a theory are concerned first with signs themselves and only later with application to the world. Thus, it is not necessary that the terms refer to specific ideas, but Berkeley requires that they be "referred to things" when applied. In

Berkeley and the Calculus: The *Analyst*

this way, a Berkeleyan could construct a general theory of forces, treating the term "force" merely as a sign and deriving theorems from general maxims taken to apply to any force whatever.[29]

Although the physics thus developed will treat "force" as lacking a referent when it appears in a calculation or demonstration, Berkeley still requires that it be given some empirical content in terms of actual or possible experience. In this case, even without a distinct abstract idea of force, one understands forces by considering experiences of actual objects, and such reference renders the notion of force useful and significant. Berkeley concludes that:

> force and number, taken in concrete, with their adjuncts, subjects, and signs, are what every one knows; and considered in abstract, so as making precise ideas of themselves, they are what nobody can comprehend. That their abstract nature, therefore, is not the foundation of science is plain (*Alciphron*, dialogue 7, §11)

Thus, only those terms that have observable content "taken in concrete" are, in Berkeley's view, properly scientific: "In illuminating nature it is vain to adduce things which are neither evident to the senses nor intelligible to reason. Let us therefore see what sense, what experience, and lastly what reason resting upon them recommend" (*De Motu*, §21). Clearly, Berkeley's instrumentalism does not extend so far as to permit theories whose terms lack all experiential content.

Berkeley retains this account of the relationship between formal theories and their interpretations in the *Analyst*, where he insists that a notation for fluxions is insufficient if the fluxions denoted are incomprehensible. He grants that the notation for fluxions and infinitesimals is clearly conceived but insists that the attempt to "consider the things them selves which are supposed to be expressed or marked thereby" reveals confusion and contradictions. Similarly, he complains that "the exponents of fluxions or notes representing fluxions are confounded with the fluxions themselves" and that the attempt to find something represented by these signs yields only mysteries (*Analyst*, §8). Berkeley thus makes an important distinction between the physical term "force" and such mathematical terms as "fluxion":

---

29. It is in this sense that Berkeley claims that the science of mechanics has "very evident propositions or theorems relating to force, which contain useful truths" (*Alciphron*, dialogue 7, §7). Elsewhere he argues that dismissing the doctrine of abstract ideas does not require one to give up general truths in the sciences: "when it is said *the change of motion is proportional to the impressed force*, . . . [i]t is only implied that whatever motion I consider, whether it be swift or slow, perpendicular, horizontal or oblique, or in whatever object, the axiom concerning it holds equally true" (Introduction, §10).

the former can be straightforwardly interpreted to refer to objects of our experience, "in the concrete," while the latter cannot.

Although Berkeley grants that we can gain a nonabstract idea of force by experiencing the behavior of bodies in collision, he allows of no analogous case for fluxions. In other words, Berkeley sees the language of physics as capable of being "referred to things" in a way that the language of the calculus cannot. Moreover, it is the availability of a straightforward interpretation of physical theories which legitimizes the use of its main terms. The calculus, however, is subject to the metaphysical criticism that its object is literally inconceivable.

There is one fairly obvious way in which the doctrine of fluxions could seemingly be assimilated within Berkeley's philosophy of physics. Recall that the kinematic conception of magnitudes treats geometric magnitudes as produced by motion and that the doctrine of fluxions defines the fluxion of a fluent as the velocity with which it is produced. It would seem that, by defining a fluxion as a velocity, the doctrine of fluxions could be brought within the purview of the Berkeleyan treatment of motion and the term "fluxion" would gain experiential content in the same way that the fundamental terms of physics are "referred to things." The problem here is that only the notion of *instantaneous* velocity can underwrite the theory of fluxions, and Berkeley insists that this is a paradigmatically abstract idea and therefore not a proper object of mathematical investigation. I will leave this matter until I consider Maclaurin's response to the *Analyst*. The important point for present purposes is that Berkeley's denial of the intelligibility of fluxions ultimately depends on his critique of abstraction.

The *Analyst* Evaluated

Having surveyed the central arguments in the *Analyst*, I will attempt to evaluate its success. Berkeley's challenge to the mathematicians of the eighteenth century contains both philosophical and mathematical elements which must be considered separately. His main mathematical thesis is that the methods of proof employed in the calculus are vulnerable to simple logical criticisms. A subordinate claim is that the key results of the calculus can be explained by the compensation of errors thesis. Philosophically, Berkeley contends that the basic concepts of the calculus are obscure and unintelligible. Although the results are mixed, I think that Berkeley has made a pretty strong case.

The most incontestable mathematical thesis Berkeley advances in the *Analyst* is his objection to Newton's proof, in the *Principia*, of the product rule. Newton's procedure here is clearly inadmissible, and Berkeley is quite probably correct in charging that the "proof" is a piece of mathematical sleight of hand designed to disguise the use of infinitesimals.

More important is Berkeley's claim that the whole of the calculus (both Newtonian and Leibnizian) depends upon fallacious forms of argument. His lemma that inconsistent suppositions are not admissible in a mathematical demonstration is clearly correct, but Berkeley has shown only that there is the *appearance* of inconsistency in the handling of infinitesimals and evanescent increments. To postulate a quantity greater than zero and subsequently treat it as equal to zero hardly seems a consistent method of reasoning, and both the Newtonian and Leibnizian formulations of the calculus appear to employ these inconsistent assumptions. The way of avoiding this criticism is to show that, when properly interpreted, the calculus does not require contradictory assumptions. In 1734, the basic procedures of the calculus were unquestionably in need of more careful justification than they had previously received. Berkeley held out no hope for such a project, but his criticisms precipitated a flurry of efforts to show that the Newtonian calculus of fluxions was immune to this logical criticism.

A third significant mathematical criticism is Berkeley's claim that the Newtonian method of fluxions does indeed require the supposition of infinitesimal magnitudes. Newton was anxious to show that his theory of prime and ultimate ratios could avoid infinitesimals, but he offers no account of what ultimate ratios of evanescent quantities are, except to define them as the ratios with which disappearing magnitudes vanish. Such a characterization does nothing to determine whether these are, indeed, ratios between infinitesimal quantities. Berkeley remarks that it requires "marvelous sharpness of Discernment" to distinguish between an infinitesimal and an evanescent increment. And, certainly, there is plenty of obscurity in the claim that a ratio can exist after the quantities which stand in that ratio have vanished—that, although the quantities have been reduced to nothing and not simply to infinitely small parts of finite magnitudes, their ratio is not a "ratio of nothings." Berkeley's interpretation of Newton can be challenged, but only by admitting that Newton's pronouncements on the nature of evanescent magnitudes are too vague to be clearly committed to anything. Thus, we must acknowledge the justice of Berkeley's main mathematical criticisms.

The compensation of errors thesis fares rather poorly, so Berkeley's mathematical work in the *Analyst* is not altogether free from defect. He succeeds in showing how classical methods can replace the calculus in certain problems in the theory of conic sections but falls far short of proving the generalized thesis. In fact, his attempts to prove the generalized thesis are as logically faulty as the methods of the calculus. Thus, the standard of rigor demanded in the *Analyst* is one which Berkeley is unable to meet consistently in his own work.

The philosophical arguments in the *Analyst* attack the intelligibility of infinitesimals and ultimate ratios. Berkeley claims that neither is conceivable and that proponents of the calculus overstated human conceptual capacities in claiming that either can be the object of knowledge. His objection depends largely on the claim that the capacity of the mind is finite and that anything infinitely small is therefore beyond comprehension. His claim is stated in terms of our inability to frame an idea of anything infinitely small, without invoking the doctrine of the minimum sensible. There is indeed no mention of the minimum sensible in the *Analyst* and no argument that infinitesimal magnitudes are inconsistent with the doctrine of minima. I take the absence of reference to the minimum sensible in the *Analyst* to be another indication of a significant change in Berkeley's views on the nature of mathematics—especially his views on the nature of geometrical reasoning. Berkeley was at one time prepared to make the minimum sensible the basis for a revised geometry which would reject almost all of the traditional results, and he considered a similar revision of the calculus in the *Commentaries*. He ultimately abandoned his project for a revision of geometry, relying instead upon the theory of representative generalization to accommodate the results of classical geometry while denying that finite magnitudes are infinitely divisible. The minimum sensible, prominent in the plan for a revision of geometry, fades into the background when the project is shelved.

Curiously, the doctrine of the minimum sensible reappears when Berkeley attacks the theory of infinitesimals in the *Principles*, its only mention in any of Berkeley's publications other than the *New Theory of Vision*. In the *Principles*, Berkeley briefly discusses infinitesimals and the calculus after detailing his objections to the doctrine of infinite divisibility (*Principles*, §§130–32). There is little in these sections that adds to what we have already seen, except that the doctrine of the minimum sensible is used to discredit the theory of infinitesimals. Berkeley declares:

> If it be said that several theorems undoubtedly true are discovered by methods in which infinitesimals are made use of, which could never have been if their existence included a contradiction in it, I answer that upon a thorough examination it will not be found that in any instance it is necessary to make

use of or conceive infinitesimal parts of finite lines, or even quantities less than the *minimum sensibile;* nay, it will be evident this is never done, it being impossible. (*Principles,* §132)

Here Berkeley argues that infinitesimals are inconceivable because it is impossible to conceive of anything less than a minimum sensible, his failure to include such an argument in the *Analyst* suggests that the doctrine of minima is no longer central to his views on analysis in 1734.[30]

Rather than depending upon the questionable doctrine of minima, Berkeley's position in the *Analyst* endorses classical methods in analysis which allow only finite magnitudes. On Berkeley's later view, problems in analysis are not to be solved by the crude method of counting up minima but by reasoning about particular perceived figures regarded as representatives of "all possible finite figures . . . of all sensible and imaginable extensions or magnitudes of the same kind."[31] Exhaustion proofs fit this epistemological scheme well in that they require only that we consider finite magnitudes and their finite differences. Treating the figures in the proof as representatives of "all sensible and imaginable figures" endows the results with the requisite generality—a perceived line can represent lines of any imaginable size, and the proportions which lie at the heart of classical geometry can be approximated to any degree of accuracy by taking ratios between successively larger lines. Berkeley is unwilling, however, to treat finite perceivable lines as representatives of infinitesimal lines. He insists that infinitesimal magnitudes are flatly inconceivable and violate the classical standards of rigor. Again, he was by no means alone in his hostility toward the doctrine of infinitesimals, so his brief dismissal of the infinitesimal as inconceivable is hardly a weakness in his case against the calculus. As I have noted, many British mathematicians of the eighteenth century agreed that infinitesimals were inconceivable, and Berkeley did not need to make a very strong case on this point to convince his intended audience.[32]

---

30. This is not to say that Berkeley had repudiated the doctrine of the minimum sensible by the time he published the *Analyst.* The *New Theory of Vision* and the *Principles* were both reissued in 1734 without deletion of passages which set out the doctrine.

31. Berkeley takes this position explicitly in query 6 of the *Analyst:* "Whether the Diagrams in a Geometrical Demonstration are not to be considered as Signs, of all possible finite Figures, of all sensible and imaginable Extensions or Magnitudes of the same kind?"

32. See, for example, Benjamin Robins's review of his *Discourse Concerning the Nature and Certainty of Sir Isaac Newton's Method of Fluxions, and of Prime and Ultimate Ratios.* He complains that the method of indivisibles "is a manner of expression hardly accompa-

The more intriguing philosophical thesis in the *Analyst* is Berkeley's claim that Newton's ultimate ratios of evanescent quantities are no more conceivable than infinitesimals. One cannot, Berkeley insists, conceive a ratio of increments where there are no increments. To claim that one can conceive the ultimate ratios assumed by the Newtonian theory of fluxions is to claim—unintelligibly, on Berkeley's account—that one can frame an abstract idea of an instantaneous velocity which "prescinds from" perceivable space and time. Here Berkeley goes beyond reporting his own inability to frame the idea of an ultimate ratio to argue that the doctrine involves logical inconsistencies.

In the end, I think that Berkeley's philosophical challenges to the calculus are effective, not in showing that the calculus is fundamentally incoherent or essentially beyond repair, but in demonstrating a need for clarification of its important concepts. Berkeley places the burden of proof on those who defend the methods of the calculus as mathematically rigorous and epistemologically unproblematic. Responses to his mathematical and philosophical critique of the calculus were not long in coming. Indeed, Cajori calls the publication of the *Analyst* "the most spectacular event of the century in the history of British mathematics" and aptly characterizes its arguments as "so many bombs thrown into the mathematical camp" (Cajori 1919, 57).

---

nied with any ideas; since it is not at all intelligible to speak of the collective number of lines or planes, where their number is wholly undetermined and infinite" (Robins 1735b, 246).

CHAPTER SEVEN

The Aftermath of the *Analyst*

A long and intense controversy among British mathematicians followed the publication of the *Analyst*. In the course of this controversy various and conflicting defenses of the Newtonian calculus were proposed, all with the aim of meeting Berkeley's objections. My purpose in this chapter is to outline the most important replies to the *Analyst* and to assess how successfully these proposals answer Berkeley's objections. Given the number of responses to the *Analyst*, I cannot make a detailed study of all the anti-Berkeleyan mathematical writings from the mid-eighteenth century. By my own count more than a dozen publications in the period 1734–50 responded to Berkeley's criticisms of the calculus in one way or another, and it is difficult to find a text on the calculus of fluxions from the period which does not contain at least a veiled reference to Berkeley's *Analyst*.[1]

Interestingly enough, the *Analyst* occasioned no serious response on the Continent, unless we count a summary review in the *Bibliothèque Brittanique* which also reviews James Jurin's attack on Berkeley, *Geometry no Friend to Infidelity*. The reviewer does little more than reproduce extracts from both works and conclude that Berkeley's concerns are

---

1. See Cajori (1919) and Guicciardini (1989, chapters 3 and 4) for an overview of the publications for this period. Even textbooks of the period made veiled replies to Berkeley: "it will easily be perceived, that a first *Fluxion* is not *Incomprehensible*, and that it is not so difficult to form a just Idea of it in the Mind, as some have imagined" (Hodgson 1736, x). In addition to the works discussed here, see Blake (1741), Emerson (1743), Martin (1739), and Simpson (1737).

hardly worth addressing.[2] Berkeley's reputation on the Continent had sunk so low by 1740 that Comte Georges-Louis Leclerc de Buffon, in the preface to his translation of Newton's *Method of Fluxions and Infinite Series*, characterizes him as an "enemy of science" whose polemics against the calculus are the product of ambition, vanity and a small mind.[3] The lack of a significant Continental response to Berkeley suggests that concern with the foundations of the calculus was less intense outside Britain in the 1730s and 1740s. But the *Analyst* is addressed to a British audience and is principally concerned with the Newtonian calculus, so one would expect a proportionately greater interest in refuting Berkeley among adherents of the Newtonian calculus of fluxions. In this chapter, therefore, I shall be concerned exclusively with works which aim to defend Newton's doctrines against Berkeley's criticisms.

For my purposes the significant replies will be loosely grouped into two categories: those which drew a response from Berkeley and those which he ignored. Within these crude categories there is a considerable variation—in particular, some authors asserted the essential correctness of the Newtonian formulation of the calculus, while others sought to answer Berkeley's charges with alternative presentations of the key concepts in the calculus of fluxions. I will begin my survey with an account of Berkeley's polemical exchanges with James Jurin and John Walton,[4] then consider other attempts to defend the New-

2. The reviewer confidently misidentifies the author of Jurin's pseudonymous response, concluding: "The reputation of Mr. Berkeley has obliged us to give an extended extract of his work, and of the response to it, in order that the reader not imagine that the accusations and objections of such an able man were well founded. The reader will judge from the extracts, where we have translated and abridged the two pieces, which of them understands the subject. For the rest, we are assured that Mr. Middleton, whose name has already appeared in our journal many times, is the author of this response, which he composed with the co-operation of Mr. Smith, professor of Astronomy at Cambridge" *Bibliothèque Brittanique* 4 (1735):430.

3. "All was quiet for several years, until in the breast of England itself there arose a learned enemy of science [un Docteur ennemi de la Science] who declared war on the mathematicians. This learned man rose to become a bishop [montre en Chaire] by informing the faithful that geometry is contrary to religion; he told them to be on guard against the geometers, who are, according to him, blind and troublesome people who neither know, nor reason, nor believe. . . . I have read his book carefully, searching for the motives which led him to insult mathematicians, and I found that it is not religious zeal, but rather vanity which guided his pen. This learned man has a mind too narrow for mathematics" (Newton 1740, xxiv–xxv).

4. Guicciardini (1989) identifies this author as "Jacob" or "Jakob" Walton, but the *National Union Catalog* and the British Library's *Catalogue of Printed Books* identify him as John Walton. The title pages of his works list the author as simply "J. Walton."

tonian calculus and attempt to reach some conclusions concerning the mathematical and philosophical effects of the *Analyst*.

Berkeley's Disputes with Jurin and Walton

The first response to the Berkeleyan critique of the calculus came almost immediately, in a pamphlet entitled *Geometry no Friend to Infidelity* (1734) by James Jurin, writing under the pompous pseudonym "Philalethes Cantabrigiensis." Berkeley responded to Jurin in 1735 with his *Defence of Free-thinking in Mathematics*, adding an appendix directed against John Walton's *Vindication of Sir Isaac Newton's Principles of Fluxions* (1735a). Berkeley's tract drew two answers, Jurin's *the Minute Mathematician; or, the Free-Thinker no Just-Thinker* (1735a) and Walton's *The Catechism of the Author of the "Minute Philosopher" Fully Answer'd* (1735b). Berkeley took no notice of Jurin's second effort, but he concluded his part in this battle of books in 1735 with his last publication on the calculus: *Reasons for not replying to Mr. Walton's Full Answer*. Predictably, Walton replied in a second edition of his *Catechism* (1735c), adding an appendix which addressed Berkeley's *Reasons for not replying*. The ratio of argument to invective in these pieces is not high, but some important substantive issues were addressed in the course of the controversy. In particular, both of Berkeley's opponents argue that the notion of a fluxion is fully intelligible, and their replies highlight the basic philosophical problems posed by Berkeley's critique of the calculus. These two disputes were sufficiently complex to demand separate treatment, so I will first survey Berkeley's controversy with Jurin and then turn to his exchanges with Walton. I will not try to cover every detail of the many charges and countercharges leveled in the course of these disputes. Instead, I will try to bring out the significant points debated by Berkeley and his opponents. In both cases the responses to Berkeley were generally inadequate and served more to shore up Berkeley's charges than to refute them. Thus, a large part of the interest in these two controversies is the extent to which they ratify Berkeley's contention that the foundations of the calculus were poorly understood by Newton's followers.

The Dispute with Jurin

James Jurin (1684–1750) was a well-known physician, Fellow and Secretary of the Royal Society from 1721 to 1727, and an ardent Newtonian. His pamphlet *Geometry no Friend to Infidelity* concedes nothing to Berkeley and proposes to refute all of the principal points in the *Analyst* while counterattacking with objections to Berkeley's metaphysics and epistemology. Jurin reads Berkeley as leveling three charges against the British mathematical community, the first two being "infidelity with regard to the Christian Religion" and "endeavouring to make others Infidels" (Jurin 1734, 6). He goes on at great length in response to these two charges, complaining about "*odium theologicum*" and an inquisition of British mathematicians. Such complaints are not of interest here, but it is worthwhile to observe that in questioning the motivation behind the *Analyst* Jurin notes the important change in Berkeley's attitude toward classical mathematics between the 1709 and 1732 editions of the *New Theory of Vision*. He proposes to "take the liberty of considering a little what has been your conduct with regard to Mathematicians," and observes that

> About [five and twenty years ago] you published an *Essay towards a new Theory of Vision*, wherein you were pleased to insert a great many severe censures upon Mathematicians, relating to their ignorance of the fundamental principles of their own science. Had you then heard of their being Infidels? If so, why were many of those censures greatly mollified, upon farther consideration, in a new edition of that piece published last year? (Jurin 1734, 22)

I argued in chapter 2 that this change in view was prompted by motives other than the covetousness of fame which Jurin ascribes to Berkeley. In particular, I argued that Berkeley had developed and applied his theory of representative generalization after the publication of the *New Theory of Vision*, and this change in his epistemological principles was sufficient to admit more of classical mathematics. Nevertheless, it is interesting to see that one of Berkeley's contemporaries took note of the dramatic change in his attitude toward classical mathematics, even if he was mistaken about its motivation.

Jurin's reading of the *Analyst* discerns a third charge against the practitioners of the calculus, namely that of "errour and false reasoning" in their own science. He breaks this down into three specific points:

> Your objections against this method [of fluxions] may, I think, all of them be reduced under these three heads.
> 1. Obscurity of this doctrine.
> 2. False reasoning used in it by Sir *Isaac Newton*, and implicitly received by his followers.
> 3. Artifices and fallacies used by Sir *Isaac Newton*, to make this false reasoning pass upon his followers. (Jurin 1734, 30)

I will concentrate on Jurin's reply to the first two objections, his attempt to show that the doctrine of fluxions is coherent and its principles adequately demonstrated, as his answer to the third is the occasion for largely irrelevant expressions of admiration for Newton and contempt for Berkeley.

Responding to Berkeley's charge that the doctrine of fluxions is incomprehensible, Jurin retorts that he finds its key terms fully intelligible and has no difficulty understanding the theory. He confesses that the doctrine is "not without its difficulties, and will doubtless seem obscure to every unqualified or inattentive reader of Sir *Isaac Newton*" (Jurin 1734, 31), but he insists that the relevant test is whether those versed in the subtleties of higher mathematics find the doctrine obscure and mysterious. As I have noted, one of Berkeley's objections amounts to little more than a report of his own failure to frame an adequate notion of fluxions, so Jurin's own report may be an adequate response on this specific point. He also complains that Berkeley's presentation of the Newtonian doctrine has prejudiced the case by making fluxions appear needlessly confusing. Jurin particularly objects to Berkeley's characterization of second and third fluxions as "velocities of velocities, the second, third, fourth, and fifth velocities, &c." He rightly contends that Newton never used such expressions and finds Berkeley's gloss an unacceptable attempt to persuade the unwary reader into thinking that Newtonian fluxions are difficult to conceive. Again, there is some measure of truth in this charge against Berkeley. The Newtonian account of second-order fluxions, for example, can be read as considering the *rate of acceleration* of a point; and it is not too far-fetched to describe the rate of acceleration as the velocity of a velocity. This may be a linguistic barbarism, but the notion of the rate of change of a rate of change of position is hardly inconceivable, and this is what the second fluxion of a moving point would amount to.[5]

In his response to the claim that Newton has used false reasoning in the calculus, Jurin first undertakes to defend the indefensible by showing that the Newtonian proof of the product rule in the *Principia*

---

5. Benjamin Robins noted this very point when he observed that "all the absurdity of expression, and all the inconsistency with himself charged on him by the author of the *Analyst*, arises wholly from misrepresentation. For example, it has been asserted, that there is as little sense in the phrase, fluxion of a fluxion, as to speak of the velocity of a velocity. This objection supposes, that the simple word velocity can always be substituted in the room of the word fluxion. But by Sir *Isaac Newton*'s description of the fluxions of magnitudes, it is evident, that the single words can never be used promiscuously: for the fluxion of any quantity is not the velocity of that quantity, but the velocity, wherewith it at all times augments or diminishes; for instance, the fluxion of a line is not the velocity, wherewith that line moves, but the velocity of the point, by whose motion the line is described" (Robins 1736a, 330–31).

Chapter Seven

is "rigorously geometrical."[6] He begins by claiming that the quantity *ab* by which Berkeley's calculation of the increment of *AB* differs from Newton's calculation of the moment of the rectangle *AB* is inconsiderable and could make no difference in practice. Jurin then attempts to argue that the quantity *ab* is properly rejected by an exegesis of Newton's remarks in the *Principia*, reaching the remarkable conclusion that "what he endeavours to obtain by these suppositions, is no other than the increment of the rectangle $(A - \frac{1}{2}a) \times (B - \frac{1}{2}b)$, and you must own that he takes *the direct and true method to obtain it*" (Jurin 1734, 44). The argument leading to this conclusion is worth quoting at length, since it reveals much about the general outline of Jurin's defense of Newton. He begins by observing that the quantities $aB + bA + ab$ and $aB + bA - ab$ are both properly called moments of the flowing quantity *AB*, the former being positive or incremental moments and the latter being negative or decremental moments. He continues:

> Now, Sir, I would humbly beg leave to inquire of you, who see so much more clearly in these matters than Sir *Isaac Newton* or any of his followers; which of these two Quantities, $aB + bA + ab$ and $aB + bA - ab$, you will be pleased to call the moment of the rectangle *AB? The case is indeed difficult*, the difference between them is no less than $2ab$, just the double of that same *ab*, which has given us all so much trouble; and yet each of them plead an equal right to the title of moment. So equal a one, that, though I am very sensible of your *address and skill*, yet there seems to be no possibility of deciding the controversy between them *by legitimate reasoning*. I see but two ways of doing it. One is that they should toss up cross or pile for the title:[7] Or if that be thought too boyish and unbeseeming the Gravity of Mathematical quantities, they must even end the dispute in an amicable manner, and without claiming any preference one of another, agree that they make two moments between them. Then, Sir, I apprehend the case will stand thus: $aB + bA + ab + aB + bA - ab$ making twice the moment of the rectangle *AB;* it follows that $aB + bA$ will make the single moment of the same rectangle. (Jurin 1734, 45–46)

---

6. Recall that Newton's procedure for finding the moment of the product *AB* takes the difference between the products $(A - \frac{1}{2}a) \times (B - \frac{1}{2}b)$ and $(A + \frac{1}{2}a) \times (B + \frac{1}{2}b)$. Berkeley insists that he must take the difference between $(A + a) \times (B + b)$ and *ab*.

7. The *Oxford English Dictionary* reports that "cross or pile" refers to the markings on opposite signs of a coin; in particular it can mean "'tossing up' to decide a stake, or anything doubtful, by the side of a coin which falls uppermost."

The Aftermath of the *Analyst*

This argument is nothing more than a restatement of Newton's original "proof," except that Jurin substitutes the products $(A + a) \times (B + b)$ and $(A - a) \times (B - b)$ where Newton had used $(A + \frac{1}{2}a) \times (B + \frac{1}{2}b)$ and $(A - \frac{1}{2}a) \times (B - \frac{1}{2}b)$. Newton thereby avoids having to divide the result by two, but otherwise the reasoning is unaltered. Jurin attempts to bolster this sorry case by arguing that it is consistent with Newton's declaration that the moment of the rectangle *AB* is proportional to the velocity with which it is increased or diminished. Once again, he argues that because the moment of *AB* can be either positive or negative, we must

> ... not suppose *AB* as lying at either extremity of the moment; but as extended in the middle of it; as having acquired the one half of the moment, and as being about to acquire the other; or as having lost one half of it, and being about to lose the other. And this is the method Sir *Isaac Newton* has taken in the demonstration you except against. (Jurin 1734, 50)

The effect of this defense is to concede all of Berkeley's objections. The increment or "affirmative moment" of the quantity *AB* is computed by taking the increment of the quantity $(A - \frac{1}{2}a) \times (B - \frac{1}{2}b)$, and this for the remarkable reason that the incremental quantities must be "extended in the middle" of *AB*. But this contradicts the claim that the moment of a quantity is its "momentaneous increment," and the whole procedure remains irredeemably mysterious.

Jurin goes on to consider Berkeley's argument that Newton fallaciously shifts hypotheses by first assuming an increment *o* and then contradicting this assumption by letting *o* vanish (*Analyst*, §§13–16). His defense is to invoke the theory of evanescent increments, and he objects specifically to Berkeley's claim that Newton's procedure requires us to assume that the increments vanish or become nothing. Jurin engages in some detailed exegesis, concentrating on the proper interpretation of the Latin phrase *"Evanescant jam argumenta illa et eorum ratio ultima erit,"* from Newton's *Quadrature of Curves*. Jurin accuses Berkeley of viciously misinterpreting Newton:

> Must not therfore the meaning of [this expression] be to consider the last proportion between the evanescent augments, in the point of their evanescence, or of their ceasing to exist? Ought it not to be thus translated, Let the augments now become evanescent, let them be upon the point of evanescence? What then must we think of your interpretation *Let the increments be nothing, let there be no increments?* Do not the words *ratio ultima* stare us in the face, and plainly tell us that though there is a last proportion of evanescent increments,

yet there can be no proportion of increments which are nothing, of increments which do not exist? I believe, Sir, every thinking person will acquit Sir *Isaac Newton* of the gross oversight you ascribe to him, and will acknowledge that it is your self alone, who have been guilty of a most palpable, inexcusable, and unpardonable blunder. (Jurin 1734, 57–58)

Once again, Jurin's response to Berkeley hardly comes up to the mark. The point at issue is whether it is legitimate to retain a quantity expressing a ratio, once the quantities which form the ratio have been supposed to vanish. Berkeley insists that this procedure involves nothing more than two inconsistent assumptions: first that the increment o is positive, and then that it is nothing or zero. Jurin replies that evanescent increments are "on the point of evanescence" and so are not truly nothing; but then it is hard to see how the terms in o can be dropped from the equation which expresses the fluxion or how, if they are dropped, we can distinguish between increments which are evanescent and those which are truly nothing.

Jurin also attacks Berkeley's compensation of errors thesis, by far the weakest part of the *Analyst,* but even here makes little headway in undermining Berkeley's position.[8] He deals only with the first example of compensating errors, where Berkeley argues that the calculus-based value for the subtangent to the parabola comes out correctly because two neglected quantities compensate. Jurin's strategy is to compute the value for the subtangent twice, each time *retaining* one of the errors identified by Berkeley.

As before, the equation $y^2 = px$ determines the parabola ABN (fig. 7.1), the subtangent is TP while $AP = x$, $PB = y$, $PM = BR = dx$, $RN = dy$, and $NL = z$. Berkeley argues that the first error is committed when the value RN was taken for RL, so Jurin proposes to avoid that error by using RL in place of RN while retaining the "second error" of supposing $dy = pdx/2y$. The result is:

$$S = TP = \frac{RB \cdot PB}{RL} = \frac{PM \cdot PB}{RL} = \frac{dx \cdot y}{dy + z} = \frac{dx \cdot y}{dy + dy^2/2y}$$

$$= \frac{2x \cdot dy}{y} \cdot \frac{y}{dy + dy^2/2y} = \frac{2x}{1 + dy/2y} = \frac{2x \cdot 2y}{2y + dy}$$

$$= 2x \cdot \frac{2y}{2y + dy}. \qquad (7.1)$$

---

8. Breidert (1989, 110–11) has studied Jurin's argumentation on this point, and my analysis follows his.

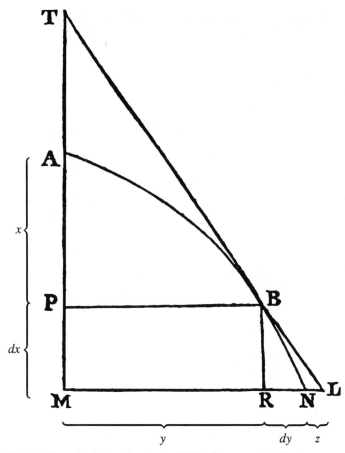

Fig. 7.1 From Berkeley, *The Analyst* (1734). Lowercase labels mine.

The second error claimed by Berkeley involved discarding the second-order infinitesimal dy^2 in the differentiation of the equation $y^2 = px$. If we retain the quantity dy^2 in the expression of the subtangent we have

$$dy = \frac{pdx - dy^2}{2y}. \tag{7.2}$$

Solving equation (7.2) for dx yields

$$dx = \frac{2ydy + dy^2}{p} = \frac{x(2ydy + dy^2)}{y^2}.$$

Now, using this expression for dx in (7.1) but reintroducing the "first error" of using RN for RL, we get

$$S = TP = \frac{PB \cdot RB}{RN} = \frac{PB \cdot PM}{RN} = \frac{dx \cdot y}{dy} = \frac{x(2ydy + dy^2)y}{y^2 dy}$$

$$= \frac{x(2y + dy)}{y} = 2x \cdot \frac{2y + dy}{2y}.$$

Jurin concludes that both of these values are correct, since there will be no sensible difference between them:

> Since therefore these errors are wholly insignificant, my conclusion when reduced to numbers, coming out exactly the same, whether the first, or second, or neither, or both of these errors be committed; and since by committing both these errors, the calculus, which would otherwise, especially in the higher operations, be exceedingly tedious and laborious, is now rendered surprisingly expeditious and easy; it seems to me that this is so far from being any defect in the method of Fluxions, that on the contrary it is one of the greatest advantages and excellencies of that invention. (Jurin 1734, 69)

Here we have a quite remarkable account of the virtues of the calculus. The quantity dy is not equal to zero, and yet Jurin asserts that the three values

$$2x, \quad 2x \cdot \frac{2y}{2y + dy}, \quad \text{and} \quad 2x \cdot \frac{2y + dy}{2y}$$

are all "true and equally exact." If we take this to mean that the three values are all literally equal, then dy is equal to zero, and the calculus is inconsistent. But if the exactness of the results means only that they differ from one another by an "insignificant" amount, then the pretense to rigorous demonstration has been abandoned by allowing (*contra* Newton) small errors to be neglected in mathematics.[9]

Perhaps the most interesting part of Jurin's reply to Berkeley is his defense of the doctrine of abstract ideas. Jurin saw quite clearly the connection between Berkeley's attack on abstraction and his views on

---

9. Breidert voices a similar conclusion: "Jurin seems not to understand that Berkeley can always respond: either dy is not equal to zero, in which case the values are not equal, or $dy = 0$, in which case $RN = 0$ and then there is division by zero, which is not permitted" (Breidert 1989, 111).

mathematics and announces early on that "as your mistake about [Mr. Locke] has a near relation to Geometry, before I close this letter, I shall take up a little of your time rectifying that likewise" (Jurin 1734, 24). Later, he insists that "though I have carefully perused what you have written on the subject, I am one of those who still adhere to the vulgar, or rather universal error of all Mankind, that neither Geometry nor any other general Science can subsist without general ideas" (Jurin 1734, 73–74). Jurin thus shares my conclusion that Berkeley's attack on abstraction is central to his critique of the calculus, and he hopes to vindicate the Newtonian methods by refuting Berkeley's case against abstraction.

But although Jurin claims to have "carefully perused" Berkeley's writings on the subject of abstract ideas, he responds only to Berkeley's assertion, discussed in chapter 1 (*Theory*, §125), that Locke's description of the abstract general idea of a triangle (*Essay* IV, vii, 9) and his account of abstraction are unintelligible. Berkeley there cites Locke's dictum that ideas which combine inconsistent ideas are inconceivable, arguing that this contradicts his account of the abstract idea of a triangle. Jurin represents Berkeley as entirely mistaken about the intent of Locke's comments on the abstract idea of a triangle. On his analysis, Locke's aim here is not to describe the abstract general idea of a triangle but to show the difficulties involved in abstracting from particulars.[10] He further argues that Locke actually characterizes the abstract general idea of a triangle "three lines, including a space" (*Essay* II, xxxi, 6), which Jurin takes to be clearly and easily comprehensible. He further argues that the two Lockean passages which Berkeley finds contradictory are, in fact, consistent:

> Here, Sir, I strongly apprehend you are fallen into one of those traps, which this Great Man would sometimes divert himself with setting to catch unwary cavillers. . . . Had his first proposition run thus, It is an Idea wherein several different and inconsistent Ideas are put together, it would undoubtedly have been contradictory to the second. But that is not the case: pray observe the words of this cautious and accurate Writer. It is an Idea, wherein SOME PARTS OF several different and inconsistent ideas are put together. (Jurin 1734, 82)

---

10. "Is it not plain to any one who attentively reads the passage you refer to, that his intention there was not to *describe* the general idea of a Triangle, but only to shew from the seeming inconsistencies in that Idea, supposed to be already known, that it required some pains and skill to form it, as well as other abstract ideas?" (Jurin 1734, 81).

Jurin's point here is that the idea of any particular triangle is a compound with other ideas as parts,[11] some of which will differ from those in the compound idea of a second triangle. But the abstract general idea of a triangle will consist of all and only those parts of the compound ideas which are common to all triangles, and such an idea can be described quite consistently as the idea of a plane figure with three sides. Jurin confesses that this idea is "something imperfect which cannot exist" precisely because, containing all and only those component ideas which apply equally to any triangle, the abstract general idea of a triangle is insufficiently determined to be the idea of any particular existing triangle. It is worth observing that this conception of abstraction does not overcome the argument from impossibility, Berkeley's principal criticism of the doctrine of abstract ideas. As discussed in chapter 1, Berkeley argues that whatever is impossible is inconceivable, so that an "impossible object" would be one of which we can have no idea. On Jurin's account of Locke, it would seem that the abstract idea of a triangle is the idea of a thing which he admits cannot actually exist.[12] Thus, Jurin accepts precisely the controversial premise which Berkeley needs in order to make his argument from impossibility go through.

Berkeley responded to Jurin with *A Defence of Free-Thinking in Mathematics,* published in 1735. The work adds relatively little to Berkeley's case against the calculus but is interesting in so far as it helps to clarify Berkeley's objections and adds some detail to his conception of abstraction and mathematics. In response to Jurin's claim that he and his colleagues can clearly conceive fluxions, Berkeley appeals to common sense, insisting that the comprehensibility of fluxions is a matter in which "every reader of common sense may judge as well as the most profound mathematician" (*Defence,* §18). He argues that the basic concepts of the calculus are so obscure as to be incomprehensible to anyone:

> Every reader of common sense, that will but use his faculties, knows as well as the most profound analyst what idea he frames or can frame of velocity without motion, or of motion

---

11. We can think of these parts as Lockean "simple ideas" which may be combined to form "complex ideas." The idea of a particular triangle is thus a complex idea composed of simple ideas.

12. "For every individual Triangle, every Triangle that can exist, must be something more than a space included by three lines, it must also have the characteristick mark of some one of the particular species of Triangles; without which it would be *imperfect, it could not exist,* which is what Mr. *Locke* here says of a Triangle in general." (Jurin 1734, 81).

The Aftermath of the *Analyst*

> without extension, of magnitude which is neither finite nor infinite, or of a quantity having no magnitude which is yet divisible, of a figure where there is no space, of proportion between nothings, or of a real product of nothing multiplied by something. He need not be far gone in geometry to know that obscure principles are not to be admitted in demonstration: That if a man destroys his own hypothesis, he at the same time destroys what was built upon it: That errour in the premises, not rectified, must produce errour in the conclusion. (*Defense*, §20)

Here Berkeley moves beyond simple reports of his own failure to conceive fluxions to portray the calculus as depending upon an ability to separate mentally ideas that are necessarily bound together. His challenges to frame an idea of "velocity without motion" or of "figure where there is no space" seem to be demands to do the impossible. Certainly, if we allow that there can be no velocity where there is no motion and no figure without space, and if we admit that what we can conceive depends upon what could possibly be, then Berkeley's case is convincing indeed. It should be clear that Berkeley's remarks are an application of his argument from impossibility against various abstractions which are supposed to be the fundamental concepts of the calculus. In this respect the *Defence* presents a stronger case for the incomprehensibility of fluxions than did the *Analyst*.

Berkeley also responds to Jurin's attempt to defend Newton's proofs of the rule for finding the fluxion of the product AB and the general rule for finding the fluxion of any power x^n. As noted above Jurin's defense of the first point is quite weak and Berkeley wastes few words in showing the inadequacy of his "defence of what [he does] not understand." The details may be left aside here, since the weakness of Jurin's case on this point is sufficient to do Berkeley's work for him. One interesting issue does emerge from this discussion, however, when Berkeley considers Jurin's claim that the error arising from the rejected term ab in the computation of the fluxion of AB is insignificant in practice. Berkeley takes this as another opportunity to insist that the rigor of a demonstration cannot be judged by the practical application of the results. Indeed, he even seems to acknowledge the existence of "quantities less than sensible," which cannot be neglected in a rigorous demonstration, although they are irrelevant to "gross practice."[13] This is further evidence, if any is needed, of the

---

13. Berkeley writes: "And, although quantities less than sensible may be of no account in practice, yet none of your masters, nor will even you yourself venture to say

extent to which Berkeley's criteria for rigorous demonstration lead him away from instrumentalism with regard to the calculus. Furthermore, his apparent acknowledgement of quantities "less than sensible" indicates that he has largely repudiated his old doctrine of sensible minima by 1735, since the entire doctrine depends upon the idea that there can be no quantity less than that perceivable by sense. The ontological status of such imperceptible quantities does conflict with the metaphysical framework of the *Principles,* because the doctrine of *esse* is *percipi* would obviously rule out such quantities.

Berkeley next attacks Jurin's attempt to employ the theory of evanescent increments, and again he finds no problem in demolishing Jurin's defense of Newton. He notes that Newton's procedure seems to demand that the quantity o first be considered as positive and then as zero:

> The next point you undertake to defend is that method for obtaining the rule to find the fluxion of any power of flowing quantity, which is delivered in the Introduction to the Quadratures, and considered in the *Analyst.* And here the question between us is, whether I have rightly represented the sense of those words, *evanescant jam argumenta illa,* in rendering them, let the increments vanish, *i.e.* let the increments be nothing, or let there be no increments? This you deny, but, as your manner is, instead of giving a reason you declaim. I, on the contrary, affirm, the increments must be understood to be quite gone, and absolutely nothing at all. My reason is, because without that supposition you can never bring the equality or expression
>
> $$nx^{(n-1)} + \frac{(n^2 - n)}{2}ox^{(n-2)} + \ldots$$
>
> down to
>
> $$nx^{(n-1)}$$
>
> the very thing aimed at by the evanescence. Say whether this be not the truth of the case? Whether the former expression is not to be reduced to the latter? and whether this can pos-

---

they are of no account in theory and in reasoning. The application in gross practice is not the point in question but the rigour and justness of the reasoning. And it is evident that, be the subject ever so little or ever so inconsiderable, this doth not hinder but that a person treating thereof may commit very great errours in logic, which logical errours are in no wise to be measured by the sensible or practical inconveniences thence arising, which, perchance, may be none at all" (*Defence,* §26).

sibly be done so long as *o* is supposed to be a real quantity? (*Defence*, §20)

Berkeley clinches his case by noting that Newton himself regarded the evanescent increment *o* as reduced to nothing, since in the Newtonian treatise *Analysis per Æquationes Numero Terminorum Infinitas*, the word *evanescere* is used equivalently with *esse nihil* when the rule for quadrature of simple curves is demonstrated.[14] Jurin's efforts to distinguish nonexistent or zero quantities from those which had simply vanished (the task which Berkeley had found to require "marvellous sharpness of Discernment") is thus brought to grief and his critique of Berkeley collapses. It is no understatement to say that Newton deserved a more competent defender than Jurin; although he makes some small headway against some of Berkeley's less telling arguments, the bulk of the case against the calculus remains unanswered.

Berkeley's reply to Jurin's defense of the Lockean doctrine of abstract ideas contains, as I discussed in chapter 1, an explicit presentation of Berkeley's characteristic "argument from impossibility" as the basis of his case against the doctrine of abstract ideas. He ignores Jurin's insistence that Locke's account of abstraction can be made consistent, and instead attempts to show how the doctrine of abstract ideas contradicts the principle that what is impossible is inconceivable. As might be expected, his case is greatly helped by Jurin's confession that the abstract general idea of a triangle is indeed the idea of something which cannot exist, and his application of the argument from impossibility proceeds accordingly.

Berkeley includes an interesting gloss of his theory of representative generalization, which he contrasts with the Lockean view that each general name stands for a distinct abstract idea:

> It is Mr. Locke's opinion that every general name stands for a general *abstract* idea, which prescinds from the species or individuals under it. Thus, for example, according to him, the general name *colour* stands for an idea which is neither blue, red, green, nor any other particular colour, but somewhat distinct and abstracted from them all. To me it seems the

14. The relevant passage reads: "If we suppose [the increment of the curve] to be diminished infinitely and to vanish, or *o* to be nothing . . ." (Newton 1964–67, 1:341). Berkeley's observation would not necessarily show his analysis of the situation to be correct if it could be shown that Newton had changed his views between the writing of this tract and the writing of the *Quadrature of Curves*. Jurin, however, insists that Newton had always worked with the same conception of fluxions and ultimate ratios, so Berkeley's comment is indeed a decisive objection to Jurin's interpretation of Newton.

word *colour* is only a more general name applicable to all and each of the particular colours; while the other specific names, as blue, red, green, and the like, are each restrained to a more limited signification. The same may be said of the word triangle. Let the reader judge whether this be not the case; and whether he can distinctly frame such an idea of colour as shall prescind from all the species thereof, or of a triangle which shall answer Mr. Locke's account, prescinding and abstracting from all the particular sorts of triangles, in the manner aforesaid. (*Defence*, §47)

The Berkeleyan case against abstract ideas is not advanced or significantly changed in the *Defence of Free-Thinking in Mathematics*, where Berkeley upholds the same general view as in the *Principles*. He repeatedly asserts that the doctrine of abstract ideas depends upon the claim that we can frame ideas of things that are impossible, and he takes this to be a sign of the doctrine's incoherence. He also repeats his claim that abstractions are a source of confusion in all branches of learning and is particularly interested in repudiating the abstractionist account of mathematics, underscoring the key role of his critique of abstraction in his philosophy of mathematics, although adding little to his attack on the calculus.

Jurin's second response to Berkeley, *The Minute Mathematician*, is heavy on invective but does raise substantive issues concerning the foundations of the method of fluxions. In particular, his efforts to set forth the first principles of the method by definition and example deserve attention. The case for the comprehensibility of fluxions in this piece begins with a postulate to the effect that mathematical quantities "may be described, and in describing may be generated or destroyed, may increase or decrease, by a continued motion" (Jurin 1735a, 18). Jurin appends several definitions to this postulate, the most important of which concern fluxions and nascent or evanescent increments. These read:

> 2. The velocity with which such flowing quantity increases or decreases, is called the fluxion of that flowing quantity.
> 4. A nascent increment is an increment just beginning to exist from nothing, or just beginning to be generated, but not yet arrived at any assignable magnitude how small soever. An evanescent increment is the same thing as a nascent increment, but only considered in a different manner, as by a continual diminution becoming less than any assignable quantity, and at last vanishing into nothing, or ceasing to exist. (Jurin 1735a, 18–19)

The Aftermath of the *Analyst*

Fig. 7.2 From Jurin, *The Minute Mathematician* (1735, 19).

This attempt to elucidate the mysterious of fluxions is extended by an example. Imagine the line *AB* (fig. 7.2) to be described by the motion of a point, then consider the points *C* and *c* between *A* and *B*:

> When the generating point, in describing the line *AB*, is arrived at the point *C*, and proceeds from thence towards *B*: At the instant of time that it sets out or departs from the point *C*, at that very instant of time an increment begins to be generated, or begins to exist, which therefore is properly called a nascent increment. And as the generating point at that instant of time is supposed to be just setting out, and not as yet to have moved to the least imaginable distance from the point *C*, nor consequently to have generated the least imaginable increment, it is plain that the nascent increment here considered will be less than any quantity that can be assigned. In like manner when the generating point returns back from *c* to *C*, in order to annihilate the increment *cC*, that increment will continually grow less and less, will become less than any assignable quantity, and will at last entirely vanish and become nothing by the return of the generating point to the point *C*. At that instant of time therefore that the generating point returns to *C*, at that very instant I say the increment vanishes, and is then properly called an evanescent increment. (Jurin 1735a, 20–21)

Despite the confident and mocking tone maintained throughout his treatise, Jurin has in effect conceded defeat.[15] He maintains that a nascent increment is a positive quantity less than any assignable real quantity, that it is generated in a durationless instant of time, and that it is not an infinitesimal magnitude. Such pronouncements from such a staunch and vocal defender of Newton are more than sufficient to establish Berkeley's claim that the foundations of the calculus were, in the 1730s, shrouded in "much Emptiness, Darkness, and Confusion."

In reply to Berkeley's charge that the method of evanescent incre-

---

15. The general tone of Jurin's defense can be gathered from the pronouncement he delivers shortly after the above passage, when he declares "Behold good Reader, the *difficult*, the *obscure*, the *mysterious*, the *incomprehensible* principles of Fluxions! I am much mistaken if a little attention do not enable thee clearly to conceive them" (Jurin 1735a, 21–22).

ments requires two inconsistent assumptions, Jurin takes on the task of showing that the required assumptions are in fact consistent. His strategy is to argue that there is no inconsistency in assuming that an increment first exists and then vanishes, because the assumptions are made at different times:

> I forbear making any remarks upon your interpretation of the word vanish. I admit it to be as you are pleased to make it, that the first supposition is, there are increments; and that the second supposition is, there are no increments. What do you infer from this? The second supposition, say you is contrary to the former, and destroys the former, and in destroying the former it destroys the expressions, the proportions, and every thing else derived from the former supposition. Not too fast, good Mr. *Logician*. If I say, the increments now exist, and, the increments do not now exist; the latter assertion will be contrary to the former, supposing now to mean the same instant of time in both assertions. But if I say at one time the increments now exist; and say an hour after, the increments do not now exist; the latter assertion will neither be contrary, nor contradictory to the former, because the first now signifies one time, and the second now signifies another time, so that both assertions may be true. The case therefore in your argument does not come up to your *Lemma*, unless you will say Sir *Isaac Newton* supposes that there are increments, and that there are no increments, at the same instant of time. Which is what you have not said, and what, I hope, you will not dare to say. (Jurin 1735a, 96–97)

This approach, however, does nothing to validate Newton's procedure.[16] Recall that Newton had obtained the ratio of increments

$$o : \left(nx^{n-1}o + \frac{(n^2 - n)}{2}o^2 x^{n-2} + \ldots \right) \tag{7.3}$$

---

16. Giorello (1985, 236–37) has also commented on this response by Jurin, but reaches a somewhat more sympathetic conclusion, namely that Jurin and Berkeley disagree about the role of kinematic concepts in geometric demonstrations: "This type of justification, which undoubtedly derives from the same 'genetic' conception of mathematical objects, is set forth by Newton in his *Quadrature of Curves* and elsewhere. It implicitly changes the conception of 'demonstration': here, indeed, the *argumenta illa* arise and evanesce, not only in the phase of the *solution* of the problems, but also in the phase of the *proof* of the proposition. . . . The intuition of time here takes the place of the Leibnizian law of continuity and realizes that which in the eyes of the bishop is merely 'amusement or a miracle for the reader'." As I explain below, this is a rather too charitable interpretation of Jurin.

and then divided through by the common term o. We can assume that the division is carried out at some time t_0 when o is positive. The division results in the reduced ratio

$$1 : \left(nx^{(n-1)} + \frac{(n^2 - n)}{2} ox^{(n-2)} + \ldots \right) \tag{7.4}$$

Then we dismiss all terms containing o from (7.4) to get the ratio

$$1 : (nx^{(n-1)}) \tag{7.5}$$

Since the fluxion of the power x^n is the ratio of the increment of the power to the increment of the root, we obtain $nx^{(n-1)}$ as the fluxion of the power after the terms containing o have been dismissed. Let us stipulate that the dismissal of the terms occurs at a time t_1 (after t_0), when the increment o has vanished. There is certainly no contradiction in assuming that the quantity o is positive at t_0 and zero at t_1, but the reduced ratio (7.4) was obtained by a step which necessarily implies that the increment o is positive. We can therefore only retain the term $nx^{(n-1)}$ as the limiting value of the ratio at time t_1 if we have a guarantee that, in effect, the limit of a ratio is the ratio of the limits. But in Newton's presentation there is no guarantee that this is the case, and talk of "vanishing quantities" implies that the move from ratio (7.3) to ratio (7.4) requires an assumption which is contradicted when we move from (7.4) to (7.5).

The net effect of Jurin's response to Berkeley is to leave the foundations of the calculus shrouded in mystery. Although Berkeley did not reply to Jurin's *Minute Mathematician*, it is clear that his silence does not indicate that he had been convinced by the arguments advanced against him. If anything, Jurin's performance solidified Berkeley's case against the calculus. After the publication of the *Minute Mathematician* Jurin went on to engage in a prolonged dispute with Benjamin Robins over the proper formulation of the Newtonian theory which raised fundamental points in the theory of limiting ratios. I will consider this debate later on but would emphasize at this point that Jurin was clearly not the man to be entrusted with the task of clarifying and defending the calculus.

Jurin's love of controversy subsequently brought him into conflict with Berkeley over the medicinal virtues of tar-water. Jurin, a representative of the British medical establishment, opposed Berkeley's *Siris* in an anonymous *Letter to the Right Reverend Bishop of Cloyne*, oc-

casion'd by His Lordship's treatise on the virtues of tar-water (1744). Berkeley was aware of his authorship and identified him in a letter to Thomas Prior, dated 19 June 1744, that included a satirical verse directed at Jurin (*Works* 8:271–72). It is unclear whether Berkeley knew that Jurin was the author of the anti-*Analyst* pamphlets attributed to "Philaletes Cantabrigiensis." I will ignore this aspect of Berkeley's battle against Jurin, as it is not relevant to the mathematical issues under consideration here, and will proceed instead to consider Berkeley's exchanges with John Walton.

The Dispute with Walton

Little is known about John Walton, except that he was a teacher of mathematics in Dublin at the time the *Analyst* was published. Berkeley refers to him condescendingly as a "Dublin professor" who "gleans after the Cantabrigian" (*Appendix*, §1) and, except for his publications against Berkeley, Walton is a mathematical unknown. His *Vindication of Sir Isaac Newton's Principles of Fluxions, against the Objections contained in the "Analyst"* tries to show that Newton's results are underwritten by consistent and coherent principles. Walton focuses on three familiar points: the doctrine of evanescent magnitudes, the proof of the product rule for finding the fluxion of the rectangle AB, and the proof of the rule for determining the fluxion of the power x^n. Walton's defense of Newton is similar in some ways to Jurin's, but their approaches differ in several respects.

In defending the Newtonian conception of nascent and evanescent magnitudes Walton relies heavily upon extracts from Newton's presentation, which he summarizes and attempts to explain. In doing so he conflates the Newtonian doctrine of moments with the doctrine of evanescent magnitudes, then claims that the entire theory is a model of clarity and rigor. Thus, after quoting the Introduction to Newton's *Quadrature of Curves*, in which fluxions are defined as the velocities with which fluents are generated or destroyed, Walton asserts that "The momentaneous Increments or Decrements of flowing Quantities, he elsewhere calls by the Name of *Moments*." He continues with an explication of the doctrine which does little to overcome Berkeley's objections:

> The Magnitudes of the momentaneous Increments or Decrements of Quantities are not regarded in the Method of Fluxions, but their first or last Proportions only; that is, the Proportions with which they begin or cease to exist: These are not their Proportions immediately before or after they begin

The Aftermath of the *Analyst*

or cease to exist, but the Proportions with which they begin to exist, or with which they vanish. . . . Now the ultimate Ratio of those Increments is that Ratio with which they vanish and become nothing; or the Ratio with which they cease to be: And the first Ratio of them is the Ratio with which they begin to exist. (Walton 1735a, 7)

This leaves open the crucial question of how one quantity that lacks determinate magnitude can stand in a ratio to another. Walton shifts to the language of limits in his attempt to elucidate this mystery, but he is quite far from anything like the modern notion of a limit:

The ultimate Ratios with which synchronal Increments of Quantities vanish, are not the Ratios of finite Increments, but Limits which the Ratios of those Increments attain, by having their Magnitudes infinitely diminish'd: The Proportions of Quantities which grow less and less by Motion, and at last cease to be, will, in most Cases, continually change, and become different in every successive Diminution of the Quantities themselves: And there are certain determinate Limits to which all such proportions perpetually tend, and approach nearer than by any assignable Difference, but never attain before the Quantities themselves are infinitely diminish'd; or till the Instant they evanesce and become nothing. These Limits are the last Ratios with which such Quantities or their Increments vanish or cease to exist; and they are the first Ratios with which Quantities or the Increments of Quantities, begin to arise or come into Being. (Walton 1735a, 8–9)

Although Walton speaks of ultimate ratios as limits here, it is clear that his conception of limits is far from the modern theory of limits and convergence. Walton's "ultimate ratio" is the limit of vanishing quantities, or the value of the ratio after both of the quantities have been infinitely diminished. But the modern theory of limits defines the limiting values of a sequence as a value which can be approached within any desired degree of accuracy,[17] and the limit of a ratio between two sequences as the ratio of the limits of the separate sequences. To distinguish this from Walton's account, consider the ratio $(1 + (1/n)) : (1 - (1/n))$. The limit of this ratio as n gets large is clearly 1. The modern theory of limits proceeds by separately considering the two sequences $\{s_n\} = (1 + (1/n))$ and $\{s_m\} = (1 - (1/m))$, finds that both converge to the limit 1, since for any $\varepsilon > 0$ each can be brought within ε of 1, then invokes the theorem that the limit of a ratio is the

---

17. In technical terms, the sequence $\{s_n\}$ converges to the limit L if and only if for any $\varepsilon > 0$ there is some term s_τ in the sequence such that for any $k > \tau$, $|s_k - L| < \varepsilon$.

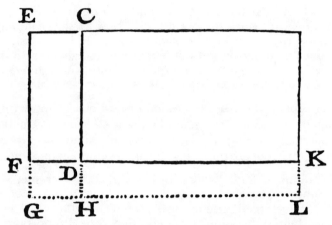

Fig. 7.3 From Walton, *A Vindication of Sir Isaac Newton's Principles of Fluxions* (1735, 14).

ratio of its limits.[18] The limit of the ratio $(1 + (1/n)) : (1 - (1/m))$ will be the ratio of the limits, that is, 1 : 1, or simply 1. Walton would require the quantities $(1/n)$ and $(1/m)$ to be infinitely diminished in order to compute the ultimate ratio, which implies that the quantities m and n would become infinitely large. The modern theory, however, does not assume the diminution of the quantities $(1/n)$ and $(1/m)$ to infinitely small magnitudes.

Walton's defense of Newton's proof of the product rule invokes this theory of evanescent magnitudes and proceeds by taking the rectangle *CK* (fig. 7.3) to represent the product, with the sides *DK* and *DC* for *A* and *B* respectively. If we then take *DF* and *DH* to represent the moments *a* and *b*, the rectangle *EL* will correspond to the product $(A + a) \times (B + b)$ and the difference between *AB* and $(A + a) \times (B + b)$, which Walton calls the "gnomon" of the rectangle, will be the area *CGK*. Walton argues that the gnomon is not the moment of the rectangle and that Newton properly computes it as $Ab + Ba$:

> For whatever be the Magnitudes of *a* and *b*, when *F* and *H* first begin to move back towards *D*, the Gnomon *CGK* and the Sum of the Rectangles *LD* and *FC*, will be as $Ab + Ba + ba$ and $Ab + Ba$. . . . Hence it appears, that under a constant Diminution of the Increments *a* and *b*, by the Motion of the Points *F* and *H* towards *D*, the Gnomon *CGK* and the Sum of the Rectangles *CF* and *DL*, constantly tend toward an

---

18. The theorem can be easily proved by exploiting the definition of convergence to a limit. If two sequences $\{s_m\}$ and $\{s_n\}$ converge to limits L_m and L_n, with $L_n \neq 0$, then the limit of the quotient $\{s_m\}/\{s_n\}$ must be L_m/L_n.

The Aftermath of the *Analyst*

Equality by a continual Diminution of their Difference *FG*; and that they become equal, and their Ratio becomes a Ratio of Equality, in the Instant that the Difference vanishes and the Points *F* and *H* coincide with *D*. . . . Hence, the Gnomon *CGK* . . . is not the Moment or Fluxion of the Rectangle *AB*, except in the very Instant when it begins or ceases to exist. (Walton 1735a, 13–15)

The problematic step here, of course, is the assumption that the points *F* and *H* coincide with *D*, while the products *Ab* and *Ba* remain positive and the product *ab* is neglected as a zero quantity. To assume that *F* and *H* coincide with *D* requires that the quantities *a* and *b* are zero, which also makes the products *Ab* and *Ba* zero; but Walton explicitly denies this, even as he discards the product *ab*.

Walton uses a similar line of reasoning in defending Newton's proof of the rule for determining the fluxion of a power. He argues that the result is not obtained by an illegitimate shifting of hypotheses, but rather by the computation of legitimate ultimate ratios:

In this Computation, Sir *Isaac* endeavours to collect the Proportion with which the isochronal Increments of x and x^n, begin or cease to exist: Their Proportion obtained on supposition that o is something, is allowed to be the same with that of 1 and

$$nx^{(n-1)} + \frac{(n^2 - n)}{2} o x^{(n-2)} + \&c.$$

And it must be acknowledg'd that this Ratio has a Limit it cannot attain before the Increments are infinitely diminish'd and become evanescent; and when, by an infinite Diminution, they become evanescent, no other Terms of their Ratio will be affected, so as to vanish with 'em, but such as are govern'd or regulated by them: In the Instant therefore that o vanishes,

$$\frac{(n^2 - n)}{2} o x^{(n-2)}$$

and all ensuing Terms of the Series absolutely vanish altogether; but the Terms 1 and $nx^{(n-1)}$ remain invariable under all possible Changes of the Increments, from any finite Degrees of Magnitude whatever, even till they become evanescent: They therefore express the last Ratio, under which the isochronal Increments of x and x^n vanish, or the Proportion of the Velocities with which those Increments cease to exist: Sir *Isaac Newton* then rightly retain'd 'em for the Measures of the Ratio of the Fluxions of x and x^n, tho' got in virtue of his first Supposition; and the Fallacy, the Inconsistency, lies on the Side of this Author, who wou'd have them rejected on the Authority of a Lemma not to the Purpose. (Walton 1735a, 23–24)

This response takes no notice of Berkeley's objections but simply asserts that supposing o to vanish (through infinite diminution) is in no way contradictory to the prior assumption that o was a positive increment of the flowing quantity x. Moreover, Walton assumes exactly what is in question when he claims that the ultimate ratio of 1 to $nx^{(n-1)}$ will persist after the increments (from which this ratio has been derived) have vanished. Walton shows a similar failure to appreciate Berkeley's point when he tries to defend the Newtonian proof of the product rule, arguing essentially that when the quantities a and b are evanescent, the product ab will be nothing, but Ab and bA will still be positive quantities. One might, of course, read Walton's vindication of Newton as an attempt to treat evanescent magnitudes as infinitesimals and thereby to place Newton's fluxional calculus on the same footing as the Continental *calculus differentialis*, but Walton explicitly contrasts the Leibnizian presentation of the calculus with the Newtonian account of fluxions, claiming that his interest is only in defending Newton.[19]

Berkeley responds to Walton's *Vindication* in a brief appendix to the *Defence of Free-Thinking in Mathematics*, restating his reservations about the comprehensibility of the calculus. He finds Walton's efforts worth little explicit attention, claiming that "the foregoing defence contains a full and explicit answer to Mr. Walton, as he will find, if he thinks it worth his pains to read what this gentleman hath written, and compare it therewith" (*Appendix*, §1). Nevertheless, he finds certain points to merit specific attention and asks that Walton be catechized on fundamental concepts. Among other things, Berkeley asks whether Walton can "conceive velocity without motion, or motion without extension, or extension without magnitude?" If Walton claims

---

19. Thus, he declares: "The Dispute between the Followers of Sir *Isaac Newton*, and the Author of the *Analyst*, is not about the Principles of the *differential Calculus*, but about those of Fluxions; and it is whether these Principles in themselves are clear or obscure, and whether the Inferences from them are just or unjust, true or false, scientific or otherwise: We are not concerned about Infinitesimals or minute Differences, but about the Ratios with which mathematical Quantites begin or cease to exist by Motion; and to consider the first or last Proportions of Quantities, does not imply that such Quantities have any finite Magnitudes: They are not the Proportions of first or last Quantities, but Limits of Ratios; which Limits the Ratios of Quantities attain *only* by an infinite Diminution of their Magnitudes, by which infinite Diminution of their Magnitudes they become evanescent and cease to exist" (Walton 1735a, 29–30).

The Aftermath of the *Analyst*

this ability, Berkeley asks that he teach others to do the same. He also asks for an explanation of the difference "between a magnitude infinitely small and a magnitude infinitely diminished" and for a way to avoid treating the ultimate ratios of flowing quantities as "ratios of nothings." Although the *Appendix* falls short of a full-scale attack on Walton's *Vindication* and need not consume much of our time, one point is worth noticing: Berkeley demands that Walton frame abstract ideas of impossible objects—velocity without motion, motion without extension, and extension without magnitude—again explicitly tying his criticism of the calculus (at least his "metaphysical" objections) to his rejection of abstract ideas.

Walton replied with another pamphlet—*The Catechism of the Author of the "Minute Philosopher" Fully Answer'd*. This "full answer" does little to clear up the issues, but it does put the question of abstraction into clearer focus. Walton insists that there is no difficulty in conceiving motion in a point and undertakes to demonstrate that the principles of the calculus are fully consistent and conceivable. The fundamental point in dispute here is the notion of instantaneous velocity. Walton maintains that acceleration can only be understood by taking the concept of instantaneous velocity seriously, and he insists that at every point in its trajectory an accelerating body will have a different velocity. From this he concludes that there must be a motion in each point of the trajectory of the accelerated body, and that the abstraction of motion from space is unproblematic:

> I can conceive Velocity and Motion in a *Point of Space;* that is, without any assignable Length or extension described by it, and so might he too if he had understood and consider'd the Nature of Motion. For Motion is an Effect of some Cause acting on the thing moved; which Effect, setting aside all Resistance, will ever be proportional to the whole Action of the generating Cause: And therefore if a Cause acts continually upon a given thing without any Interruption of its Velocity: the Velocity cannot be the same in any two different Points of the Space described, however near those Points may be to each other. (Walton 1735b, 9–10)

To Berkeley's demand "when ab is nothing, whether $Ab + Ba$ be not also nothing?" Walton responds with an absurd attempt to show that the vanishing of the gnomon means that it becomes nothing, but still something:

> I agree with him that nothing is the Product of nothing multiply'd by something; but must know what he means by the

vanishing of the Gnomon and Sum of the two rectangles. If by vanishing he means that they vanish and become nothing as Areas, I grant they do; but absolutely deny, upon such an Evanescence of the Gnomon . . . that nothing remains. For there still remain the moving Sides, which are now become the sides of the Rectangle. (Walton 1735b, 12–13)

But by treating the evanescent gnomon as a line, Walton has retreated from the Newtonian foundation of the calculus to something very much like the Leibnizian calculus of infinitesimal magnitudes. To treat an evanescent surface as a line and such lines as increments of surfaces is to adopt the infinitesimalist viewpoint, for infinitesimal magnitudes are characterized as magnitudes which stand to finite magnitudes in the ratio of a point to a line or a line to a surface.[20]

Walton tries to refute Berkeley's charge that fluxions are merely "ratios of nothings" by engaging in a subtle abstraction of a ratio from the quantities which form the ratios:

Neither Sir *Isaac Newton* nor I have said, that *Fluxions* are measured by the Proportions of Magnitudes infinitely small, nor by the Proportions of any Magnitudes whatever generated in equal Times; but that *they are measured by the first or last Proportions of isochronal Increments generated or destroy'd by Motion;* which Proportions are the *Ratios* with which such Increments begin to exist before they have acquired any Magnitude, or with which they cease to exist and vanish after they have lost all Magnitude. These *Ratios* subsist when the isochronal Increments have no Magnitude, for as much as the Motions subsist with which those Increments, *just now, in this very Instant,* begin or cease to exist; to which Motions these *Ratios* are proportional. (Walton 1735b, 20–21)

In other words, a fluxion or ultimate ratio is not a "ratio of nothings," but rather a ratio which subsists after the magnitudes which form it have disappeared. Whatever else they may show, Walton's remarks clearly suggest that Berkeley's characterization of ultimate ratios as "Ghosts of departed Quantities" is on the mark.

Berkeley's rejoinder was his last publication in the *Analyst* controversy: *Reasons for Not Replying to Mr. Walton's "Full Answer," in a Letter*

---

20. Berkeley noted precisely this point in his rejoinder to Walton's *Full Answer:* "But, saith, Mr. Walton, the sides of the given rectangle still remain, which two sides according to him must form the increment of the flowing rectangle. But in this he directly contradicts Sir Isaac Newton, who asserts that $Ab + Ba$ and not $A + B$ is the increment of the rectangle AB. And, indeed, how is it possible that a line should be the increment of a surface?" (*Reasons*, §10).

The Aftermath of the *Analyst*

to P. T. P. In it he returns to the points upon which Walton was to be catechized and insists that the obscurity of his answers indicate that he is secretly an opponent of the Newtonian doctrine. After all, he reasons, "be a man's assertions ever so strong in favour of a doctrine, yet if his reasonings are directly levelled against it, whatever question there may be about the matter in dispute, there can be none about the intention of the writer" (*Reasons*, §2).

I will not work through Berkeley's response in detail but will note some of Berkeley's points which will be of interest later. Berkeley begins his critique of Walton's claim that motion is conceivable in a point with a telling parody of Walton's reasoning:

> Pray, Sir, consider his reasoning. The same velocity cannot be in two points of space; therefore velocity can be in a point of space. Would it not be just as good reasoning to say the same man cannot be in two nutshells; therefore a man can be in a nutshell? Again, velocity must vary upon the least change of space; therefore there may be velocity without space. Make sense of this if you can. What have these consequences to do with their premises? Who but Mr. Walton could have inferred them? Or how could even he have inferred them had it not been in jest? (*Reasons*, §4)

The problem here is a serious confusion over how to understand the claim that a body is in motion at an instant. Berkeley denies that a body can be said to be in motion if it covers no space, and it is certainly hard to see how a body can cover any space in a durationless instant. Thus the notion of instantaneous velocity, taken to assert that a body can cover some space in no time, is ruled out on Berkeleyan principles. But Walton's view that the continuous acceleration of a body will make its velocity differ at any two points in its trajectory also seems evident.

One way to overcome this paradox is through a counterfactual definition of instantaneous velocity: the instantaneous velocity of a body at an instant is the distance per unit time that the body would cover if it continued unaccelerated from that point. This allows us to define the expression "velocity at time t_0" without assuming that the body covers an infinitesimal distance at the instant t_0. In fact, Berkeley adopts a course very similar to this:

> Suppose the center of a falling body to describe a line, divide the time of its fall into equal parts, for instance, into minutes. The spaces described in those equal parts of time will be unequal. That is, from whatsoever points of the described line

you measure a minute's descent, you will still find it a different
space. This is true. But how or why from this plain truth a
man should infer, that motion can be conceived in a point, is
to me as obscure as any the most obscure mysteries that occur
in this profound author. (*Reasons*, §5)

On Berkeley's account, to say that an accelerated body has a different velocity at any two points in its descent is to say that if we measure the distance traveled in one unit time from any two points, the results will differ. Equivalently, uniform velocity is that in which equal distances are covered in equal periods of time. It is a short step from this to the counterfactual definition of velocity at a point as the distance that *would* be traveled from the point in one unit of time if the motion were to continue unaccelerated. Thus, Berkeley does not banish the concept of instantaneous velocity as irredeemably abstract, and it seems that an acceptable account of the calculus could be formulated with the fluxion of a point counterfactually defined as its instantaneous velocity and the rest of the theory of fluxion similarly based on kinematic considerations. However, when Colin Maclaurin proposes just such an account of the theory of fluxions, Berkeley does not take his efforts seriously.

Walton responded to Berkeley's *Reasons for not Replying* in an appendix to the second edition of his *Catechism of the Author of the "Minute Philosopher" Fully Answer'd,* but this piece makes no serious contribution to the debate. Although Berkeley did not respond to it, John Hanna, an Irish Newtonian and minor astronomer, critiqued it in his *Some Remarks on Mr. Walton's Appendix, which he Wrote in Reply to the Author of the "Minute Philosopher"; concerning Motion and Velocity.*[21] On balance, Walton's critique of the *Analyst* is only marginally more effective than Jurin's failed attempt to vindicate Newton. In crucial respects the foundations of the calculus of fluxions remained without a proper clarification, and Walton's attempt to define the key concepts of the calculus in terms of limits and instantaneous velocity was confused and lacking in rigor. This is not to say that the state of the calculus was irremediable in the 1730s. As we will see, several authors proposed interpretations which went a long way toward overcoming Berkeley's objections. However, the two authors with whom Berkeley most actively contested the foundations of the calculus did little by way of vindicating Newton.

21. Hanna (1736) deals less with the mathematical issues raised by the *Analyst* than with notions of velocity and time, so will not be considered here. He rejects Walton's analysis of motion, but does not endorse Berkeley's critique of the calculus.

The Aftermath of the *Analyst*

Other Responses to Berkeley

Aside from the works to which Berkeley responded, numerous other publications aimed at countering his critique of the calculus in the decade after the *Analyst*. Six works are of particular interest to my investigation because they represent a wide variety of approaches to the calculus and raise some significant issues for the interpretation of Berkeley: Benjamin Robins's *Discourse Concerning the Nature and Certainty of Sir Isaac Newton's Methods of Fluxions, and of Prime and Ultimate Ratios* (1735a), John Colson's commentary to his translation of Newton's *Method of Fluxions and Infinite Series* (1736), Thomas Bayes's *Introduction to the Doctrine of Fluxions* (1736), James Smith's *New Treatise of Fluxions* (1737), Colin Maclaurin's *Treatise of Fluxions* (1742), and Roger Paman's *Harmony of the Ancient and Modern Geometry Asserted* (1745). Some of these works attempt to place the Newtonian methods on the same footing as the classical method of exhaustion, while others introduce completely new principles to justify the methods of the calculus. Some authors retain the kinematic conception of magnitudes, while others seek to banish the concepts of motion and velocity from the science of geometry. As might be expected, these varied efforts meet with different degrees of success. Some authors—Robins, Maclaurin, and Paman among them—offer ingenious and mathematically sound proposals for rigorizing the calculus, while others—notably Colson and Smith—fail to make any headway against Berkeley.

Benjamin Robins and the Theory of Limits

Benjamin Robins (1707–51) made important contributions in applied mathematics, notably in ballistics. But he was also a key figure in the battle against Berkeley. His *Discourse* stands firmly in the Newtonian tradition and is designed to make good on Newton's claim that his work is "agreeable to the Geometry of the Ancients." Robins is not interested in merely repeating Newton's claims for the rigor of his theory, but instead seeks to found the calculus of fluxions on the method of exhaustion. Thus, he opens his *Discourse* by declaring that, while the Newtonian calculus is in no way obscure or unreliable, Newton's concise manner of expressing himself necessitates a more extended treatment of the doctrine of fluxions and the theory of prime and ultimate ratios.[22] However, he then proceeds to develop an ac-

---

22. "For though Sir Isaac Newton has very distinctly explained both these subjects, the first in his treatise on the Quadrature of curves, and the other in his Mathematical

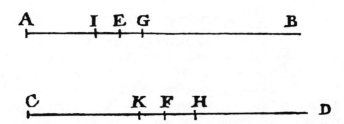

Fig. 7.4 From Robins, *A Discourse Concerning the Nature and Certainty of Sir Isaac Newton's Principles of Fluxions, and of Prime and Ultimate Ratios* (1735, 7).

count of fluxions which is an original and important contribution to the foundations of the calculus.

Robins takes his point of departure from Newton's kinematic conception of geometric magnitudes, defining the fluxion of a quantity as the velocity with which it is produced. He is at pains to stress that the key to the doctrine of fluxions is to be found in the proportions which fluxions bear to one another:

> And as different fluents may be understood to be described together in such manner, as constantly to preserve some one known relation to each other; the doctrine of fluxions teaches, how to assign at all times the proportion between the velocities, wherewith homogeneous magnitudes, varying thus together, augment or diminish. (Robins 1735a,6)

This leads Robins to his first fundamental theorem: "that the proportion between the fluxions of magnitudes is assignable from the relation known between the magnitudes themselves" (Robins 1735a, 7). The proof is by exhaustion and worth analyzing in some detail because it shows how clearly Robins understands Berkeley's challenge to the calculus and how markedly his approach differs from Newton's casual treatment of foundational questions.

Robins's proof proceeds as follows:[23] Given two lines, AB and CD (fig. 7.4), generated by two moving points, assume that the velocity of the point describing CD bears some ratio to that of the point describing AB. Let AE be denoted by the variable x, and assume that

$$CF = x^n.$$

---

principles of natural philosophy; yet as the author's great brevity has made a more diffusive illustration not altogether unnecessary; I have here endeavoured to consider more at large each of these methods; whereby, I hope, it will appear, that they have all the accuracy of the strictest mathematical demonstrations." (Robins 1735a, 2)

23. I have simplified Robins's case slightly but leave the essentials of his presentation intact. Robins actually considers the case where $CF = x^n/a^{(n-1)}$.

The Aftermath of the *Analyst*

Robins desires to show that the ratio between the fluxion of *CD* and the fluxion of *AB* is assignable at every point. To do this, he considers increments on both lines, so that as the increment *EG* is generated in the line *AB*, a corresponding increment *FH* arises in *CD*. Denote *EG* by *e*. Then, by hypothesis:

$$CH = (x + e)^n$$

which expands to yield

$$CH = x^n + nx^{(n-1)}e + \frac{n(n-1)}{2}x^{(n-2)}e^2 + \ldots \qquad (7.6)$$

Subtracting the equal quantities *CF* and x^n from the requisite sides of (7.6) yields the increments

$$FH = nx^{(n-1)}e + \frac{n(n-1)}{2}x^{(n-2)}e^2 + \ldots$$

Assume for purposes of illustration that $n > 1$, in which case the point describing *CF* is uniformly accelerated. Robins asserts that

> [N]o line whatever, that shall be greater or less than the line represented by the second term of the foregoing series (*viz.* $nx^{(n-1)}e$) will bear to the line denoted by *e* the same proportion, as the velocity, wherewith the point moves at *F*, bears to the velocity of the point moving in the line *AB*; but that the velocity at *F* is to that at *E* as $nx^{(n-1)}e$ to *e*, or as $nx^{(n-1)}$ to 1. (Robins 1735a, 9)

In other words, the fluxion of x^n is $nx^{(n-1)}$.

Robins approached this proof with a clear understanding of Berkeley's objections. Rather than follow Newton's strategy of dividing by the increment and then dismissing terms which contain it (a procedure Berkeley dubbed the *fallacia suppositionis*), Robins undertakes an exhaustion proof. He first recalls that, under the hypothesis that $n > 1$, the point describing the line *CD* is uniformly accelerated. Thus, the ratio *FH* : *EG* (being a ratio of increments) is greater than the ratio of the fluxion at *F* to the fluxion at *E*. This is because fluxions are defined as velocities, and the increments generated by continually accelerated points will increase with each unit time. As in any exhaustion proof, he proceeds *reductio ad absurdum:* assume that the ratio of the fluxion at F to the fluxion at E is $p : q$, where

$$(p : q) > (nx^{(n-1)} : 1). \tag{7.7}$$

Observe, in equation (7.6) above, that the quantity e in the series expansion can be taken so small that any term in the series will be greater than the sum of all subsequent terms.[24] That is, for any term τ_i there is an e sufficiently small that

$$\tau_i > \sum_{n=1}^{\infty} \tau_{i+n}.$$

Adding τ_i to each side of the inequality yields

$$2\tau_i > \sum_{n=0}^{\infty} \tau_{i+n}. \tag{7.8}$$

Furthermore, for sufficiently small e, we can make the ratio of the second term to double the third term in the series expansion at (7.6) arbitrarily large, since as e diminishes the ratio $(nx^{(n-1)}e) : (n(n-1)x^{(n-2)}e^2)$ grows without bound. But inequality (7.8) assures that there is an e small enough that double the third term in (7.6) is greater than the sum of all the terms after the second, that is,

$$\left(n(n-1)x^{(n-2)}e^2\right) > \left(\frac{n(n-1)}{2}x^{(n-2)}e^2 + \frac{n(n-1)(n-2)}{6}x^{(n-3)}e^3 + \ldots\right). \tag{7.9}$$

Because e is positive, we are guaranteed that

$$(nx^{(n-1)}e + n(n-1)x^{(n-2)}e^2) > nx^{(n-1)}e. \tag{7.10}$$

---

24. This follows from the fact that successive terms in the series contain higher powers of e; as e gets smaller, each successive power of e diminishes more rapidly than any of its predecessors.

The Aftermath of the *Analyst*

By dividing inequality (7.10) through by e, we can construct an inequality between ratios, namely

$$((nx^{(n-1)}e + n(n-1)x^{(n-2)}e^2) : e) > (nx^{(n-1)}e : e). \qquad (7.11)$$

But as e is diminished, (7.10) approaches equality, and the ratios in (7.11) become arbitrarily close. Eventually, then, we can construct a chain of inequalities from (7.7) and (7.11):

$$(p : q) > ((nx^{(n-1)}e + n(n-1)x^{(n-2)}e^2) : e) > (nx^{(n-1)}e : e). \qquad (7.12)$$

Adding the term $nx^{(n-1)}e$ to both sides of inequality (7.9), we get

$$(nx^{(n-1)}e + n(n-1)x^{(n-2)}e^2) > \left(nx^{(n-1)}e + \frac{n(n-1)}{2}x^{(n-2)}e^2 + \frac{n(n-1)(n-2)}{6}x^{(n-3)}e^3 + \ldots \right).$$

Dividing by e yields an inequality of ratios:

$$((nx^{(n-1)}e + n(n-1)x^{(n-2)}e^2) : e) > \left(\left(nx^{(n-1)}e + \frac{n(n-1)}{2}x^{(n-2)}e^2 \right. \right.$$

$$\left. \left. + \frac{n(n-1)(n-2)}{6}x^{(n-3)}e^3 + \ldots \right) : e \right) \qquad (7.13)$$

Then, inequalities (7.12) and (7.13) imply that

$$(p : q) > \left(\left(nx^{(n-1)}e + \frac{n(n-1)}{2}x^{(n-2)}e^2 \right. \right.$$

$$\left. \left. + \frac{n(n-1)(n-2)}{6}x^{(n-3)}e^3 \ldots \right) : e \right)$$

In other words, the ratio $p : q$ exceeds $FH : EG$, or the fluxion at F exceeds the fluxion at H, which is absurd, since we have already established that the ratio $FH : EG$ (as a ratio of finite increments) is greater than the ratio of the fluxion at F to the fluxion at E.

There are other cases to be considered in this exhaustion proof, such as that where $p : q$ is less than the ratio of fluxions and that where

$n \leq 1$; nevertheless, the general idea of Robins's procedure should be tolerably clear. Note that he makes no reference to evanescent magnitudes or infinitesimal quantities. The whole procedure rests upon the binomial theorem for the expansion of the power $(x + e)^n$ and the theory of ratios.

Robins's plan to defend Newton extends beyond showing that fluxions can be defined and determined using exhaustion techniques. As the full title of his *Discourse* suggests, he wants to consider the "nature and certainty" of the Newtonian methods of fluxions and of prime and ultimate ratios. There are two important reasons for Robins to defend the doctrine of prime and ultimate ratios: first, because Newton himself used the method and, second, because it considerably shortens demonstrations which would otherwise require complicated exhaustion proofs for every case. As he did with Newton's account of fluxions, Robins aims to ground the Newtonian doctrine of prime and ultimate ratios in the classical theory of exhaustions. He insists that the Newtonian method is nothing more than an abbreviation of the classical procedure, thought obscure by some only because of Newton's concise way of expressing himself:

> The concise form, into which Sir Isaac Newton has cast his demonstrations, may very possibly create a difficulty of apprehension in the minds of some unexercised in these subjects. But otherwise his method of demonstrating by the prime and ultimate ratios of varying magnitudes is not only just, and free from any defect in itself; but easily to be comprehended, at least by those who have made these subjects familiar to them by reading the ancients. (Robins 1735a, 47)

Robins starts from Newton's introduction of the doctrine of prime and ultimate ratios in the *Principia:*

> Quantities, and the ratios of quantities, which in any finite time converge continually to equality, and before the end of that time approach nearer to each other than by any given difference, become ultimately equal. (*Principia* I, 1, 1)

Newton's proof of this proposition is surprisingly simple:

> If you deny it, suppose them to be ultimately unequal, and let D be their ultimate difference. Therefore they cannot approach nearer to equality than by that given difference D; which is contrary to the supposition. (*Principia* I, 1, 1)

Robins's task is to provide an underlying theory for these Newtonian pronouncements that will avoid Berkeley's objections. In particular,

The Aftermath of the *Analyst*

Robins must show that no contradiction is involved in first introducing a finite increment and then dismissing it when an ultimate ratio is calculated.

Toward this end Robins begins with a series of definitions and lemmas strikingly similar to the modern theory of limits. He defines the expression "ultimate magnitude," then proves a series of theorems which lead to a definition of "ultimate ratio." In each case he proves the existence and uniqueness of limits and ratios of limits and that the limit of a ratio is the ratio of limits. Although it would be an exaggeration to credit Robins with fulfilling modern standards of rigor for the calculus, his response certainly meets Berkeley's objections. It is therefore worthwhile looking briefly into his procedure.

The fundamental definition in Robins's account of the calculus is of the term "ultimate magnitude." It reads:

> We shall in the first place define an ultimate magnitude to be the limit, to which a varying magnitude can approach within any degree of nearness whatever, though it can never be made absolutely equal to it. (Robins 1735a, 53)

We can write $x \to B$ when B is the ultimate magnitude of the varying quantity x. Note that Robins's definition departs in one important respect from the modern definition of a limit: he insists that a variable *never* attains its limit, while the modern theory of limits allows variables to attain their limits.[25] Robins takes his model of ultimate magnitudes from the theory of exhaustions, which presumably accounts for his doctrine that variables do not attain their ultimate magnitudes (or limits). The circle is the ultimate magnitude of a sequence of circumscribed polygons, but there is no polygon in the sequence equal in area to the circle. And in the general geometric case where a curvilinear figure is "exhausted" by sequences of rectilinear figures, the approximations will never ultimately be equal to the curvilinear figure they approach. In such contexts as these Robins is correct to insist that the varying quantities not attain their limits, but his theory is thereby restricted to special cases of limiting processes.

This definition of ultimate magnitudes leads to the fundamental

---

25. For example, the sequence $\{s_n\} = (n - n)$ is simply the sequence $\{0, 0, 0, \ldots\}$. This sequence has the limit 0, and clearly attains its limit. More significantly, in such limits as $\lim_{x \to 0} \sin(x)$ there is no difficulty in allowing the variable to reach the limiting value. Such cases cannot be easily accommodated within Robins's account of ultimate magnitudes. As Cajori observes, "Here for the first time is the stand taken openly, clearly, explicitly, that a variable (say the perimeter of a polygon inscribed in a circle) can never *reach its limit* (the circumference). . . . According to Robins's definition, Achilles never caught the tortise" (Cajori 1919, 97).

proposition that when varying magnitudes remain in a constant ratio to one another their ultimate magnitudes stand in the same ratio. We can express this symbolically as follows: If $x \to B$ and $y \to C$ and $x/y = k$ (with k constant), then $B/C = k$. The *reductio ad absurdum* proof proceeds by considering four cases. The details need not concern us here,[26] except to note that Robins avoids any talk of infinitesimal or vanishing differences, basing his procedure entirely upon the algebra of inequalities.

"Ultimate ratios" can then be defined to include the case where variable quantities may not bear the same ratio to one another throughout their convergence to their ultimate magnitudes:

> If there be two quantities that are (one or both) continually varying, either by being continually augmented, or continually diminished; and if the proportion, they bear to each other, does by this means perpetually vary, but in such manner that it constantly approaches nearer and nearer to some determined proportion, and can also be brought at last in its approach nearer to this determined proportion than to any other, that can be assigned, but can never pass it: this determined proportion is then called the ultimate proportion, or the ultimate ratio of those varying quantities. (Robins 1735a, 57)

This definition of ultimate ratios shares with the definition of ultimate magnitudes the restriction that the ultimate ratio is never actually attained by the varying quantities, and the additional restriction that the quantities x and y approach their ultimate magnitudes monotonically. Within these restrictions, however, Robins establishes a version of the theorem that the limit of a ratio $x : y$ exists, is unique and is identical with the ratio $A : B$ of the limits, where $x \to A$, $y \to B$, and $B \neq 0$. Having defended the Newtonian doctrine of prime and ultimate ratios, Robins works through the proofs Berkeley had attacked, indicating that the Berkeleyan criticisms can be avoided by reading Newton's method as an exercise in the theory of exhaustion. In essence, Robins defends Newton's use of ultimate ratios on the grounds that the neglected terms in the binomial expansion of the power $(x + o)^n$ can be eliminated by an exhaustion proof.[27]

Robins contrasted his and Jurin's responses to Berkeley in his essay "A Review of some of the Principal Objections that have been made to the Doctrine of Fluxions and Ultimate Proportions; with some Re-

---

26. Proofs of the four cases can be found in (Robins 1735a, 53–56). Sageng (1989, 270–81) analyzes Robins's argumentation.

27. See Robins (1735a, 69–72) for the details.

marks on the different Methods that have been taken to obviate them" (Robins 1735c). His statement is concise and to the point and clearly indicates how seriously he took the *Analyst:*

> These objections have been particularly levell'd at the expression of Sir *Isaac Newton, Fluxiones sunt in ultima ratione decrementorum evanescentium vel prima nascentium.* Which being usually thus translated, that fluxions are in the ultimate ratio of the evanescent decrements, or in the first ratio of the nascent augments, it has from hence been ask'd, What these nascent or evanescent augments are? If of any magnitude, then it will be confess'd by the espousers of this doctrine, that their ratio is not the same with the ratio of the fluxions. If it is answered, that they are of no magnitude; it is then said, that to talk of the ratio of nothings, is such a strain of language, as it is supposed the warmest followers of the inventor will scarce undertake to defend. (Robins 1735c, 437)

After presenting Jurin's attempted defense of Newton's doctrine of ultimate ratios, he offers his own solution to the problem. This amounts to admitting that Newton's presentation is unclear, but maintaining that it can be reinterpreted in a way that avoids Berkeley's objections:

> [I have] endeavoured to shew, that this objection is founded on an erroneous hypothesis; for that by the ultimate proportion of varying quantities was only meant the limit of their varying proportion, and not a proportion that these varying quantities could ever exist under during their variation; and consequently that the true explication of this passage should be, Fluxions are in that proportion, which is the ultimate to all those varying proportions that the decrements bear to each other, whilst they are vanishing or diminishing; that is, the limit of the proportion that the decrements bear to each other as they diminish, is the true proportion of the fluxions. By this interpretation, which is supported by Sir *Isaac Newton's* own Words the above-mentioned objection immediately falls to the ground; since it is altogether founded on the supposition, that the decrements in their imagin'd evanescent state did really bear to each other the proportion of the fluxions; whereas this passage, when truly understood, does not suppose that the decrements can, in any circumstance whatever, bear to each other that proportion; but asserts, on the contrary, that the proportion of the fluxions is only a proportion limiting all the varying proportions that these decrements have to each other in their various degrees of diminution. (Robins 1735c, 438)

Robins's preference for treating ultimate ratios as limits which ratios of variables can never actually attain brings him into conflict with Jurin, who holds that the ultimate ratio of evanescent quantities is attained. In a reply to Robins, Jurin insists, "By arriving at a limit I understand Sir *Isaac Newton* to mean, that the variable quantity, or ratio, becomes absolutely equal to the determinate quantity, or ratio, to which it is supposed to tend" (Jurin 1735b, 380). The result of this difference in views was a long and acrimonious controversy between Jurin and Robins which filled literally hundreds of pages of the *Present State of the Republick of Letters* and its successor, the *History of the Works of the Learned*. There is no need to examine this vicious dispute, but it is interesting to observe that Berkeley's critique of the calculus should have provoked such an outpouring of mathematical vituperation.[28] The length and intensity of this controversy shows, first, that Berkeley was correct to characterize the doctrine of fluxions as a technique whose foundations were inadequately understood and, second, that there was no accepted theory of limits upon which to ground the calculus in the 1730s. It is therefore inaccurate to argue that Berkeley was uncharitable in portraying British mathematicians as confused about the foundations of their science. Although modern presentations of the calculus can evade Berkeley's critique, it is an exercise in anachronism to overcome his reservations about the calculus by appealing to the doctrine of limits.

Johnston has argued that not only was Robins successful in reinterpreting the theory of fluxions to avoid Berkeley's criticism, but that Berkeley himself recognized this and accepted Robins's theory as the correct approach to problems in analysis. The evidence is scanty and inconclusive. Johnston observes that Robins's *Discourse* was published in 1735, the last year in which Berkeley published anything on the calculus, then interprets Berkeley's withdrawal from the *Analyst* controversy as indicating satisfaction with Robins's account of fluxions and defense of Newton's theory.[29]

There are several reasons for doubting this interpretation. First, we

28. See Cajori (1919) for a summary of the dispute, which began between Robins and Jurin, but later became a battle between Jurin and Henry Pemberton. The texts in the controversy include Jurin (1735b; 1736a–c; 1737a–d), Robins (1736a–b), and Pemberton (1737a–f). Sageng (1989, 331–42) also studies this controversy.

29. Johnston makes his case for this interpretation of Berkeley's views as follows: "It is noticeable that Berkeley participated vigorously in the controversy *until* Robins's book appeared. After that he says not a word. The reason is, as we have said, that Robins showed that infinitesimals were not essential to the calculus. Berkeley must have been convinced by his arguments, and therefore realised that it was no longer possible, from his point of view, to take part in the controversy" (Johnston 1923, 266).

should note that Berkeley became bishop of Cloyne in 1735 and turned his attention to ecclesiastical matters in Ireland. Thus, his failure to respond to Robins's theory may well be due to his having been occupied with other business. Second, the same reasoning would lead to the absurd conclusion that Berkeley accepted the account of the calculus in Jurin's 1735 *Minute Mathematician*. Third, Johnston fails to note Berkeley's reference in *Siris* (1744) to the mathematicians' confused and much-disputed doctrine of fluxions.[30] Nevertheless, Robins's account of fluxions is a very good response to Berkeley. He clearly saw the problem that Berkeley had presented and worked out a method of overcoming it that adheres quite closely to the classical conception of rigor which Berkeley explicitly accepts in the *Analyst*. Berkeley apparently found this presentation wanting but made no clear statement of his reasons for disregarding it.

John Colson and Newton's *Method of Fluxions*

John Colson (1680–1760) published a translation of Newton's *Method of Fluxions and Infinite Series* that included a preface and commentary which attempted to rebut the charges in the *Analyst*. Although he later assumed the prestigious Lucasian chair in mathematics at Cambridge (a chair once held by Newton), Colson was evidently a man of quite modest mathematical abilities.[31] To judge from his response to the *Analyst*, Colson was ill-equipped to handle Berkeley's objections to the Newtonian calculus. His defense of Newton is of interest chiefly because it validates all of Berkeley's principal points. Colson's edition of Newton was taken seriously by the British mathematical community,[32]

---

30. Berkeley writes, in a footnote in *Siris* (§271), "Our judgement in these matters is not to be overborne by a presumed evidence of mathematical notions and reasonings, since it is plain the mathematicians of this age embrace obscure notions and uncertain opinions, and are puzzled about them, contradicting each other and disputing like other men: witness their doctrine of fluxions, about which, within these ten years, I have seen published about twenty tracts and dissertations, whose authors being utterly at variance with each other, instruct bystanders what to think of their pretensions to evidence" (*Works* 5:127). Such a comment would hardly be expected from a man who accepted Robins's theory as a conclusive answer to his earlier doubts about the intelligibility of fluxions.

31. Cajori (1919, 130) represents him as "a man of great industry but only ordinary ability." This judgement conforms with that of Thompson Cooper in the *Dictionary of National Biography*, who reports that "he was a plain, honest man, of great industry and assiduity, but the university was much disappointed in its expectations of a professor that was to give credit to it by his lectures."

32. John Eames's review in the *Philosophical Transactions* speaks highly of it, insisting that Colson "takes the Author's Demonstration into strict Examination, endeavours far-

but the confusions and inconsistencies in his work testify to the aptness of Berkeley's claim that Newton's followers had no clear understanding of the foundations of the calculus.

Colson devotes much of his preface to a rebuttal of Berkeley's critique of the calculus. He mentions that Newton's principles "have been scrupulously sifted and examin'd, have been vigorously opposed and (we may say) ignominiously rejected as insufficient, by some Mathematical Gentlemen" and cites the *Analyst* as a work whose objections are to be answered (Colson 1736, ix). Indeed, Colson suggests that his whole purpose in translating and editing Newton's *Treatise* was to put an end to the disputes over the calculus.[33]

Colson proposes a distinction between "relative" and "absolute" conceptions of the infinite and, on this basis, simultaneously embraces the infinitesimal as a foundation for the Newtonian theory of fluxions and rejects the method of indivisibles and the Leibnizian *calculus differentialis* as based upon extravagant hypotheses. Colson characterizes evanescent magnitudes as infinitesimals which arise from the infinite divisibility of geometric magnitudes; the calculus of fluxions, on his account, supposes that

> Quantity is infinitely divisible, or that it may (mentally at least) so far continually diminish, as at last, before it is totally extinguish'd, to arrive at Quantities that may be call'd vanishing quantities, or which are infinitely little, and less than any assignable Quantity. Or it supposes that we may form a Notion, not indeed of absolute, but of relative and comparative infinity. (Colson 1736, xi)

He contrasts this theory of the relative infinite with the illegitimate procedures of others:

> 'Tis a very just exception to the Method of Indivisibles, as also to the foreign infinitesimal Method, that they have recourse at once to infinitely little Quantities, and infinite orders and gradations of these, not relatively but absolutely such. They assume these Quantities *simul & semel*, without any ceremony,

---

ther to illustrate and enforce its Evidence, and to clear it from all the Objections that either have or may be urged against it" (Eames 1738, 87).

33. "For it was now become highly necessary, that at last the great Sir *Isaac* himself should interpose, should produce his genuine Method of Fluxions, and bring it to the test of all impartial and considerate Mathematicians; to shew its evidence and simplicity, to maintain and defend it in his own way, to convince his Opponents, and to teach his Disciples and Followers upon what grounds they should proceed in vindication of the Truth and Himself" (Colson 1736, x).

The Aftermath of the *Analyst*

as Quantities that actually and obviously exist, and make Computations with them accordingly; the results which must needs be as precarious, as the absolute existence of the Quantities they assume. And some late Geometricians have carry'd these Speculations, about real and absolute Infinity, still much farther, and have raised imaginary Systems of infinitely great and infinitely little Quantities, and their several orders and properties; which, to all sober Inquirers into mathematical Truths, must certainly appear very notional and visionary. (Colson 1736, xi–xii)

The distinction here is, to say the least, obscure. Colson may have in mind the Aristotelian distinction between potential and actual infinity, or perhaps Locke's distinction between infinity and infinite. But in speaking of evanescent magnitudes as genuine magnitudes less than any assignable quantity, yet positive, he goes beyond anything that Aristotle intended.[34] However we choose to interpret Colson's pronouncements on this issue, it is clear that he has granted Berkeley a very important point: evanescent magnitudes are infinitesimals.

In attempting to answer Berkeley's objections to the Newtonian proof of the product rule, Colson not only misses Berkeley's point entirely but manages to contradict himself.[35] His attempt to vindicate the Newtonian method of finding the fluxion of any power x^n is more interesting, although it also fails to establish the desired result. Colson argues that the Newtonian procedure involves nothing more than moving from the general case of a proposition to a particular case contained under it. He observes that as the quantity o in the ratio $1 : (nx^{(n-1)} + (n(n-1)/2)x^{(n-2)}o + \ldots)$ is diminished, the ratio approaches that of $1 : nx^{(n-1)}$. Colson insists that the ultimate ratio $1 :$

---

34. Colson's attempts to elucidate this distinction are every bit as obscure as his first pronouncements: "Absolute Infinity, as such, can hardly be the object either of our Conceptions or Calculations, but relative Infinity may, under a proper regulation. Our Author observes this distinction very strictly, and introduces none but infinitely little Quantities that are relatively so; which he arrives at by beginning with finite Quantities, and proceeding by a gradual and necessary progress of diminution. His Computations always commence by finite and intelligible Quantities; and then at last he inquires what will be the result in certain circumstances, when such or such Quantities are diminish'd *in infinitum*" (Colson 1736, xii). It remains obscure why diminution *in infinitum* is a legitimate way of obtaining infinitesimals.

35. See Colson (1736, xii–xiii) and Cajori (1919, 151–2) for the details. In essence, Colson accepts Newton's proof as valid but then insists that the moments a and b will become infinitesimal and can be regarded as equal to zero, so that "$A - \frac{1}{2}a$, A, and $A + \frac{1}{2}a$, are to be taken indifferently for the same Quantity . . . and the want of this Consideration has occasion'd not a few perplexities." But, as Berkeley asks, how can these quantities be taken indifferently for each other if a is not equal to zero?

$nx^{(n-1)}$ is attained when the quantity o becomes an infinitesimal and vanishes, and that attainment of the ultimate ratio is a particular case included under the general case when the increment o is finite:

> This therefore is their ultimate Ratio, the Ratio of their moments, fluxions, or velocities by which x and x^n continually increase or decrease. Now to argue from a general theorem to a particular case contain'd under it, is certainly one of the most legitimate and logical, as well as one of the most usual and useful ways of arguing, in the whole compass of the Mathematicks. (Colson 1736, xvii)

This would be a perfectly fine response given a proof that the limiting case is always covered in reasoning which concerns limits. But this is not generally the case. For example,

$$\lim_{x \to \infty} \left(\frac{1}{x}\right) = 0.$$

Here, however, the quotient is positive at every value but the limit is not. The principle that Colson invokes would allow us to infer that the limit of a sequence of positive quantities is necessarily positive.

Colson hopes to overcome this difficulty by introducing considerations from algebra. He claims that Berkeley's charge of inconsistent assumptions:

> is not an Objection against the Method of Fluxions, but against the common Analytics. This Method only adopts this way of arguing, as a constant practice in the vulgar Algebra, and refers us thither for the proof of it. If we have an Equation any how compos'd of the general Numbers $a, b, c, \&c.$ it has always been taught, that we may interpret these by any particular Numbers at pleasure, or even by 0, provided that the Equation, or the Conditions of the Question, do not expressly require the contrary. (Colson 1736, xviii)

Here Colson adds the condition that the interpretation of variables must be consistent with "the Equation, or the Conditions of the Question," which is precisely the point at issue.[36] Berkeley charges that the calculus gives two inconsistent interpretations to the symbol "o" and

36. Recall Berkeley's demand "Whether because, in stating a general Case of pure Algebra, we are at full Liberty to make a Character denote either a positive or negative Quantity, or nothing at all, we may therefore, in a geometrical Case, limited by Hypotheses and Reasonings from particular Properties and Relations of Figures, claim the same Licence?" (*Analyst*, query 27).

The Aftermath of the *Analyst*

Colson's defense amounts to the irrelevant claim that the calculus would be vindicated if its assumptions were consistent.

Colson makes other responses to Berkeley's objections in his Commentary to Newton's *Method of Fluxions* but raises very few new points. However, he makes one particularly candid admission when he likens vanishing quantities to such "impossible" magnitudes as imaginary numbers. He argues that there is no idea of quantity corresponding to the symbol "$\sqrt{-1}$," yet it is useful for purposes of algebraic calculation:

> These impossible quantities . . . are so far from infecting or destroying the truth of these Conclusions, that they are the necessary means and helps of discovering it. And why may we not conclude the same of other species of impossible quantities, if they must needs be thought and call'd so? . . . Therefore the admitting and retaining these Quantities . . . 'tis enlarging the number of general Principles and Methods, which will always greatly contribute to the Advancement of true Science. In short, it will enable us to make a much greater progress and proficiency, than we otherwise can do, in cultivating and improving what I have elsewhere call'd The Philosophy of Quantity. (Colson 1736, 338–39)

This is certainly a possible response to Berkeley, but it gives up the ground that was to be defended. Berkeley insists upon a purely geometric foundation for the calculus and explicitly denied the adequacy of a formalistic foundation. Algebra is a science, on Berkeley's account, but a science of signs only; the calculus must treat of its proper object, extension, which cannot be properly investigated by a mere algebraic formalism.

Thomas Bayes

The Reverend Thomas Bayes (1702–61) was a dissenting minister and Fellow of the Royal Society, best known for his posthumously published work in the theory of probability. But he also authored an anonymous response to the *Analyst* in 1736. His *Introduction to the Doctrine of Fluxions* attempts to place the calculus on an axiomatic foundation, complete with postulates, definitions, axioms, corollaries, and remarks. He achieves mixed results at best, since he leaves some of Berkeley's principal questions unanswered. I will examine the key postulates and definitions of Bayes's project, with an eye toward determining their effectiveness as responses to Berkeley.

The foundation of Bayes's approach is the familiar notion of continuous variation. He bases his first postulate upon the kinematic conception of magnitudes, remarking that "[t]he notion of Fluxions was originally gained by the observation of quantities being described by a continual motion" and postulating that *"Quantities may be supposed as continually changing, so as every distinct instant of time to be different from what they were before"* (Bayes 1736, 9). Bayes is clearly aware that the concept of continuity is not without metaphysical difficulties but postulates that the concept of continuous variation is intelligible, whether or not quantities in nature are actually continuous.[37]

With this postulate (and a redundant second postulate declaring that the notion of fluxions is intelligible) as a basis, Bayes sets out the foundation of his theory of fluxions in five key definitions. These read:

> 1. A Flowing quantity is one that continually increases or decreases, and in such a manner that some time is requisite to make any increment or decrement.
> 2. The Fluxion of a flowing quantity is its rate or swiftness of increase or decrease.
> 3. The change of a flowing quantity is the difference between the flowing quantity itself, and its value at a particular instant of time.
> 4. The time in which a change is made, is the time the flowing quantity takes to alter from a prior to a subsequent value, whose difference is the change.
> 5. That change is said to vanish at a given instant, which is the difference between the flowing quantity before that instant and its value then; and that change is said to begin to arise at a given instant, which is the difference between the flowing quantity after that instant, and its value at that instant. (Bayes 1736, 12)

The third and fifth definitions are more than a little obscure, but Bayes's remarks indicate that he is working with a conception of variable magnitudes where the differences between magnitudes must always be *positive*. Thus, his definitions must be stated in such a way that the differences between two values of a variable are never negative.[38]

---

37. In a revealing remark, he declares: "it is not the business of the Mathematician to dispute whether quantities do in fact ever vary in the manner that is supposed, but only whether the notion of their so doing be intelligible; which being allowed, he has a right to take it for granted, and then to see what deductions he can make from that supposition" (Bayes 1736, 9).

38. Elsewhere Bayes indicates his preference for treating all quantities as positive when he writes that "the supposition of a negative quantity, or a quantity less than nothing, is an absurdity" (Bayes 1736, 44). Later, he adds that "An Algebraist never scruples to subtract a greater quantity from a less; but if he really designs to do this, he

The Aftermath of the *Analyst*

Definition 3 introduces the term "change" for what others would call differences or increments. Assuming that the variable quantity x attains the value A, the change of x (relative to A) will be the difference $x - A$ for values of x greater than A, and $A - x$ for values of x less than A. In general, then, the change of x relative to A will be $|x - A|$. Definition 5 attempts to introduce the concepts of nascent and evanescent magnitudes. Again, assuming that x attains the value A at some time t_0, the change of x (relative to the value A) will vanish at t_0 and arise immediately after t_0. In other words, the increment of x will be either nascent or evanescent in the neighborhood of t_0.

Bayes then introduces ultimate ratios of vanishing magnitudes in a prolix definition which does little to overcome Berkeley's objections:

> *Def.* 6 If there be two permanent quantities A and B, and two other flowing Quantities a and b, and the ratio of a to b be always, during a given time, that of the sum or difference of the first permanent quantity A, and another flowing quantity x to the sum or difference of the second permanent quantity B, and another flowing quantity y, and at the end of the given time all the flowing quantities vanish; then the ratio of the permanent quantities A and B, is the last or ultimate ratio of the vanishing quantities a and b; which I thus express, ult.$a : b :: A : B$. *i.e.* if $a : b :: A \mp x : B \mp y$ always, during the time T, and at the end of that time, a, b, x, y all vanish; then ult.$a : b :: A : B$. (Bayes 1736, 13)

Bayes's idea is to define the ultimate ratio of the vanishing quantities a and b in terms of the ratio between the constant quantities A and B. Because the ultimate ratio of vanishing quantities cannot be defined directly in terms of the fixed ratio $A : B$, except in the degenerate case where the ratio $a : b$ is constant throughout the evanescence of the quantities a and b. Bayes introduces two additional vanishing quantities x and y to link the constants A and B with the vanishing quantities a and b in the proportion $a : b :: A \mp x : B \mp y$. He does not thereby answer Berkeley's criticism but simply postulates that there is a ratio between vanished quantities or, to use the Berkeleyan idiom, that the ghost of departed quantities is a genuine ratio. Bayes thought the possible objection could be avoided by stressing the arbitrariness of mathematical definitions. He insists that his definition is the same as Newton's and "can't be disputed; for whether a and b, properly speaking, have any proportion as they arise, or vanish, yet A and B

---

may try till his heart akes [*sic*] before he will be able to accomplish it, or to know what he is about" (Bayes 1736, 46).

have; and that I am at liberty to call by what name I please" (Bayes 1736, 13).

Bayes's approach to the foundations of the calculus is quite far removed from the modern theory of limits. In speaking of ratios between quantities that have vanished, Bayes explicitly declines to treat ultimate ratios as limiting values of continually diminished ratios. His contribution to the debate over the foundations of the calculus is certainly an advance over that provided by Colson, but it does not really overcome Berkeley's principal objections. Bayes's failure can best be seen in his response to Berkeley's objections to Newton's proof of the rule for finding the fluxion of any power x^n (*Analyst*, §§13–16). Bayes argues, as did Jurin, that Newton can consistently assume an increment o first to exist and then to vanish:

> To suppose the increments to be something and nothing at the same time, is contradictory; but to suppose them first to exist, and then to vanish, is perfectly consistent; nor will the consequences drawn from the supposition of their prior existence, if just, be any way affected by the supposition of their subsequent vanishing, because the truth of the latter supposition no ways contradicts the truth of the former. (Bayes 1736, 37)

The problem, however, is that the derivations of the value for the fluxion of a power do seem to depend upon an assumption which is contradicted. Furthermore, Bayes begs the crucial question of whether a chain of reasoning which depends upon the supposition of positive quantities yields results which hold when these quantities have vanished.

Bayes's attempt to overcome the appearance of inconsistency instead emphasizes the shortcomings of his analysis:

> To make this more plain, consider what is made out from each supposition: from the first that x has increased by o, this consequence is drawn, that the proportion between the increments of x and x^n, *so long as they exist*, may be expressed by that of
>
> $$1 \text{ to } nx^{n-1} + \frac{n^2 - n}{2} o x^{(n-2)} \; \mathit{\&c.}$$
>
> if o always express the increment of x. And this consequence is no ways affected by supposing o continually to decrease, and at length to vanish. (Bayes 1736, 38)

Bayes conveniently ignores the fact that supposing the increment o to vanish or become equal to zero means the expression for the ratio of increments will have been obtained by a step which is equivalent to division by zero.³⁹ Thus, Bayes simply overlooks Berkeley's point.

Smith (1980) argues that Bayes's *Introduction* is a successful defense of Newtonian methods against Berkeley's charges.⁴⁰ However, considering the obscurity of Bayes's pronouncements on the nature of ultimate ratios and his failure to address Berkeley's main criticisms, this conclusion overstates the case in favor of Bayes. Smith correctly notes that some of Bayes's proof strategies foreshadow standard methods used in later rigorizations of the calculus, but this does not blunt Berkeley's criticisms. We can grant that Bayes's *Introduction* is a serious work by an important mathematician, but it is hardly a conclusive answer to Berkeley.

James Smith

James Smith is an obscure figure in British mathematics, known to us only for his 1737 *New Treatise of Fluxions,* unquestionably the most bizarre attempt to answer Berkeley's objections. Where other authors thought to counter Berkeley's interpretation of evanescent increments as "ratios of nothings," Smith embraces this characterization. The *New Treatise* had no discernible effect in the British mathematical community, but it retains some interest for my investigation because it marks the outer limit of obscurity in the responses to Berkeley. Other commentators have remarked on the weaknesses of Smith's doctrine,⁴¹ and I will not attempt to summarize his general theory. Smith himself

---

39. Similarly, he elsewhere insists that "it is [no] objection to the justness of their reasoning, that an algebraical note is sometimes to be interpreted, at the end of the process, in a sense which cou'd not have been substituted for it in the beginning of it; since if quantities themselves are considered as continually changing, the sense of the mark which represents or expresses them must, in order to its doing so, continually change along with them" (Bayes 1736, 49).

40. He concludes his essay by addressing the query "did Bayes succeed?" and insists that "On grounds of the mathematical correctness and lucidity of his explanations, we cannot but answer: yes. He was able to see the essence of Newton's mathematical ideas and to observe that implicit in them was the procedure of using prime and ultimate ratios as the proof technique for the deduction of the basic results of fluxion theory. Moreover, Bayes was able to present them explicitly and lucidly. Newton's fluxions were not presented in a rigorously satisfactory manner. Bayes showed that Newton's method could be adapted to provide convincing proofs of Newton's results—'convincing' here is meant to be in the sense of Bayes' own time. He had not achieved 19th-century rigor; he could not, nor should we expect this in the 18th century" Smith (1980, 387).

41. Breidert says of Smith: "He runs, one might say, into the drawn sword, and simply places back on the table all of the points Berkeley had called into doubt" (Breidert 1989, 118). Cajori concludes his digest of the *New Treatise* with the comment: "We are tempted to make the remark that in 1737 Smith left the subject even more mysterious than he found it" (Cajori 1919, 169).

recognized that his ideas were far outside the mainstream, and it would take me too far afield to engage in a detailed exegesis of his thought.[42] I will therefore examine only those parts of his doctrine which bear directly on his response to Berkeley.

In a part of his *New Treatise* entitled "A Treatise of Nascent and Evanescent Quantities, first and last Ratios," Smith outlines his theory of vanishing quantities and their application to various problems. Toward the end of this section he raises an objection to the theory of evanescent increments which seems to be drawn from his reading of Berkeley. His idea is to clarify the theory by stating an objection and then replying, but the objection gets the better of the theory. The objection is in the form of a dilemma:

> Nascent and evanescent Quantities are Something or Nothing; for, *Enter ens & non-ens non datur medium.* If Something, then the Ratio of evanescent Quantities is the same with the Ratio before they were evanescent, or when they had any finite Magnitude.... If they are meer Nothing, or Non quanta; [the ratio of evanescent quantities] is $= 0/0 = 0$... which is absurd. (Smith 1737, 22)

The "answer" to this objection is amusing, if unenlightening:

> Evanescent Quantities are really Nothing, or Non-quanta.... And yet it does not follow that their last Ratio, or the Ratio they nihilesce with, is Nothing.... The Increments are indeed annihilated and gone, but their last Ratio remains, and is as real as any Ratio they ever had; for a Ratio is an abstract Idea, and exists only in the Mind; and tho' the *relatum* and *correlatum* are annihilated, yet it does not follow that their Ratio, that is the Idea the Mind had of them at their nihilescing, or in the very last Instant of their Existence, is annihilated. (Smith 1737, 23)

Smith's language here verges on parody. Replying to Berkeley by declaring ultimate ratios to be abstract ideas is amusing enough; but when Smith adds that they "exist only in the mind," we are invited to

---

42. Smith acknowledges the eccentricity of his views: "I have said some Things of Motion and Velocity, which, at first sight, may seem to be impossible, but, I believe, upon due Consideration, they will be found to be true. Our common Definition of Motion, *Translatio corporis de loco in locum* is certainly imperfect, and I am inclined to think, that *Aristotle's* old exploded Definition of Motion will, some time or other, come into Vogue again. *Actus entis in potentia, quatenus in potentia, est.* Motion is an Effect, and every Effect has a coinstaneous [*sic*] Existence with the Action by which it is produced" (Smith 1737, Preface).

conjecture that he was aware of other Berkeleyan doctrines and intended his reply to the objection ironically.

This suspicion is strengthened in Smith's further comments on the nature of evanescent quantities. Without specifically mentioning it, Smith clearly alludes to Berkeley's epithet "ghosts of departed quantities" in discussing these mysterious entities:

> There is, sometimes, something very strange in the Nature of these evanescing Augments, and it is literally true of them, what *Juvenal* figuratively says of Man.
> *Mors sola fatetur,*
> *Quantula sunt hominum corpuscula*
> We know nothing of them till they be dead and gone. (Smith 1737, 25)

As amusing as Smith's remarks may be, we can hardly take this seriously as a response to Berkeley's objections. It remains to be seen whether Smith can rebut Berkeley's charges against Newton's two fundamental proofs but, as might be expected, Smith makes no headway in replying to these objections either. The sixth section of his *New Treatise*, an "Analysis of *The Analyst*'s Objections," consists of long extracts from the *Analyst* with short and irrelevant rebuttals. Berkeley claims that Newton's computation of the moment $aB + bA$ for the rectangle AB should contain an additional term ab; Smith misses the point entirely and declares, "all the Objections in this Quotation are about the Magnitude of the Increment of $A \times B$. Therefore there is not one Objection in all this Quotation against Sir *Isaac Newton*, or this Lemma" (Smith 1737, 51). Berkeley alleges a *fallacia suppositionis* in the proof from Newton's *Quadrature of Curves;* Smith simply argues that Newton's two assumptions are consistent because made at different times. The inadequacy of this defense I have already examined and so will proceed to other authors.

Colin Maclaurin

Colin Maclaurin (1698–1746) was a Scottish Newtonian and the most important British mathematician in the 1730s and 1740s. His *Treatise of Fluxions* is the most highly regarded mathematical work provoked by the *Analyst*.[43] If we judge significance by length, Maclaurin's *Treatise* is a very important work indeed: its two volumes and 763 quarto

---

43. Guicciardini calls it "The most authoritative answer to Berkeley" (Guicciardini 1989, 47), while Cajori deems it "the ablest and most rigorous text of the eighteenth century" (Cajori 1919, 181).

pages dwarf anything else published on the subject at the time. The level of technical sophistication is also quite high. Rather than simply replying to Berkeley, Maclaurin extended the calculus to cover new problems and went significantly beyond what anyone else in British mathematics had achieved.[44]

Maclaurin's initial motivation in writing the *Treatise* was to counter Berkeley's charges against the calculus, as he admits at the very beginning of the Preface: "A Letter published in the Year 1734, under the title of *the Analyst,* first gave Occasion to the ensuing Treatise" (Maclaurin 1742, i). As he prepared his work, the publications of Jurin, Robins, and Colson appeared, and Maclaurin decided to delay publication until he had covered every aspect of the calculus. The fifty-page Introduction following the Preface contains an extensive study of the classical method of exhaustion and a historical account of the development of infinitesimal techniques. Maclaurin continually stresses the superior rigor of the classical methods, even quoting Locke in support of his opinion that the infinite should be banned from mathematics.[45] On Maclaurin's account, it is essential that mathematics avoid disputed and questionable points, and he draws an interesting contrast between mathematics and philosophy on this point:

> Philosophy probably will always have its mysteries. But these are to be avoided in geometry: and we ought to guard against abating from its strictness and evidence the rather, that an absurd philosophy is the natural product of a vitiated geometry. (Maclaurin 1742, 47)

Having placed the standard of rigor quite high, Maclaurin proceeds to combine the Newtonian tradition, particularly its kinematic conception of geometric magnitudes, with the classical method of exhaustion, and ultimately develops a foundation for the method of fluxions which resembles that in Robins's *Discourse*.[46]

---

44. See Tweedie (1915), Turnbull (1951), and Sageng (1989) on Maclaurin's work.

45. "For whilst Men talk and dispute of infinite Space or Duration, as if they had as compleat and positive *Ideas* of them, as they have of the Names they use for them, or as they have of a Yard, or an Hour, or any other determinate Quantity, it is no wonder, if the incomprehensible Nature of the thing, they discourse of, or reason about, leads them into Perplexities and Contradictions; and their Minds be overlaid by an Object too large and mighty, to be surveyed and managed by them" (*Essay* II, xviii, 21).

46. Cajori reports that "To what extent, if any, Maclaurin may have been influenced by Robins in the mode of treating fluxions is difficult to say" (Cajori 1919, 189). James Wilson, the editor of Robin's *Mathematical Tracts,* was convinced that Maclaurin "conformed himself entirely to Mr. Robins's sentiments in regard to Sir Isaac Newton's doctrine" and "has even expressly followed his plan in treating the subject" (Robins 1761, 312).

The Aftermath of the *Analyst*

Maclaurin's plan in the *Treatise* is to set out a philosophically unexceptionable account of motion and to use it as the means to vindicate the calculus. In particular, he is concerned that the key concepts of the calculus be clearly conceivable in order to meet Berkeley's metaphysical criterion for mathematical rigor.[47]

Maclaurin takes the concept of uniform motion to be perfectly unproblematic:

> Time is conceived to flow always in an uniform course, that serves to measure the changes of all things. When the space described by motion flows as the time, so that equal parts of space are described in any equal parts of time, the motion is uniform; and the velocity is measured by the space that is described in any given time. (Maclaurin 1742, 53)

Although this account of uniform motion makes reference to the Newtonian concept of absolute time, it could be phrased without it. Indeed, Berkeley himself accepts something very similar in his response to Walton.[48]

With the concept of uniform velocity in hand, Maclaurin defines instantaneous velocity and accelerated motion and uses them as the foundation for his calculus of fluxions. He argues that instantaneous velocity is fully comprehensible, at least in the sense that any body in motion for a given time must have some definite velocity at every instant in the interval:

> Any space and time being given, a velocity is determined by which that space may be described in that given time: And, conversely, a velocity being given, the space which would be described by it in any given time is also determined. This being evident, it does not seem to be necessary, in pure geometry, to enquire further what is the nature of this power, affection, or mode, which is called *Velocity,* and is commonly ascribed to the body that is supposed to move. It seems to be sufficient for our purpose, that while a body is supposed in motion, it must be conceived to have some velocity or other at any term of the time during which it moves, and that we can

---

47. He insists "No quantities are more clearly conceived by us than the limited parts of space and time. They consist indeed always of parts; but of such as are perfectly uniform and similar.... [B]y motion they become the measures of each other reciprocally" (Maclaurin 1742, 53).

48. There, Berkeley had insisted that the (accelerated) motion of a falling body is one in which "The spaces described in ... equal parts of time will be unequal. That is, from whatsoever points of the described line you measure a minute's descent, you will still find it a different space" (*Reasons,* §5).

demonstrate accurately what are the measures of this velocity at any term, in the enquiries that belong to this doctrine. (Maclaurin 1742, 53–54)

Given an intelligible idea of motion, Maclaurin defines instantaneous velocity, first for uniform motion and then for accelerated motion. The instantaneous velocity of a uniformly moving body is given by the distance it travels in a unit time; for an accelerated body, instantaneous velocity is defined counterfactually as the distance the body would travel in a unit time if it were to continue unaccelerated during that time.[49]

Maclaurin insists that we need not consider instantaneous velocity as the movement of a body across an infinitely small space in an infinitely small time and takes his treatment of instantaneous velocity to be consistent with nearly any philosophical treatment of motion. He then makes a quite explicit reply to Berkeley's complaint that the idea of instantaneous velocity is an illegitimate abstraction:

> When we suppose that a body has some velocity or other at any term of the time during which it moves, we do not therefore suppose that there can be any motion in a term, limit or moment of time, or in an indivisible point of space: and as we shall always measure this velocity by the space that would be described by it continued uniformly for some given finite time, it surely will not be said that we pretend to conceive motion or velocity without regard to space and time. (Maclaurin 1742, 56)

Again, this procedure seems to meet Berkeley's standards.[50] Berkeley certainly cannot deny the legitimacy of definition in terms of counter-

---

49. "As a power which acts continually and uniformly is measured by the effect that is produced by it in a given time, so the velocity of an uniform motion is measured by the space that is described in a given time. If the action of the power vary, then its exertion at any term of the time is not measured by the effect that is actually produced after that term in a given time, but by the effect that would have been produced if its action had continued uniform from that term: and, in the same manner, the velocity of a variable motion at any given term of time is not to be measured by the space that is actually described after that term in a given time, but by the space that would have been described if the motion had continued uniformly from that term" (Maclaurin 1742, 55).

50. Guicciardini claims that Maclaurin's procedure is circular: "There is a circularity in basing the calculus, a mathematical tool devised to study kinematics, on the concepts of time and velocity. It is interesting to see that this was never really felt to be a difficulty by Newtonian mathematicians" (Guicciardini 1989, 44–45). Later, he objects that Maclaurin's definition of a fluxion is "equivalent to defining the 'measure' of $y'(t)$ as $y'(t)\delta t$: a procedure that involves circularity" (Guicciardini 1989, 49). But it is unclear

factual considerations, since his own metaphysics is underwritten in part by a counterfactual definition of the term "existence." In a famous passage Berkeley declares, "The table I write on, I say, exists, that is, I see and feel it; and if I were out of my study I should say it existed, meaning thereby that if I was in my study I might perceive it, or that some other spirit actually does perceive it" (*Principles*, §3). Similarly, he analyzes the claim that the earth is in motion in terms of the possibility of certain observations under counterfactual circumstances:

> [T]he question, whether the earth moves or no, amounts in reality to no more than this, to wit, whether we have reason to conclude from what hath been observed by astronomers, that if we were placed in such and such circumstances, and such or such a position and distance, both from the earth and sun, we should perceive the former to move among the choir of the planets, and appearing in all respects like one of them: and this, by the established rules of Nature, which we have no reason to mistrust, is reasonably collected from the phenomena. (*Principles*, §58)

Taking these pronouncements seriously (and there is no reason not to do so), I conclude that Berkeley could accept Maclaurin's account of instantaneous velocity.

With this definition of instantaneous velocity, Maclaurin defines fluxions in the familiar way, as the velocities with which geometric fluents are generated.[51] Four axioms concerning accelerated and retarded motion are the starting point for Maclaurin's theory of fluxions. They read:

> AXIOM I: The space described by an accelerated motion is greater than the space which would have been described in the same time, if the motion had not been accelerated, but had continued uniform from the beginning of the time.
> AXIOM II: The space described by a motion while it is accel-

---

why Maclaurin should be accused of reasoning in a circle or why a kinematic concept such as uniform velocity is mathematically inadmissible. There is no formal flaw in Maclaurin's presentation of the calculus, however distant his conception of the subject may be from modern presentations of the material.

51. "The velocity with which a quantity flows, at any term of the time while it is supposed to be generated, is called its *Fluxion* which is therefore always measured by the increment or decrement that would be generated in a given time by this motion, if it was continued uniformly from that term without any acceleration or retardation: or it may be measured by the quantity that is generated in a given time by an uniform motion which is equal to the generating motion at that term" (Maclaurin 1742, 57).

erated, is less than the space which is described in an equal time by the motion that is acquired by that acceleration continued uniformly.
AXIOM III: The space described by a retarded motion is less than the space which would have been described in the same time, if the motion had not been retarded, but had continued uniform from the beginning of the time.
AXIOM IV: The space described by a motion while it is retarded, is greater than the space which is described in an equal time by the motion that remains after that retardation, continued uniformly. (Maclaurin 1742, 59)

Then, using laborious exhaustion proofs, Maclaurin develops an account of fluxions with limiting considerations very much like those found in Robins. Maclaurin avoids any talk of evanescent increments but instead uses sequences of approximating values to find fluxions. This method of computing instantaneous velocity by taking it as the limit of a sequence of finite intervals of space and time certainly evades Berkeley's criticism. Indeed, it seems that Maclaurin's *Treatise* has a positively Berkeleyan foundation. Maclaurin follows Berkeley in accepting the authority of the classical methods in analysis. And where Berkeley, although convinced that the calculus could be presented in a classically rigorous form, was not prepared to undertake the revision himself, Maclaurin was up to the challenge and in the end wrote a comprehensive treatise in the calculus with a Berkeleyan basis.

Although Berkeley's objections to the calculus were eventually answered by Maclaurin, his charges—that basic concepts were obscure and proofs of fundamental results less than convincing—were correct in 1734. I see Berkeley's critique as a demand for clarification of basic concepts in the calculus, in which light Maclaurin's response can hardly be treated as a refutation of Berkeley. Indeed, Maclaurin essentially grants Berkeley's case against the Newtonian and Leibnizian methods and proceeds from there. It is therefore strange that Stammler treats Maclaurin's work as destroying the Berkeleyan approach to mathematics:

> The historical significance of our Irish thinker contains a certain tragic irony. This consists in the fact that through his work one of the most outstanding mathematicians was led to consider thoroughly the foundations of his subject, in which he discovered . . . how untenable the Berkeleyan position is, however clever it must be called. The destruction of his philosophy was provoked by his theory of mathematics. (Stammler 1921, 58)

This evaluation makes sense only if we see Berkeley's approach to the philosophy of mathematics as committed to the nonsensical thesis that mathematical practice is in principle incompatible with his philosophical theory. But it is clear that Berkeley's position cannot be caricatured in this way. *The Analyst* accomplishes all that we can ask of a philosopher of mathematics. Berkeley found contemporary mathematical theories wanting, explained why he found them so, and asked that the new methods be replaced with procedures that lived up to the classical standard. In the end, his critique was taken seriously and new methods were developed. We should all be so fortunate as to have our views refuted in this manner.

Roger Paman

Roger Paman's *Harmony of the Ancient and Modern Geometry Asserted* (1745) attempts to defend the calculus without relying upon the kinematic concepts that had been standard fare for previous British writers on the subject. It began as a series of papers which Paman left in the care of Dr. David Hartley in 1740, when he embarked on the famous South Sea expedition led by Captain George Anson. Paman was one of the fortunate few to survive this voyage, returning to England in 1744. His papers on the calculus were communicated to the Royal Society in 1742 and later published. Paman's work has been largely ignored by historians of mathematics,[52] but it is of interest in this investigation because it is as compelling a response to Berkeley as that of Maclaurin but developed along significantly different lines. Paman's work on the foundations of the calculus grew out of his attempt to answer Berkeley's assertion that "all Attempts for setting the abstruse and fine Geometry" would fail unless the "Object and End" of geometry were better understood (*Analyst*, §35). He was particularly interested in Berkeley's suggestion that the concepts of instantaneous velocity and infinity had no place in geometry. Paman reports that a friend (a Reverend Frank of St. John's College, Cambridge) had proposed the project to him, and that "perhaps, never any Person was more zealous in the Cause of Mathematics in general, was more prepossessed in favour of Sir *Isaac Newton*'s Doctrines in particular, or more strongly prejudiced against the Author of the Analyst, or his Performance" (Paman 1745, xxvii).[53] Nevertheless, as he undertook to recast the calculus in the language of finite differences he found

---

52. Breidert is the only other author who has included a discussion of Paman's work; see Breidert (1989, 119–22).

53. I have supplied the page numbers for Paman's preface.

himself proposing an entirely new approach to the foundations of the method of fluxions.

Paman opposed not only the kinematic conception of magnitudes but contemporary attempts to base the calculus on a theory of limits. As he observes, the term "limit" can be interpreted either as a quantity approximated to, or as a sequence of quantities which approximate a given quantity. Although his own doctrine is ultimately an exercise in the theory of limits as we understand that subject today, Paman was wary of using the language of limiting processes in the calculus. He begins by distinguishing the "minimajus" and "maximinus" of a quantity, which correspond roughly to the modern notions of least upper bound and greatest lower bound. Because "limit" can be taken in two senses, Paman finds it inappropriate for his purposes:

> [W]hen . . . I have declared, that nothing was farther from my Intentions than to introduce Novelties, by multiplying Technical Terms or unnecessary Distinctions, and that I would gladly relinquish these Terms, if others in common Use could be found expressive of my Meaning, it has been hinted, that the Word Limit would supply their Place: But when, out of a Desire of conforming to the usual Stile of Language, I attempted to substitute the Word Limit for the Terms Maximinus and Minimajus, it quickly appeared this Substitution could not take Place. For . . . I found, upon applying the Definition of a Maximinus or Minimajus thereto, this Substitution could not be effected, without confounding the approximating Quantity with the Quantity approximated to, and advancing what was directly repugnant to the Acceptation of a Limit in Geometry, whether ancient or modern. (Paman 1745, iii–iv)

In commenting upon the doctrine of limits, Paman discusses the question, touched upon in examining the dispute between Robins and Jurin, of whether a sequence of approximations ever attains its limiting value, that is, whether a state of "ultimate equality" obtains between the approximations and the limiting quantity. Paman observes that "the moderns" have taken the term limit to designate the quantity approximated to, and this has given rise to the question of whether a limit is ever attained.[54] He proposes to sidestep the issue by taking the

---

54. "The moderns have indeed taken the Word Limit in a more extensive Sense, by making one Line or Figure the Limit of another; but in doing this the Limit is always considered as the Quantity approximated to, whereas the Maximinus or Minimajus is always considered as the approximating Quantity. . . . But Mr. *Mac Laurin* and Mr. *Robins* both expressly call the Circle the Limit of the inscribed and circumscribed Polygons; and even the cautious Sir *Isaac Newton* . . . infers, in his Doctrine of prime

maximinus and minimajus of quantities to be always expressed in terms of finite differences. On this account, the ultimate equality of a limit and an approximation could only be attained if the difference between them were diminished ad infinitum, but his doctrine is confined to the consideration of finite differences.[55]

Paman's account begins by introducing the concept of the first and last state of a variable x. His first postulate declares

> That one Character may represent a certain Quantity; and that another may represent every Quantity of the same Nature; thus let p represent any particular Quantity, and let x, y, z, represent every Quantity of the same Nature. (Paman 1745, 1)

The notion of "same Nature" here reflects Paman's interest in a geometric interpretation of his doctrine, with quantities being taken from geometric "species" such as lines, angles, and surfaces. Thus, he explains that if p represents a given line, then x, y, and z represent lines of any length. Like Bayes, Paman also implicitly confines his account to positive quantities. He then concludes that when a constant p and a variable x are given, there are important inequalities that can be established. For example, $p > x$ whenever $x < p$; or, $p > x$ for all values of x "between p and Nothing," with negative quantities left out of consideration. Similarly, $p > bx$ for all values of x between p/b and zero. This leads to a distinction between the first and last states of a variable:

> By the first State of x, I mean all the Values of x, between some certain assignable Value and Nothing.
>
> By the last State of x, I mean all the Values of x, greater than, or above some certain assignable Value. (Paman 1745, 2)

The first or last state of a variable will depend upon the choice of a "certain assignable Value." Thus, if k is a constant and x a variable, the first state of x relative to k will be the interval $(0, k)$ and the last

---

and ultimate Ratios, an ultimate Equality between an approximating Quantity and its Limit; and tho' that ultimate Equality be denied by Mr. *Robins,* and not always insisted upon by Mr. *Mac Laurin,* yet both these Gentlemen agree in making their Difference at last to become less than any assigned or assignable Quantity; whereas a Maximinus or Minimajus must ever differ from the Expression it is referred to by an assignable Quantity" (Paman 1745, v–vi).

55. "I do not therefore dissent from Sir *Isaac;* but the plain Truth of the Matter is, I have no Occasion for Sir *Isaac*'s ultimate Equality; the Distinction of the Kind serves my Purpose, and makes finite Maximinority or Minimajority equivalent, in Consequence to, and briefer in Process than infinite or ultimate Equality" (Paman 1745, vi).

state of x relative to k is the interval (k, ∞).[56] The interesting cases involve variable quantities which appear in algebraic operations. Thus, when x is a variable and k a constant, $p > kx$ whenever $x < p/k$ and the first state of x in this case will contain all of the values of x between p/k and zero.

Before defining the maximinus and minimajus of a quantity, Paman refines the notion of two quantities being "of the same kind." Paman wants to extend the notion, originally borrowed from the geometric model where various magnitudes belong to certain species, into algebraic territory and identifies the kind of a quantity with the degree of the equation which defines it:

> The better to introduce this Distinction into Analytics, whenever more indeterminate Quantities than one occur in any Expression, or Equation, I have always considered one of them as the indeterminate Unit; and, calling it the radical Quantity of the rest, have distinguished every Term by the Power of the radical Quantity it involves: Thus I call ax an x Quantity, ax^3 an x^3 Quantity; and surely there will not be thought any thing strange or metaphysical in so doing, for as to the Notion of a radical Quantity, that the foreign Mathematicians have thought it requisite, is plain from their term *Function*, and it is generally understood, if not mentioned, in Analytic Operations among ourselves, being in Effect no more than considering the Quantity so called, as the Abscisse, and the Expression involving any of its Powers as the Ordinate of a Curve. (Paman 1745, vi)

The requisite definitions follow easily:

> Now, for Brevity Sake, when one Quantity is the greatest Quantity of its Kind less than another, in the first state of x, I call the former the first Maximinus of the latter. And instead of saying, for Instance, m is the greatest determinate Quantity less than $m + ax + bx^2 + cx^3$, *&c.* in the first State of x, I call

---

56. The presence of the symbol ∞ in my characterization of Paman's definition should not lead to the conclusion that the concept of infinity is somehow illegitimately involved in the theory. Paman himself is explicit on this point: "From hence it appears, that, by the first State of x, I do not mean the nascent or evanescent State of Sir *Isaac Newton*, nor, by any of the values of x in its first State, the Minimum Magnum of Dr. *Barrow*, or the Infiniment Petit of the Marquis *de l'Hospital;* but all the finite Values of x less than a particular Value, which particular Value is assignable from the Quantities compared: And in the last State of x I do not consider any of its Values as infinitely great, or as the Maximum Magnum of Dr. *Barrow;* but I mean thereby all the Values of x greater than a particular Value, the Assignability whereof depends upon the Quantities compared" (Paman 1745, ix).

The Aftermath of the *Analyst*

> m that determinate Quantity, which is the first Maximinus of $m + ax + bx^2 + cx^3$, &c. and, in lieu of calling mx the greatest x Quantity (the greatest Quantity which is to x in a given Ratio) less than $mx + ax^2 + bx^3 + cx^4$, &c. I call it that x Quantity which is the first Maximinus of $mx + ax^2 + bx^3 + cx^4$. (Paman 1745, x)

The idea here is fairly simple and has an obvious application to the calculus of fluxions. Consider the familiar series expansion for the increment of a quantity x^n:

$$nx^{(n-1)}o + \frac{n(n-1)}{2}x^{(n-2)}o^2 + \ldots$$

In the first state of o there will be sufficiently small values of o such that $nx^{(n-1)}o$ is the greatest o quantity less than the value of the series. That is to say, no determinate o quantity ko can be found such that

$$\left(nx^{(n-1)}o + \frac{n(n-1)}{2}x^{(n-2)}o^2 + \ldots\right) < (nx^{(n-1)}o + ko)$$

throughout the first state of o. In other words, $nx^{(n-1)}o$ is the maximinus of the series in the first state of o.

Paman illustrates this theory by a geometric example, which he takes to show that "These Maximinus's and Minimajus's exist wherever variable Quantities are concerned, and are evident to the Senses, in Geometry" (Paman 1745, xi). His example is as follows:

> [T]he Angle *NAP* [fig. 7.5], which the Tangent *NA* makes with another right line *AP*, intersecting the Curve *AOM* in the Point of Contact *A*, is (when the Curve is convex to *AP*) the greatest of all those rectilinear Angles that are less than the Angle *MAP*, formed by the Curve *MOA* with the same Line AP; and when the Curve is concave to *AP*, *NAP* will be the least of all those rectilinear Angles which are greater than *MAP*; that is, in the former Case, no rectilinear Angle can be greater than *NAP*, and less than *MAP*; and in the latter no rectilinear Angle can be less than *NAP* and greater than *MAP*; the one being the Maximinus and the other the Minimajus of it. (Paman 1745, xii)

Paman defines the fluxion of a quantity as first maximinus of the increment. The best way to understand his approach is to investigate

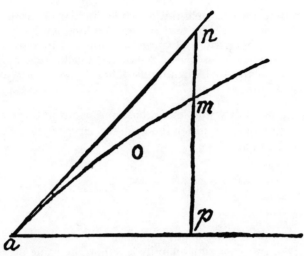

Fig. 7.5 From Paman, *The Harmony of the Ancient and Modern Geometry Asserted*, Preface (1745, xi).

his relevant definitions and to compare them against Berkeley's criticisms. The two key definitions read:

> 1. If any Expression be augmented, or diminished, by the Augmentation or Diminution of it's [*sic*] radical Quantity, I call the Increment, or Decrement of the Expression, the Difference of that Expression.
> 2. If x be the radical Quantity of any Expression represented by y; and if x be augmented, or diminished, by any indeterminate Quantity z, I call z the Increment, or Decrement of x, the Fluxion of x, and denote it by $\dot{x}$. (Paman 1745, 24–25)

The second definition equates the fluxion of the root x with its increment, and might be thought to depart rather substantially from Newton's presentation. However, Paman adds an elucidation to this second definition which, in effect, defines the fluxion of a fluent as the maximinus or minimajus of its increment:

> And I call that $\dot{x}$ Quantity, which is the first Maximinus or Minimajus of the Difference of y (arising from the Substitution of $x \mp \dot{x}$ for x in the Value of y) the first Fluxion of y, and denote it by $\dot{y}$; thus if $y = x^m$ [it follows that] $\dot{y} = mx^{(m-1)}\dot{x}$ for, $mx^{(m-1)}\dot{x}$, is that $\dot{x}$ Quantity which is the first Maximinus or Minimajus of the Difference between x^m and $(x \mp \dot{x})^m$. (Paman 1745, 25)

The Aftermath of the *Analyst*

On this account, we begin with an analytic expression for the fluent, say

$$y = ax^n \qquad (7.14)$$

with a being a constant, x and y variables. Let $\dot{x}$ be the increment of the root quantity x and o be the corresponding increment of y; then

$$(y + o) = a(x + \dot{x})^n. \qquad (7.15)$$

Expanding (7.15) gives the result

$$y + o = ax^n + anx^{(n-1)}\dot{x} + \frac{n(n-1)}{2}ax^{(n-2)}\dot{x}^2 + \ldots \qquad (7.16)$$

We can now analyze the increment of y by subtracting equation (7.14) from equation (7.16):

$$o = anx^{(n-1)}\dot{x} + \frac{n(n-1)}{2}ax^{(n-2)}\dot{x}^2 + \ldots$$

The fluxion of y, written $\dot{y}$, is defined as the first maximinus of the series on the right side of equation (7.16). In other words,

$$\dot{y} = nx^{(n-1)}\dot{x}$$

And $nx^{(n-1)}\dot{x}$ is the $\dot{x}$ quantity which is the first minimajus or maximinus of the difference of y. Further, we can assume the root quantity x to flow uniformly, in which case the increment $\dot{x}$ can be taken as a unit and the fluxion of y will be $nx^{(n-1)}$ exactly as in Newton.

Again, we have a reply to Berkeley which meets the principal objections raised in the *Analyst*. Paman accounts for fluxions without introducing infinitesimal magnitudes or engaging in the *fallacia suppositionis*, and there is essentially nothing in his treatise that Berkeley could object to. His practice of taking only finite increments is very much in the spirit of Berkeley's own failed attempt to rescue the calculus by seeking compensating errors, but the device of taking least upper bounds goes well beyond anything Berkeley could achieve. It is perhaps surprising that Paman's work had no discernible influence on British mathematics in the period, but this may be attributable to the dominance of Maclaurin's *Treatise of Fluxions*. By the time Paman's

Harmony of the Ancient and Modern Geometry was published, Maclaurin's work had already gained sufficient stature to be regarded as the definitive answer to Berkeley's objections.

The Significance of the *Analyst*

Although Berkeley's critique of the calculus provoked a flood of responses, it is still worth asking whether the *Analyst* is a significant mathematical or philosophical contribution. Some commentators take it as an important development in eighteenth-century British mathematics; others treat it and the resulting dispute as a distraction of little or no interest aside from its allegedly detrimental effects on the course of mathematical thought in Great Britain. There is no question that Berkeley had a powerful influence on mathematics in eighteenth-century Britain, but it is not immediately obvious whether he influenced it for the better or worse.

George A. Gibson was convinced that Berkeley's polemics against the calculus were salutary, and he regarded the *Analyst* as an important stimulus to British mathematics. In what remains the high-water mark of favorable reactions to the *Analyst,* Gibson declared:

> Berkeley did a great service to sound reasoning in mathematics by the publication of the *Analyst*. The rapid accumulation of results, due to the introduction of the new analysis, had tended to throw into the background the logical principles on which any truly scientific knowledge of mathematics can alone be based, and the controversy the *Analyst* called forth is favourably distinguished from that on the invention of the calculus by the comparative absence of the grosser personalities. Were it for nothing else than the *Discourse* and *Dissertation* of Robins, and the *Fluxions* of Maclaurin, Berkeley's name should be had in reverence of mathematicians. (Gibson 1899, 31–32)

Cajori (1919) takes a similar view of the controversy, as does Wisdom, who remarked that Berkeley's mathematical works are a "contribution of the utmost value, in that they influenced the whole development of pure mathematics by directing it along the path it ought to pursue" (Wisdom 1939, 3).

These favorable reactions to Berkeley's critique of the calculus share two important assumptions: first, that rigor is something desirable in a mathematical theory, and second, that a philosophical critique which arouses mathematicians' concern with the rigor of their methods is a good thing, both philosophically and mathematically.

The Aftermath of the *Analyst*

Within the context of these assumptions, Berkeley's *Analyst* and the resulting foundational debate over the calculus can be read as part of a march toward the philosophically and mathematically important goal of rigorous demonstration. Berkeley is to be congratulated for having seen that the calculus was not, in 1734, a rigorous theory, and such authors as Maclaurin, Paman, and Robins are to be hailed for raising the standard of rigor in Britain.

On the other hand, it is possible to regard the *Analyst* as a disaster for eighteenth-century British mathematics. Philip Kitcher takes such a view, when he reads the aftermath of the *Analyst* as a period of wrangling over issues irrelevant to genuine mathematical research. He writes:

> The publication of *The Analyst* precipitated a flurry of writings on the calculus, as Newton's successors rallied to his defense. Their concern contrasts sharply with the attitude of the Leibnizians when faced with similar challenges, and indicates the significance which the eighteenth-century British mathematical community attached to the problem of rigor. Berkeley's critique had unfortunate consequences. Colin Maclaurin, the most talented of Newton's successors, presented his major work on the calculus, the *Treatise on Fluxions*, in cumbersome geometrical style, explicitly responding to Berkeley's objections. In the course of his attempt at defense, Maclaurin was drawn into philosophical issues which are largely irrelevant to mathematical research, and, in an effort to make his mathematics conform to his philosophical presuppositions, he developed a style for the Newtonian calculus which widened the gap between British and Continental mathematicians. . . . Priding itself on its rigor and its maintenance of a proper geometrical approach to mathematics, the British mathematical community fell further and further behind. (Kitcher 1984, 239–40)

Other writers have expressed similar opinions of the matter, although they do not explicitly blame Berkeley for singlehandedly destroying British mathematics. Guicciardini sees a significant decline in British mathematics after the *Analyst* and regards the project of answering Berkeley as a "sterile exercise" unworthy of the historian's attention (Guicciardini 1989, 173 n. 7). Sherry (1987) does not specifically discuss the decline in British mathematics, but follows Kitcher in taking Berkeley's concern with rigorous demonstration as largely misplaced.

This reading of the *Analyst* assumes that concern with rigor is less important than the conquest of new results. In other words, the attempt to place a mathematical theory on a philosophically respectable foundation is misguided whenever that attempt hinders the expansion of the theory. In this history of mathematics the heroes are Continental mathematicians, who boldly embraced the infinitesimal calculus and extended its application to new domains while leaving the foundational questions for another, bleaker day. The villain is Berkeley, who duped the stodgy British into wasting their time on philosophical debates and employing cumbersome techniques which lacked the power of the progressive Continental methods.

I think both approaches to the *Analyst* are dubious. They assume that there is a right way for mathematics to develop, and take the history of mathematics to be a chronicle of the successes along this privileged path of development. The first takes the history of mathematics as the record of rigor triumphing over scandalously lax practices. The second focuses on the growth of more powerful and extensive theories and regards foundational debates as an unfortunate indulgence in metaphysical (as opposed to mathematical) subtleties. But surely it is gratuitous to assume that the history of mathematics or philosophy must conform to a script, and particularly a script where good consistently triumphs over evil. False starts, confusion, and recurrent "foundational crises" are all part of the history of mathematics, however much one may care to ignore them. The queen of the sciences has a risqué past, and the history of philosophico-mathematical controversy is one way of bringing it to light. Berkeley is part of this history and his *Analyst* is significant, not because it sent British mathematics down the "right" path, but because it raised questions that were of interest to his age. And the proof positive that these questions mattered is the fact that they were answered. Admittedly, these answers came in many different forms from thinkers of different abilities, but this only shows how difficult the questions were.

The issue at stake here is one of general historiography. One may reasonably ask why certain figures or events are deemed significant while others are left out of a historical study. In general, choice of topic and methodology determine what goes into a historical study and what is left out as irrelevant. In the present case my interest has been to explore Berkeley's relationship to the mathematics of the seventeenth and eighteenth centuries, and I have concentrated on the analysis and interpretation of texts from the era. As a result, I assume that any publication (such as the *Analyst*) which prompts a serious and extended debate is very significant. This is certainly not the only way to approach the history of mathematics in this period. One might hold that what really matters in the history of mathematics is the

The Aftermath of the *Analyst*

project of tracing the lineage of today's mathematical theories back through the ages. In this project, Berkeley's work and responses to it are not of great significance because the calculus as we know it today derives from nineteenth-century Continental studies in the theory of limits and convergence. This approach to the history of mathematics is certainly the dominant view at present. Indeed, anyone today speaking of "the rigorization of the calculus" invariably intends the nineteenth-century work of Cauchy, Weierstrass, and Dedekind and not the earlier work of Maclaurin, Paman, or Robins. But if our interest is to understand British mathematics as it actually looked in the eighteenth century, we cannot ignore Berkeley. This, in part, is my interest, and consequently I take Berkeley seriously.

Conclusions

My study of Berkeley and mathematics has ranged widely in the preceding chapters, but I am now in a position to draw some general conclusions about the place of mathematics in Berkeley's thought and Berkeley's place in the history of mathematics.

The first thing that should be clear is the fact that Berkeley had an identifiable philosophy of mathematics, and his mathematical writings are not a loose collection of random thoughts on the subject. Berkeley's approach to mathematics is everywhere informed by a desire to avoid abstractions and to interpret mathematics as part of his general theory of signs. On the ontological question of what mathematical objects are, Berkeley holds that geometry and arithmetic are conversant about different kinds of signs: geometry takes perceived figures as representatives of other figures, while arithmetic uses numerical signs to represent any collections whatever. Thus, geometric signs must resemble the figures they represent, while arithmetical symbols are "perfectly arbitrary & in our power, made at pleasure" (*Commentaries*, 732).

Berkeley's account of mathematical epistemology flows naturally from these considerations. In geometric demonstrations concrete particulars, not abstract lines or figures, are the immediate objects of our reasoning, but the theory of representative generalization explains how geometric knowledge can be demonstrative and general without being abstract. Arithmetic and algebra, in contrast, have no specific object other than the symbols themselves, and our knowledge of the truths in these sciences is gained simply by understanding the rules for the combination and manipulation of symbols. Any application of algebra or arithmetic must therefore involve an interpretation in which the symbols are "referred to things." But in such circumstances it is crucial that we employ only those symbolic manipulations which correspond to the nature of the things represented by our notation: from the arithmetical fact that 20/5 = 4, it does not follow that one-fifth of a $20 bill is worth $4.

Berkeley's criteria for rigorous demonstration reflect this mathematical epistemology. A rigorous demonstration in algebra or arithmetic requires only that we understand and correctly apply the computational rules for the symbolic system. Geometry, however, must proceed "from the distinct Contemplation and Comparison of Figures, [whose] Properties are derived, by a perpetual well-connected chain of Consequences, the Objects being still kept in view, and the attention ever fixed upon them" (*Analyst*, §2). Thus, the calculus must be judged by criteria of rigor which go beyond the simple requirements implied by Berkeley's formalistic treatment of algebra and arithmetic.

These considerations can help us outline Berkeley's conception of the hierarchy of the sciences. Natural philosophy is concerned with uncovering the laws of nature, which Berkeley characterizes as a kind of grammar, knowledge of which allows us to make reliable predictions about what kinds of ideas we can expect to follow others. In Berkeleyan metaphysics, bodies are causally inert and only minds are active; nevertheless, physical theories which purport to uncover causes can be accepted for their predictive success. At the level of physical theory, then, Berkeley endorses a fairly strong version of instrumentalism. But Berkeley does not allow for a thoroughgoing instrumentalism with regard to mathematics. Most notably, the calculus cannot be justified purely on the basis of its problem-solving success: because geometry has a proper object, the algebraic notation of the calculus must be interpreted to refer to geometric objects; thus a formalistic or instrumental justification is ruled out. Standing above mathematics in this hierarchy is metaphysics, as Berkeley makes clear:

> Whether there be not really a *Philosophia prima*, a certain transcendental Science superior to and more extensive than Mathematics, which it might behove our modern Analysts rather to learn than despise? (*Analyst*, query 49)

Presumably, Berkeley regards metaphysics as a science which admits of neither falsehood in its claims nor instrumentalism in its interpretation.

Conclusions

Another point that should be clear by now is that mathematics and its philosophical interpretation have a central place in Berkeley's epistemology, and no adequate treatment of Berkeley's philosophy can treat his mathematical writings as peripheral. The critique of abstraction and the theory of representative generalization are the foundation of Berkeley's philosophy of mathematics and are rightly regarded as central to his philosophy as a whole. Perhaps surprisingly, Berkeley's philosophy of mathematics does not have a direct connection with his immaterialism. To whatever extent immaterialism rests on the denial of abstraction, there will be a connection between it and the Berkeleyan philosophy of mathematics. But it is not obvious that Berkeley's antiabstractionism mandates the rejection of material substance. A Hobbesian materialist, for example, could endorse the critique of abstract ideas and might well accept much or all of the Berkeleyan philosophy of mathematics without embracing immaterialism.[1]

As for Berkeley's place in mathematical history, I have shown that the *Analyst* and the disputes it engendered deserve a prominent place in any serious or extended treatment of eighteenth-century mathematics. Berkeley was not a crank or a mathematical incompetent, nor were his objections irrelevant to the British mathematical community; he rightly argued that the calculus failed to meet a standard of rigor that its proponents acknowledged as applicable. His case against the calculus ultimately depends upon peculiarly Berkeleyan doctrines, particularly the critique of abstract ideas, but his fundamental charges were taken seriously and set much of the agenda for British mathematics for more than two decades. This is not to say that Berkeley's critique and the responses to it were decisive for the development of the mathematical theories we use today. As it happens, British mathematics and its calculus of fluxions did not flourish in the second half of the eighteenth century. The problems addressed in British mathematics and the techniques applied to their solution led eventually to a significant divergence from the dominant Continental school of analysis. But even if the mathematical work in Britain in the post–*Analyst* era came to a dead end, the role that Berkeley played in the history of mathematics is remarkable and worthy of further scholarly investigation.

---

1. Hobbes himself would not have embraced Berkeley's formalistic treatment of arithmetic and algebra, since he endorsed Barrow's view that geometry is the true foundation of all mathematics. Nevertheless, someone who accepts Hobbes' antiabstractionism and his radical materialism could still find Berkeley's philosophy of mathematics congenial.

Conclusions

Finally, I must also acknowledge some problems confronting Berkeley's philosophy of mathematics. His central argument against abstract ideas is not wholly decisive, and the theory of signs which underpins his positive account of mathematical reasoning does not seem to be fully free of abstractions. Furthermore, he fails to establish the compensation of errors thesis at the center of his attempted explanation of the success of the calculus. But such difficulties do not entirely vitiate the Berkeleyan project. Berkeley's genius lies in his ability to ask interesting questions.[2] Some of the most important questions Berkeley raised can be found in the fifty-two queries at the end of the *Analyst*. That he was unable adequately to answer all of them to our satisfaction takes nothing away from the importance of the questions. They were of intense interest to the thinkers of his age and they still merit our attention.

2. Indeed, Berkeley's *Querist* is a work consisting entirely of questions dealing with issues in economics and social policy.

Bibliography

Aaron, R. I. 1971. *John Locke.* 3d ed. Oxford: Clarendon Press of Oxford University Press.
Andersen, Kirsti. 1985. Cavalieri's Method of Indivisibles. *Archive for History of the Exact Sciences* 24:292–367.
Annas, Julia. 1987. Die Gegenstände der Mathematik bei Aristoteles. In *Mathematics and Metaphysics in Aristotle,* ed. Andreas Graeser, 131–47. Bern and Stuttgart: Paul Haupt.
Apollonius of Perga. 1952. *Conics.* Trans. R. Catesby Taliaferro. Reprinted in *Great Books of the Western World* 11:593–804. Chicago: Encyclopaedia Britannica.
Apostle, Hippocrates George. 1952. *Aristotle's Philosophy of Mathematics.* Chicago: University of Chicago Press.
Aquinas, Saint Thomas. 1948. *Thomas von Aquin In Librum Boethii de Trinitate Quæstiones Quinta et Sexta, Nach dem Autograph Cod. Vat. lat. 9850.* Ed. Paul Weyser. Fribourg: Société Philosophique; Louvain: Editions E. Nauvelaerts.
———. 1964. *Summa theologica: Latin text and English translation, introductions, notes and appendicies.* 60 vols. Cambridge: Blackfriars; New York: McGraw-Hill.
Ardley, G. W. R. 1962. *Berkeley's Philosophy of Nature.* University of Auckland Bulletin, no. 63, Philosophy Series, no. 3. Auckland, New Zealand: University of Auckland.
Aristotle. 1984. *The Complete Works of Aristotle: The Revised Oxford Translation.* Ed. Jonathan Barnes. 2 vols. Bollingen Series, no. 71. Princeton, N.J.: Princeton University Press.
Armstrong, David. 1961. *Berkeley's Theory of Vision.* Melbourne: Melbourne University Press.
Armstrong, Robert L. 1969. Berkeley's Theory of Signification. *Journal of the History of Philosophy* 7:163–76.

Bibliography

Arnauld, Antoine, and Pierre Nicole. 1981. *La logique, ou l'art de penser*. 2d ed., ed. Pierre Clair and François Girbal. Paris: J. Vrin. Familiarly known as the *Port-Royal Logic*.

Ashe, George. 1707(?). Concerning the Squaring of the Circle &c. by Mr. George Ashe. British Library Additional MS. 4811 (Sloane Papers), fol. 53r–57r.

———. 1684. A new and easy way of demonstrating some propositions in Euclide. *Philosophical Transactions* 14:672–76.

Atherton, Margaret. 1987. Berkeley's Anti-Abstractionism. In *Essays on the Philosophy of George Berkeley*, ed. Ernest Sosa, 45–60. Dordrecht: D. Reidel.

———. 1990. *Berkeley's Revolution in Vision*. Ithaca and London: Cornell University Press.

Ayers, Michael. 1981. Locke's Doctrine of Abstraction: Some Aspects of its Historical Purpose and Significance. In *John Locke*, ed. Reinhardt Brandt, 5–24. Proceedings of the Symposium at the Herzog-August Bibliothek in Wolfenbüttel, 10–18 July 1979. Berlin: Walter de Gruyter.

Baron, Margaret E. 1969. *The Origins of the Infinitesimal Calculus*. Oxford: Pergamon Press.

Barrow, Isaac. [1860] 1973. *The Mathematical Works of Isaac Barrow*. Ed. W. Whewell. 2 vols. bound as one. Cambridge: Cambridge University Press. Reprint. Hildesheim: Georg Olms Verlag.

———. 1670. *Lectiones Geometricæ In quibus (præsertim) Generalia Curvarum Linearum Symptomata Declaruntur*. London: John Dunmore.

———. 1683. *Lectiones Mathematicæ XXIII; In quibus Principia Matheseôs generalia exponuntur: Habitæ Cantabrigiæ A.D. 1664, 1665, 1666*. London: George Wells.

———. [1734] 1970. *The Usefulness of Mathematical Learning explained and demonstrated: Being Mathematical Lectures read in the Publick Schools at the University of Cambridge*. Trans. John Kirkby. London: Stephen Austen. Reprint. London: Cass and Company.

———. 1916. *The Geometrical Lectures of Isaac Barrow*. Ed. and trans. J. M. Child. Chicago and London: Open Court.

Baum, Robert J. 1969. *George Berkeley's Philosophy of Mathematics*. Ann Arbor, Mich. University Microfilms.

———. 1972. Instrumentalist and Formalist Elements in Berkeley's Philosophy of Mathematics. *Studies in History and Philosophy of Science* 3:119–34.

[Baxter, Andrew.] 1733. *An Inquiry into the Nature of the Human Soul; wherein the Immateriality of the Soul is evinced from the Principles of Reason and Philosophy*. 2 vols. London: G. Strahan.

[Bayes, Rev. Thomas.] 1736. *An Introduction to the Doctrine of Fluxions, and Defence of the Mathematicians Against the Objections of the Author of "The Analyst," so far as they are Designed to Affect their General Methods of Reasoning*. London: J. Noon.

Bayle, Pierre. 1697. *Diccionaire historique et critique*. Rotterdam: Leers.

———. 1965. *Historical and Critical Dictionary: Selections*. Ed. and trans. Richard H. Popkin. Indianapolis: Bobbs-Merrill.

Bibliography

Belfrage, Bertil. 1985. The Order and Dating of Berkeley's Notebooks. *Revue Internationale de Philosophie* 154:196–214.

———. 1986. The clash on semantics in Berkeley's *Notebook A*. In *George Berkeley: Essays and Replies*, ed. David Berman, 117–26. Dublin-Irish Academic Press.

Berkeley, George. Additional MS. 39304, British Library, London.

———. 1930. *Berkeley's Commonplace Book*. Edited with Introduction, Notes, and Index by G. A. Johnston. London: Faber and Faber.

———. 1948–57. *The Works of George Berkeley, Bishop of Cloyne*. Ed. A. A. Luce and T. E. Jessop. Edinburgh and London: Nelson.

———. 1987. *George Berkeley's Manuscript Introduction*. Ed. Bertil Belfrage. Oxford: Doxa.

———. [1976] 1989. *Philosophical Commentaries*. Transcribed from the manuscript and edited by George H. Thomas, with explanatory notes by A. A. Luce. Alliance, Ohio: Mount Union College. Reprint. New York: Garland.

———. 1992. *"De Motu" and "The Analyst": A Modern Edition with Introductions and Commentary*. Ed. and trans. Douglas M. Jesseph. New Synthese Historical Library, no. 42. Dordrecht and Boston: Kluwer Academic Publishers.

Berlioz, Dominique. 1988. Berkeley et la polémique du calcul infinitésimal. In *Entre forme et histoire: la formation de la notion de développement à l'âge classique*, ed. Olivier Bloch, Bernard Balan, and Paulette Carrive, 71–85. Paris: Meridiens Klincksieck.

Berman, David, ed. 1986. *George Berkeley: Essays and Replies*. Dublin: Irish Academic Press.

[Blake, Francis.] 1741. *An Explanation of Fluxions, in a Short Essay on the Theory*. London: W. Innys.

Blay, Michel. 1986. Deux Moments de la critique du calcul infinitésimal: Michel Rolle et George Berkeley. *Revue d'histoire des sciences* 39:223–53.

Bos, H. J. M. 1974. Differentials, higher-order differentials and the derivative in the Leibnizian calculus. *Archive for History of the Exact Sciences* 14:1–90.

———. 1980. Newton, Leibniz, and the Leibnizian Tradition. In *From the Calculus to Set Theory, 1630–1910: An Introductory History*, ed. I. Grattan-Guinness, 49–93. London: Duckworth.

Bosmans, H. 1927. André Tacquet (S. J.) et son traité d'"Arithmétique théorique et practique." *Isis* 9:64–82.

Boyer, Carl B. [1949] 1959. *The History of the Calculus and its Conceptual Development*. New York: Dover.

———. [1968] 1989. *A History of Mathematics*. 2d ed., rev. Eta Merzbach. New York: John Wiley & Sons.

Bracken, H. M. 1974. *Berkeley*. New York and London: St. Martin's.

———. 1977–78. Bayle, Berkeley, and Hume. *Eighteenth Century Studies* 11:227–45.

———. 1984. Hume on the "Distinction of Reason." *Hume Studies* 10:89–108.

Breidert, Wolfgang. 1969. 'Momentum' und 'Minimum': Zu zwei Notizen in Berkeley's "Philosophischen Kommentaren." *Archiv für Begriffsgeschichte* 13:76–78.

―――. 1986a. Berkeley's *De Ludo Algebraico* and Notebook B. *Berkeley Newsletter* 9:12–14.

―――. 1986b. Berkeley's Kritik an der Infinitesimalrechnung. In *300 Jahre "Nova methodus" von G. W. Leibniz (1684-1984)*, ed. Albert Heinekamp, 185–90. Symposium of the Leibniz-Gesellschaft, *Studia Leibnitiana* Sonderheft no. 14. Stuttgart: Franz Steiner Verlag.

―――. 1989. *George Berkeley, 1685–1753*. Vita Mathematica, no. 4. Basel, Boston, and Berlin: Birkhäuser.

Brook, Richard J. 1973. *Berkeley's Philosophy of Science*. Archives Internationales d'Histoire des Idees, no. 65. The Hague: Martinus Nijhoff.

Browne, Peter. [1728] 1976. *The Procedure, Extent, and Limits of Human Understanding*. London: William Innys. Reprint. British Philosophers and Theologians of the 17th and 18th Centuries. New York: Garland.

Buchdahl, Gerd. 1969. *Metaphysics and the Philosophy of Science; The Classical Origins, Descartes to Kant*. Oxford: Basil Blackwell.

Cajori, Florian. 1917. Discussion of Fluxions from Berkeley to Woodhouse. *American Mathematical Monthly* 24:145–54.

―――. 1919. *A History of the Conceptions of Limits and Fluxions in Great Britain from Newton to Woodhouse*. Chicago and London: Open Court.

―――. 1925. Indivisibles and "Ghosts of Departed Quantities" in the History of Mathematics. *Scientia* 37:301–6.

―――. 1929. Controversies Between Wallis, Hobbes, and Barrow. *Mathematics Teacher* 2:146–51.

Cantor, Geoffrey. 1984. Berkelcy's *Analyst* Revisited. *Isis* 75:668–83.

Cantor, Moritz. 1892–1908. *Vorlesungen Über die Geschichte der Mathematik*. 4 vols. Leipzig: Teubner.

Carnot, L. M. N. 1797. *Réflexions sur la Métaphysique du Calcul Infinitésimal*. Paris: Duprat.

Cassirer, Erich. 1914. *Berkeleys System: Ein Beitrag zur Geschichte und Systematik des Idealismus*. Giessen: Töplemann.

Cavalieri, Buonaventura. 1635. *Geometria indivisibilibus continuorum nova quadam ratione promota*. Bologna: Clement Ferroni.

―――. 1647. *Exercitationes Geometricae Sex . . .* Bologna: Jacobi Monti.

Cheyne, George. 1703. *Fluxionum Methodus Inversa: sive Quantitatum Fluentem leges generaliores*. London: J. Matthew.

―――. 1705. *Philosophical Principles of Natural Religion: containing the Elements of Natural Philosophy, and the Proofs for Natural Religion, arising from them*. London: G. Strahan.

Cicovacki, Predrag. 1990. Locke on Mathematical Knowledge. *Journal of the History of Philosophy* 28:511–24.

Claussen, Friedrich. 1889. *Kritische Darstellung der Lehren Berkeleys über Mathematik und Naturwissenschaften*. Halle: Erhardt Karras.

Clavius, Christoph. 1612. *Christophori Calvii Bambergensis e Societate Iesu Operum Mathematicorum Tomus primo-quintus*. 5 vols. Munich: A. Hierat.

Cleary, John Joseph. 1982. *Aristotle's Theory of Abstraction: A Problem About the Mode of Being of Mathematical Objects*. Ann Arbor, Mich. University Microfilms.

Bibliography

[Collins, Anthony.] 1713. *A discourse of free-thinking, occasion'd by the rise and growth of a sect call'd free-thinkers.* London.

Colson, John, ed. and trans. 1736. *The Method of Fluxions and Infinite Series: with its Application to the Geometry of Curve-Lines. By the Inventor Sir Isaac Newton, K^t. Late President of the Royal Society. Translated from the Author's Latin Original not yet made publick, To which is subjoin'd, A Perpetual Comment upon the whole Work, Consisting of Annotations, Illustrations, and Supplements, In order to make this Treatise a Compleat Institution for the use of Learners.* London: J. Nourse.

[Colson, John?]. 1737. Review of *The Method of Fluxions and Infinite Series: with its Application to the Geometry of Curve-Lines,* in two parts. *The Present State of the Republick of Letters* 18:223–36, 444–57.

Dancy, Johnathan. 1987. *Berkeley: An Introduction.* Oxford and New York: Blackwell.

Crapulli, Giovanni. 1969. *Mathesis Universalis: Genesi di un'Idea Nel XVI Secolo.* Lessico Intellettuale Europeo 2. Rome: Edizioni dell'Ateneo.

Crombie, A. C. 1977. Mathematics and Platonism in Sixteenth-Century Italian Universities and in Jesuit Educational Policy. In *Prismata: Naturwissenschaftsgeschichtliche Studien,* ed. Y. Maeyama and W. G. Saltzer, 63–94. Wiesbaden: Franz Steiner Verlag.

De Gandt, François. 1991. Cavalieri's Indivisibles and Euclid's Canons. In *Revolution and Continuity: Essays in the History and Philosophy of Early Modern Science,* ed. Peter Barker and Roger Ariew. Studies in Philosophy and the History of Philosophy, no. 24, 157–82. Washington, D.C.: Catholic University of America Press.

De Morgan, Augustus. 1852. On the Early History of Infinitesimals in England. *London, Edinburgh and Dublin Philosophical Magazine and Journal of Science* 4:321–30.

Dear, Peter. 1988. *Mersenne and the Learning of the Schools.* Cornell History of Science Series. Ithaca and London: Cornell University Press.

Dechales, Claude François Milliet de. 1674. *Cursus; seu mundus mathematicus.* 3 vols. Lyon.

Devaux, Philippe. 1953. Berkeley et les mathématiques. *Revue internationale de philosophie* 7:101–33.

Digby, Sir Kenelm. [1644] 1978. *Two Treatises: In the one of which, The Nature of Bodies; in the other, The Nature of Mans Soule; is looked into: In way of Discovery, of the Immortality of Reasonable Selves.* British Philosophers and Theologians of the 17th and 18th Centuries. New York and London: Garland.

Dijksterhuis, E. J. 1987. *Archimedes.* Trans. C. Dikshoorn, with a bibliographic essay by Wilbur Knorr. 2d ed. Princeton, N.J.: Princeton University Press.

Doney, Willis, ed. 1988. *Berkeley on Abstraction and Abstract Ideas.* New York and London: Garland.

Duhem, Pierre. 1985. *Medieval Cosmology: Theories of Infinity, Place, Time, Void, and the Plurality of Worlds.* Ed. and trans. Roger Ariew. Chicago: University of Chicago Press.

Eames, John. 1738. A brief account by Mr. John Eames F. R. S. of a work entitled, *The Method of Fluxions and Infinite Series, with its Application to the*

Bibliography

Geometry of Curve Lines, by the inventor Sir Isaac Newton, Kt. &c. translated from the author's Latin original not yet made publick. To which is subjoin'd a perpetual comment upon the whole &c. by John Colson. Philosophical Transactions of the Royal Society 39:87–89.

Edwards, C. H. 1979. *The Historical Development of the Calculus.* New York and Berlin: Springer Verlag.

Emerson, William. 1743. *The Doctrine of Fluxions: not only Explaining the Elements thereof, but also its Application and Use in the Several Parts of Mathematics and Natural Philosophy.* London: Bettenham.

Euclid. [1925] 1956. *The Thirteen Books of Euclid's "Elements" Translated from the Text of Heiberg.* Ed. and trans. T. L. Heath. 3 vols. Cambridge: Cambridge University Press. Reprint. New York: Dover.

Evans, W. D. 1914. Berkeley and Newton. *Mathematical Gazette* 7:418–21.

Fauvel, John, and Jeremy Gray, eds. 1987. *The History of Mathematics: A Reader.* Basingstoke, Hampshire: MacMillan.

Feingold, Mordechai, ed. 1990. *Before Newton: The Life and Times of Isaac Barrow.* Cambridge: Cambridge University Press.

Flage, Daniel. 1987. *Berkeley's Doctrine of Notions: A Reconstruction Based on his Theory of Meaning.* New York: St. Martin's.

Fogelin, Robert. 1988. Hume and Berkeley on the Proofs of Infinite Divisibility. *The Philosophical Review* 97:47–70.

Fontialis, Jacobus. 1740. *Opera Posthuma.* Namur.

Foster, John, and Howard Robinson, eds. 1985. *Essays on Berkeley: A Tercentenary Celebration.* Oxford: Clarendon Press of Oxford University Press.

Frege, Gottlob. [1884] 1962. *Die Grundlagen der Arithmetik: Eine logisch mathematische Untersuchung über den Befriff der Zahl.* Breslau: Koebner. Reprint. Hildesheim: Georg Olms.

———. [1893–1903] 1962. *Die Grundgesetze der Arithmetik, begriffschriftlich abgeleitet.* 2 vols. Jena: H. Pohle. Reprint. Hildesheim: Georg Olms.

Garrison, James W. 1987. Newton and the Relation of Mathematics to Natural Philosophy. *Journal of the History of Ideas* 48:609–27.

Gibson, George A. 1898. Vorlesungen über die Geschichte der Mathematik von Moritz Cantor. Dritter Band, Dritte Abteilung. A Review: with special reference to the *Analyst* controversy. *Proceedings of the Edinburgh Mathematical Society* 17:9–32.

———. 1899. Berkeley's *Analyst* and its critics: an episode in the development of the doctrine of limits. *Bibliotheca Mathematica* 13:65–70.

Gillispie, Charles Coulston. 1971. *Lazare Carnot Savant: A Monograph treating Carnot's Scientific Work, with Facsimile Reproductions of his Unpublished Writings on Mechanics and on the Calculus, and an Essay concerning the Latter by A. P. Youschkevitch.* Princeton: Princeton University Press.

Giorello, Giulio. 1985. *Lo Spettro e il Libertino: Teologia, Matematica, Libero Pensiero.* Milan: Mondadori.

Giusti, Enrico. 1980. *Bonaventura Cavalieri and the Theory of Indivisibles.* Milan: Edizioni Cremonese.

Grabiner, Judith. 1981. *The origins of Cauchy's rigorous calculus.* Cambridge, Mass.: MIT Press.

Bibliography

Graeser, Andreas, ed. 1987. *Mathematics and Metaphysics in Aristotle.* Bern and Stuttgart: Paul Haupt.

Grattan-Guinness, I. 1969. Berkeley's Criticism of the Calculus as a Study in the Theory of Limits. *Janus* 56:215–27.

Grattan-Guinness, I., ed. 1980. *From the Calculus to Set Theory, 1630–1910: An Introductory History.* London: Duckworth.

Gray, Robert. 1978. Berkeley's Theory of Space. *Journal of the History of Ideas* 16:415–34.

Grayling, A. C. 1986. *Berkeley: The Central Arguments.* London: Duckworth.

Guicciardini, Niccolò. 1989. *The development of the Newtonian calculus in Britain: 1700–1800.* Cambridge: Cambridge University Press.

Guldin, Paul. 1635–41. *Centrobaryca, seu de Centro Gravitatis . . .* 4 vols. Vienna: G. Gelbhaar.

Hall, A. R. 1980. *Philosophers at War: The Quarrel Between Newton and Leibniz.* Cambridge: Cambridge University Press.

Hanna, John. 1736. *Some Remarks on Mr. Walton's Appendix, which he Wrote in Reply to the Author of the "Minute Philosopher"; concerning Motion and Velocity.* Dublin: Fuller.

Harris, N. G. E. 1988. Locke's Triangle. *Canadian Journal of Philosophy* 18:31–41.

Hayes, Charles. 1704. *A Treatise of Fluxions; or, an Introduction to Mathematical Philosophy; Containing a Full Explication of that Method by which the most celebrated Geometers of the Present Age have made such Vast Advances in Mechanical Philosophy.* London: Midwinter.

Heine, Eduard. 1871. Die Elemente der Funktionenlehre. *Journal für die reine und angewandte Mathematik* 74:172–90.

Heinekamp, Albert, ed. 1986. *300 Jahre "Nova methodus" von G. W. Leibniz (1684–1984).* Symposium of the Leibniz-Gesellschaft, *Studia Leibnitiana,* Sonderheft, no. 14. Stuttgart: Franz Steiner Verlag.

Hobbes, Thomas. [1845] 1966a. *Thomæ Hobbes Malmesburiensis Opera Philosophica Quæ Latine Scripsit Omnia in Unum Corpus Nunc Primum Collecta.* Ed. William Molesworth. 5 vols. Aalen, Germany: Scientia Verlag.

———. [1845] 1966b. *The English Works of Thomas Hobbes of Malmesbury, now First Collected and Edited by Sir William Molesworth.* Ed. William Molesworth. 11 vols. Aalen, Germany: Scientia Verlag.

Hodgson, James. 1736. *The Doctrine of Fluxions, Founded on Sir Isaac Newton's Method, Published by Himself in his Tract upon the Quadrature of Curves.* London: T. Wood.

Imlay, Robert. Berkeley on Abstract General Ideas. *Journal of the History of Philosophy* 9:321–28.

Jesseph, Douglas M. 1989. Philosophical Theory and Mathematical Practice in the Seventeenth Century. *Studies in History and Philosophy of Science* 20:215–44.

———. 1990. Berkeley's Philosophy of Geometry. *Archiv für Geschichte der Philosophie* 72:301–22.

Jessop, T. E. 1963. Berkeley's Philosophy of Science. *Hermathena* 97:23–35.

Johnston, G. A. 1916. The Influence of Mathematical Conceptions on Berkeley's Philosophy. *Mind* 25:177–92.

———. 1918. Berkeley's Logic of Mathematics. *The Monist* 28:25–45.

———. 1923. *The Development of Berkeley's Philosophy*. London: Macmillan.

Jones, J. F., III. 1983. Intelligible Matter and Geometry in Aristotle. *Apeiron* 17:94–102.

Jurin, James [Philalethes Cantabrigiensis, pseud.]. 1734. *Geometry no Friend to Infidelity; or, A Defence of Sir Isaac Newton and the British Mathematicians, in a Letter to the Author of the "Analyst": Wherein it is Examined how far the Conduct of such Divines as Intermix the Interest of Religion with their Private Disputes and Passions, and Allow Neither Learning nor Reason to those they differ from, is of Honour or Service to Christianity, or Agreeable to the Example of our Blessed Savior and his Apostles*. London: T. Cooper.

———. 1735a. *The Minute Mathematician; or, the Free-Thinker no Just-Thinker. Set forth in a Second Letter to the Author of the "Analyst"; Containing a Defence of Sir Isaac Newton and the British Mathematicians, against a late Pamphlet, entitled "A Defence of Free-Thinking in Mathematicks."* London: T. Cooper.

———. 1735b. Considerations upon some passages contained in two Letters to the Author of the *Analyst*, written in defence of Sir Isaac Newton, and the British Mathematicians. By Philalethes Cantabrigiensis. *The Present State of the Republick of Letters* 16:369–96.

———. 1736a. Considerations occasioned by a Paper in the last *Republick of Letters*, concerning some late Objections against the Doctrine of Fluxions, and the different Methods that have been taken to obviate them. By Philalethes Cantabrigiensis. *The Present State of the Republick of Letters* 17:72–91.

———. 1736b. Considerations upon some passages of a Dissertation concerning the Doctrine of Fluxions, published by Mr. Robins in the *Republick of Letters* for April last. By Philalethes Cantabrigiensis. *The Present State of the Republick of Letters* 18:45–82.

———. 1736c. The Remainder of the Paper begun in our last, entitled, Considerations upon some passages of a Dissertation concerning the Doctrine of Fluxions, published by Mr. Robins in the *Republick of Letters* for April last. By Philalethes Cantabrigiensis. *The Present State of the Republick of Letters* 18:111–80.

———. 1737a. The Contents of Dr. Pemberton's Observations published the last Month. *The History of the Works of the Learned* 20:230–39.

———. 1737b. A Reply to Dr. Pemberton's Observations published in the *History of the Works of the Learned* for the Month of April. *The History of the Works of the Learned* 20:385–97.

———. 1737c. A Reply to Dr. Pemberton's Observations published in the *History of the Works of the Learned* for the Month of June. *The History of the Works of the Learned* 21:66–79.

———. 1737d. The Conclusion and Postscript to the last Reply of Philalethes Cantabrigiensis to Dr. Pemberton, published in our *History* for July. *The History of the Works of the Learned* 21:235–36.

———. 1744. *A letter to the Right Reverend the Bishop of Cloyne, occasion'd by His Lordship's treatise on the virtues of tar-water. Impartially examining how far that*

Bibliography

medicine deserves the character His Lordship has given of it. London: Jacob Robinson.

Keill, John. 1739. *Introductiones ad Veram Physicam et Veram Astronomium* . . . Leyden: Verbeek.

Keisler, Gerald. 1976. *Elementary Calculus: an Infinitesimal Approach*. New York: Academic Press.

Keough, Andrew. 1934. A Catalogue of George Berkeley's Gift to Yale's Library. *Yale University Library Gazette* 34:1–26.

Keynes, Geoffrey. 1976. *A Bibliography of George Berkeley, Bishop of Cloyne: His works and his Critics in the Eighteenth Century*. Oxford: Clarendon Press of Oxford Unviersity Press.

Kirkby, John. 1748. *The Doctrine of Ultimators. Containing a New Acquisition to Mathematical Literature, Naturally Resulting from the Consideration of an Equation, as Reducible from its Variable to its Ultimate State; or, A Discovery of the True and Genuine Foundation of what has Hitherto Mistakenly Prevailed under the Improper Names of Fluxions and the Differential Calculus*. London: Hodges.

Kitcher, Philip. 1973. Fluxions, limits, and infinite littleness: a Study of Newton's presentation of the calculus. *Isis* 44:33–49.

———. 1981. Mathematical Rigor: Who Needs It? *Noûs* 15:469–93.

———. 1984. *The Nature of Mathematical Knowledge*. Oxford and New York: Oxford University Press.

Klein, Jacob. 1968. *Greek Mathematical Thought and the Origin of Algebra*. Trans. Eva Brann. Cambridge, Mass.: MIT Press.

Lasswitz, Kurd. 1890. *Geschichte der Atomistik von Mittlealter bis Newton*. Hamburg: Leopold Voss.

Lear, Johnathan. 1982. Aristotle's Philosophy of Mathematics. *Philosophical Review* 91:161–92.

Leclerc, Jean. 1711. Review of Berkeley's *New Theory of Vision*. *Bibliothéque choisie* 22:58–88.

Leibniz, G. W. [1848–53] 1962. *G. W. Leibniz Mathematische Schriften*. Ed. C. I. Gerhardt. 7 vols. Hildesheim: Georg Olms.

———. 1989. *Philosophical Essays*. Ed. and trans. Roger Ariew and Daniel Garber. Indianapolis, Ind. and Cambridge, Mass.: Hackett Publishing Company.

Leroy, André-Louis. 1956. Valeur exemplaire des erreurs mathématiques de Berkeley. *Revue de Synthèse* 77:155–69.

L'Hôpital, G. F. A. 1696. *Analyse des infiniment petits pour l'intelligence des lignes courbes*. Paris: Imprimiere Royale.

Lindberg, David, ed. 1978. *Science in the Middle Ages*. Chicago: University of Chicago Press.

Locke, John. 1975. *An Essay Concerning Human Understanding*. Ed. P. H. Nidditch. Oxford: Clarendon Press of Oxford University Press.

Lokken, Roy N. 1980. Discussions on Newton's Infinitesimals in Eighteenth-Century Anglo-America. *Historia Mathematica* 7:142–55.

Lorenz, Theodor. 1904. Weitere Beiträge zur Lebensgeschichte George Berkeleys. *Archiv für Geschichte der Philosophie* 17:159–70.

Luce, A. A. [1934] 1967. *Berkeley and Malebranche: A Study in the Origins of Berkeley's Thought*. Oxford: Clarendon Press of Oxford University.

———. 1963. *The Dialectic of Immaterialism*. London: Hodder and Staughton.
Maclaurin, Colin. 1742. *A Treatise of Fluxions, in Two Books*. 2 vols. Edinburgh: Ruddimans.
[Maclaurin, Colin]. 1744a. An account of a book intituled, *A Treatise of Fluxions, in two books. Philosophical Transactions of the Royal Society* 42:325–63.
———. 1744b. The continuation of an account of *A Treatise of Fluxions &c*, Book II. *Philosophical Transactions of the Royal Society* 42:403–15.
Mahoney, Michael S. 1978. Mathematics. In *Science in the Middle Ages*, ed. David Lindberg, 145–78. Chicago: University of Chicago Press.
———. 1990. Barrow's Mathematics: Between Ancients and Moderns. In *Before Newton: The Life and Times of Isaac Barrow*, ed. Mordechai Feingold, 179–249. Cambridge: Cambridge University Press.
Malebranche, Nicholas. 1963. *Recherche de la Vérité, ou, l'on Traite de la Nature de l'Esprit de l'Homme et de l'Usage qu'il en doit faire pour éviter l'Erreur dans les Sciences*. Ed. Geneviève Rodis-Lewis. 3 vols. Paris: Libraire Philosophique J. Vrin.
McCracken, Charles J. 1983. *Malebranche and British Philosophy*. Oxford: Clarendon Press of Oxford University Press.
McKim, Robert. 1986. The entries in Berkeley's Notebooks: A reply to Bertil Belfrage. In *George Berkeley: Essays and Replies*, ed. David Berman, 156–61. Dublin: Irish Academic Press.
Martin, Benjamin. 1739. Πανγεωμετρια; *or the Elements of all Geometry, containing . . . An Appendix, containing an Epitome of the Doctrine of Fluxions; and a Specimen of the Method de Maximis & Minimus*. London: J. Noon.
Masi, Michael. 1983. Arithmetic. In *The Seven Liberal Arts in the Middle Ages*, ed. David L. Wagner, 147–68. Bloomington, Ind.: Indiana University Press.
Meyer, Eugen. 1894. *Humes und Berkeleys Philosophie der Mathematik, vergleichend und kritisch dargestellt*. Abhandlungen zur Philosophie und ihrer Geschichte, no. 3. Halle: Max Niemeyer.
Moked, Gabriel. 1988. *Particles and Ideas: Bishop Berkeley's Corpuscularian Philosophy*. Oxford: Clarendon Press of Oxford University Press.
Mueller, Ian. 1970. Aristotle on Geometrical Objects. *Archiv für Geschichte der Philosophie* 52:156–71.
———. 1981. *Philosophy of Mathematics and Deductive Structure in Euclid's "Elements."* Cambridge, Mass.: MIT Press.
Muller, John. 1736. *A Mathematical Treatise: Containing a System of Conic-Sections; with the Doctrine of Fluxions and Fluents, applied to various Subjects*. London: W. Innys.
Nagel, Ernest. 1935. "Impossible Numbers": A Chapter in the History of Modern Logic. *Studies in the History of Ideas* 3:429–74.
Neri, Luigi. 1980. "Filling the World with a Mite": un paradosso dell'infinita divisibilità negli scritti giovanili di George Berkeley, *Revista di Filosofia* 71:67–97.
[Newton, Sir Isaac.] 1712. *Commercium Epistolicum D. Johannis Collins, et aliorum de Analysi promota*. London: J. Tonson.

Bibliography

[―――.] 1722. An Account of the Book entituled *Commercium Epistolicum Collonii & aliorum, De Analysi promota;* published by order of the *Royal Society . . . Philosophical Transactions of The Royal Society* 20:173–224.

―――. [1729] 1934. *Sir Isaac Newton's Mathematical Principles of Natural Philosophy and his System of the World.* 2 vols. Trans. Andrew Motte; rev. and ed. Florian Cajori. Berkeley, Calif.: University of California Press.

―――. 1736. *The Method of Fluxions and Infinite Series: with its Application to the Geometry of Curve-Lines. By the Inventor Sir Isaac Newton, K^t. Late President of the Royal Society. Translated from the Author's Latin Original not yet made publick, To which is subjoin'd, A Perpetual Comment upon the whole Work, Consisting of Annotations, Illustrations, and Supplements, In order to make this Treatise a Compleat Institution for the use of Learners.* Ed. and trans. John Colson. London: J. Nourse.

―――. 1740. *La Methode des Fluxions, et des Suites Infinies. Par M. Chevalier Newton.* Trans. Georges-Louis Leclerc de Buffon. Paris: DeBure.

―――. 1964–67. *The Mathematical Works of Isaac Newton.* Ed. Derek T. Whiteside. 2 vols. The Sources of Science, ed. Harry Woolf, no. 3. New York and London; Johnson Reprint Corp.

―――. 1967–81. *The Mathematical Papers of Isaac Newton.* Ed. D. T. Whiteside and M. A. Hoskins. 8 vols. Cambridge: Cambridge University Press.

Nieuwentijt, Bernard. 1694. *Considerationes circa analyseos ad quantitates infinite parvas applicatæ principia et calculi differentialis usum in resolvendis problematibus geometricis.* Amsterdam: Wolters.

―――. 1695. *Analysis infinitorum, seu curvlineorum proprietates ex polygonourm datura deductæ.* Amsterdam: Wolters.

―――. 1696. *Considerationes secundæ circa calculi differentialis principia; et responsio ad virum noblissimum G. G. Liebnitium.* Amsterdam: Wolters.

Norris, John. [1701–4] 1978. *An Essay Towards the Theory of the Ideal or Intelligible World.* 2 vols. British Philosophers and Theologians of the 17th and 18th Centuries. New York and London: Garland.

Olson, Mark A. 1988. Descartes' First Meditation: Mathematics and the Laws of Logic. *Journal of the History of Philosophy* 26:407–38.

Paman, Roger. 1745. *The Harmony of the Ancient and Modern Geometry Asserted: In Answer to the Call of the Author of the "Analyst" upon the Celebrated Mathematicians of the Present Age, to clear up what he Stiles, their Obscure Analytics.* London: J. Nourse.

Pardies, Ignace Gaston. 1746. *Short, but yet Plain Elements of Geometry: Shewing How by a Brief and Easy Method, most of what is Necessary and Useful in Euclid, Archimedes, Apollonius, and other Excellent Geometricians, both Ancient and Modern, may be Understood.* 8th ed., trans. John Harris. London: A. Ward.

Pemberton, Henry. 1737a. Some Observations on the Appendix to the *Present State of the Republick of Letters* for December 1736. *The History of the Works of the Learned* 20:155–57.

―――. 1737b. Some Observations by Doctor Pemberton on the Misrepresentations of him published in *The History of the Works of the Learned* for the last Month. *The History of the Works of the Learned* 20:305–7.

———. 1737c. Observations by Dr. Pemberton on Philalethes's Reply published in *The History of the Works of the Learned* for the last Month. *The History of the Works of the Learned* 20:438–42.

———. 1737d. Observations by Dr. Pemberton on Philalethes's Reply published in *The History of the Works of the Learned* for the last Month. *The History of the Works of the Learned* 21:124–30.

———. 1737e. Dr. Pemberton's Answer to the two Questions Put by Philalethes Cantabrigiensis in *The History of the Works of the Learned* for the last Month. *The History of the Works of the Learned* 21:285–86.

———. 1737f. An Advertisement by Dr. Pemberton concerning the Questions published in *The History of the Works of the Learned*. *The History of the Works of the Learned* 21:449–50.

Pereira, Benedictus. 1576. *De communibus omnium rerum naturalium principiis et affectionibus libri quindecim* . . . Rome.

Pitcher, George. 1977. *Berkeley*. Arguments of the Philosophers. London and New York: Routledge and Kegan Paul.

Popkin, Richard H. 1951–52. Berkeley and Pyrrhonism. *Review of Metaphysics* 5:223–46.

Proclus. 1970. *Commentary on the First Book of Euclid's "Elements."* Trans. E. Morrow. Princeton, N.J.: Princeton University Press.

Pycior, Helena M. 1987. Mathematics and Philosophy: Wallis, Hobbes, Barrow, and Berkeley. *Journal of the History of Ideas* 48:265–86.

Raphson, Joseph. 1702. *A Mathematical Dictionary; or, A Compendious Explication of all Mathematical Terms, Abridg'd from Monsieur Ozanam, and others*. London: Midwinter and Leigh.

———. 1715. *The History of Fluxions, Shewing in a Compendious Manner the first Rise of, and various Improvements made in that Incomparable Method*. London: R. Mount.

Raynor, David. 1980. "Minima Sensibilia" in Berkeley and Hume. *Dialogue* 19:196–200.

Resnik, Michael D. 1980. *Frege and the Philosophy of Mathematics*. Ithaca and London: Cornell University Press.

Rigaud, Stephen J., ed. [1841] 1965. *Correspondence of scientific men of the seventeenth century; Including letters of Barrow, Flamsteed, Wallis and Newton, printed from originals in the collection of the Right Honourable Earl of Macclesfield*. 2 vols. Hildesheim: Georg Olms.

Robins, Benjamin. 1735a. *A Discourse Concerning the Nature and Certainty of Sir Isaac Newton's Methods of Fluxions, and of Prime and Ultimate Ratios*. London: W. Innys.

[Robins, Benjamin?]. 1735b. An account of a book entituled *A Discourse Concerning the Nature and Certainty of Sir Isaac Newton's Methods of Fluxions, and of Prime and Ultimate Ratios*. *The Present State of the Republick of Letters* 16:245–270.

———. 1735c. A Review of some of the Principal Objections that have been made to the Doctrine of Fluxions and Ultimate Ratios; with some Remarks on the different Methods that have been taken to obviate them. *The Present State of the Republick of Letters* 16:436–47.

Bibliography

———. 1736a. A Dissertation shewing, that the account of the doctrines of Fluxions, and of prime and ultimate ratios, delivered in a treatise, entituled *A Discourse Concerning the Nature and Certainty of Sir Isaac Newton's Methods of Fluxions, and of Prime and Ultimate Ratios*, is agreeable to the real Sense and Meaning of their Great Inventor. *The Present State of the Republick of Letters* 17:290–335.

———. 1736b. Remarks on the Considerations relating to Fluxions, etc. that were published by Philalethes Cantabrigiensis in the *Republick of Letters* for the last month. *The Present State of the Republick of Letters* 18:87–110.

———. 1761. *Mathematical Tracts of the late Benjamin Robins, Esq. Fellow of the Royal Society, and Engineer General to the Honourable the East India Company*. Ed. James Wilson. 2 vols. London: J. Nourse.

Robinson, Abraham. 1965. *Non-Standard Analysis*. Amsterdam: North Holland.

Robles, José A. 1980. Percepcion y Infinitesimales en Berkeley. *Dianoia* 26: 151–77.

———. 1981. Percepcion y Infinitesimales en Berkeley II. *Dianoia* 27: 166–85.

———. 1984. Berkeley y su Critica a los Fundamentos del Cálculo. *Rivista Latinoamericana de Filosofia* 10:141–50.

Sageng, Erik Lars. 1989. *Colin Maclaurin and the foundations of the method of fluxions*. Ann Arbor, Mich.: University Microfilms.

Sasaki, Chikara. 1985. The Acceptance of the Theory of Proportion in the Sixteenth and Seventeenth Centuries: Barrow's Reaction to the Analytic Mathematics. *Historia Scientiarum* 29:83–116.

Sherry, David. 1987. The Wake of Berkeley's Analyst: Rigor Mathematicæ? *Studies in History and Philosophy of Science* 18:455–80.

———. 1991. The Logic of Impossible Quantities. *Studies in History and Philosophy of Science* 22:37–62.

Simpson, Thomas. 1737. *A New Treatise of Fluxions: wherein the Direct and Inverse Method are Demonstrated after a New, Clear, and Concise Manner, With their Application to Physics and Astronomy: also the Doctrine of Infinite Series and Reverting Series Universally, are Amply Explained, Fluxionary and Exponential Equations Solved: Together with a Variety of New and Curious Problems*. London: T. Gardner for the Author.

Smiglecki, Martin. 1634. *Logica Martini Smigleckii Societatis Jesu S. Theologiæ Doctoris, Selectis disputationibus & quæstionibus illustrata, Et in duos Tomor distributa*. 2 vols. Oxford: H. Crypps, E. Forrest, and H. Curteyne.

Smith, Vincent Edward. 1954. *St Thomas on the Object of Geometry*. Milwaukee, Wis.: Marquette University Press.

Smith, G. C. 1980. Thomas Bayes and Fluxions. *Historia Mathematica* 7: 379–88.

Smith, James. 1737. *A New Treatise of Fluxions*. London: printed for the author.

Stammler, Gerhardt. 1921. *Berkeleys Philosophie der Mathematik*. Kant-Studien, Ergänzungsheft, no. 55. Berlin: Reuther & Reichard.

Stock, Joseph. [1776] 1989. *An Account of the Life of George Berkeley, D. D. Late Bishop of Cloyne in Ireland. With Notes, Containing Strictures Upon his Works*.

Reprinted in *George Berkeley: Eighteenth Century Responses*, ed. David Berman, 1:5–85. New York: Garland.

Strong, Edward W. 1957. Mathematical Reasoning and its Objects. In *George Berkeley: Lectures delivered before the Philosophical Union of the University of California in honor of the two hundredth anniversary of the death of George Berkeley, Bishop of Cloyne*, ed. Steven C. Pepper, Karl Aschenbrenner and Benson Mates, 65–88. University of California Publications in Philosophy, no 29. Berkeley and Los Angeles: University of California Press.

Struik, D. J., ed. [1969] 1986. *A Source Book in Mathematics, 1200–1800*. Princeton, N.J.: Princeton University Press.

Tacquet, André. 1654. *Elementa geometriæ planæ ac solidæ, auibus accedunt selecta ex Archimede theoremata*. Antwerp: I. Meurs.

———. 1665. *Arithmeticæ Theoria et Praxis, editio secunda correctior*. Antwerp: I. Meurs.

Taylor, Brook. 1715. *Methodus Incrementorum Directa et Inversa*. London: W. Innys.

Thomae, Johannes. 1898. *Elementare Theorie der analytischen Functionen einer complexen Veränderlichen*. 2d ed. Halle: L. Nebert.

Thomason, S. K. 1982. Euclidean Infinitesimals. *Pacific Philosophical Quarterly* 63:168–85.

Tindall, Matthew. 1730. *Christianity as old as the creation; or, The gospel, a republication of the religion of nature*. London.

Toland, John. 1702. *Christianity not mysterious; or, A treatise shewing that there is nothing in the gospel contrary to reason, nor above it: and that no Christian doctrine can be properly call'd a mystery*. 2nd ed. London: Buckley.

Turnbull, Herbert W. 1951. Bi-centenary of the Death of Colin Maclaurin (1698–1746). Aberdeen University Studies, no. 127. Aberdeen: Aberdeen University Press.

Tweedie, Charles. 1915. A study of the life and writings of Colin Maclaurin. *The Mathematical Gazette* 8:133–51.

Urmson, J. O. 1982. *Berkeley*. Oxford: Clarendon Press of Oxford University Press.

Vermeulen, Ben. 1985. Berkeley and Nieuwentijt on Infinitesimals. *Berkeley Newsletter* 8:1–7.

Vermij, R. H. 1989. Bernard Nieuwentijt and the Liebnizian Calculus. *Studia Leibniziana* 1:69–86.

Wallace, William. 1984. *Galileo and His Sources: The Heritage of the Collegio Romano in Galileo's Science*. Princeton, N.J.: Princeton University Press.

Wallis, John. 1685. *A Treatise of Algebra Both Historical and Practical, Shewing the original Progress and Advancement thereof, from time to time; and by what steps it hath attained the heighth at which it now is*. London: J. Playford for R. Davis.

———. 1690. *The Doctrine of the Blessed Trinity Briefly Explained, In a Letter to a Friend*. Oxford: n.p.

———. 1693–99. *Johannis Wallis S. T. D. . . Opera Mathematica*. 3 vols. Oxford: At the Sheldonian Theater.

Bibliography

Walmsley, Peter. 1990. *The Rhetoric of Berkeley's Philosophy*. Cambridge Studies in Eighteenth-Century English Literature and Thought, no. 6. Cambridge: Cambridge University Press.

Walton, John. 1735a. *A Vindication of Sir Isaac Newton's Principles of Fluxions, against the Objections contained in the "Analyst."* Dublin: S. Powell.

———. 1735b. *The Catechism of the Author of the "Minute Philosopher" Fully Answer'd*. Dublin: S. Powell.

———. 1735c. *The Catechism of the Author of the "Minute Philosopher" Fully Answer'd; With an Appendix in Answer to the "Reasons for not Replying to Mr. Walton's Full Answer."* Dublin: S. Powell.

Warnock, G. J. 1953. *Berkeley*. London: Penguin.

Weinberg, Julius. 1965. The Nominalism of Berkeley and Hume. In *Abstraction, Relation, and Induction: Three Essays in the History of Modern Thought*, 3–60. Madison, Wis.: University of Wisconsin Press.

Whiteside, Derek T. 1960–62. Patterns of Mathematical Thought in the Later Seventeenth Century. *Archive for History of the Exact Sciences* 1: 179–338.

Winkler, Kenneth P. 1989. *Berkeley: An Interpretation*. Oxford: Clarendon Press of Oxford University Press.

Winnett, Arthur R. 1974. *Peter Browne: Provost, Bishop, Metaphysician*. London: S. P. C. K.

Wisdom, John O. 1939. The *Analyst* Controversy: Berkeley's Influence on the Development of Mathematics. *Hermathena* 29:3–29.

———. 1941. The Compensation of Errors in the Method of Fluxions. *Hermathene* 57:49–81.

———. 1942. The *Analyst* Controversy: Berkeley as Mathematician. *Hermathena* 59:111–28.

———. 1953. Berkeley's Criticism of the Infinitesimal. *British Journal for the Philosophy of Science* 4:22–25.

Wright, J. M. F. [1833] 1972. *A Commentary on Newton's "Principia" with a Supplementary Volume Designed for the Use of Students at the Universities*. The Sources of Science, ed. Harry Woolf, no. 124. New York and London: Johnson Reprint Corp.

Zurkuhlen, Heinrich. 1915. *Berkeleys und Humes Stellung zur Analysis des Unendlichen*. Berlin: Blanke.

Index

Aaron, R. I., 28n
abstraction, 3, 5, 7, 9–43, 224–25, 240–43, 245–46, 254–55, 297, 299; and geometry, 46–48, 241; and the calculus, 182–83, 215–17, 219, 230, 243, 256; Berkeley's alternative to, 33–38; Berkeley's rejection of, 3, 6, 20–33, 102–4, 246, 300; kinds of, 11–13, 22–23, 29–30
abstractionism, 5, 47–48; and infinite divisibility, 51, 53–55
Addison, Joseph, 179
algebra, 89–91, 108, 113, 114–17, 219–22, 272–73, 277, 297–98; as an extension of arithmetic, 89, 114–17; as *mathesis universalis,* 89–90, 92–93; contrasted to geometry, 222, 273; object of, 89
analysis, 123, 158, 159; nonstandard, 131, 198
Andersen, Kirsti, 133n, 134n
angle: definition of, 156, 157n; of contact, 153, 156–58
Anson, George, 285
Apollonius, 85, 202–4
Apostle, Hippocrates G., 9n
Aquinas, St. Thomas, 10–12
Archimedes, 46, 64, 123; and exhaustion method, 127; axiom of, 125–26, 157
argument from impossibility, 21–31, 242, 245–46; weaknesses of, 27–31
Aristotle, 9–11, 184, 271, 278n
arithmetic, 3, 88–122; applied, 111; contrasted to geometry, 113–14, 159, 222; objectivity of, 99; priority of, over geometry, 90–91
Armstrong, David M., 68n
Arnauld, Antoine, 17–18
Ashe, George, 165
axioms, 5–6, 84–85, 106, 124, 158–59, 181, 184–85

Barrow, Isaac, 2, 48n, 59n, 100, 123, 143–44, 288n, 299n; and abstraction, 14–17, 31n; and infinite divisibility, 49, 63–66; and infinitesimals, 169; influence on Berkeley, 63–67; on priority of geometry over arithmetic, 90–92; on rigor, 184
Baum, Robert, 39n, 106n
Bayes, Thomas, 259, 273–77
Bayle, Pierre, 51n
Belfrage, Bertil, 69n, 159n 163, 165n
Berkeley, George: *Alciphron,* 23–24, 33, 76, 83–84, 104–5, 110–13, 116, 178–79, 182n, 213, 224–26; *Analyst,* 45, 76, 84–87, 178–231, 233–36, 238–40, 247–49, 253–54, 267, 270–72, 276–77, 279, 285, 291–95; *Appendix,* 250, 254–55; *Arithmetica,* 17,

Berkeley (*continued*)
45–46, 93–95; *Commentaries*, 17, 26, 33n, 41, 45–47, 63–69, 82–83, 96, 108–9, 117, 153–62, 180, 182; *Defence*, 24, 179, 233, 242–46; *De Ludo Algebraico*, 46, 114–15, 118; *De Motu*, 76, 175–76, 213, 223–26; *De Radicibus Surdis*, 115; *Guardian* Essays, 178n; *Miscellanea Mathematica*, 45–46, 114–15; *New Theory of Vision*, 21, 27, 30–33, 51n, 57, 78–84, 96, 234, 241; "Of Infinities," 53, 162–73; *Principles*, 6–7, 22–27, 33–38, 42–43, 45, 69–78, 83, 85–86, 97, 99–100, 102–4, 110–11, 115, 117–18, 173–75, 186n, 213; *Querist*, 300; *Reasons*, 233, 256–58; *Siris*, 249–50, 269; *Three Dialogues*, 33n, 73n
Blake, Francis, 231
Blay, Michel, 192n
Boethius, 11
Bombelli, Rafael, 119
Bosmans, H., 93n
Bosses, Bartholomeus Des, 6n
Boyer, Carl B., 2n, 123n
Bracken, H. M., 38, 51n, 57
Breidert, Wolfgang, 2n, 46n, 95n, 161n, 238n, 240n, 277n, 285n
Brook, Richard J., 106n
Browne, Peter, 41–43
Buffon, George-Louis Leclerc, Comte de, 232

Cajori, Florian, 2n, 154n, 155n, 196n, 198n, 230, 231n, 265n, 268n, 269n, 271n, 277n, 279n, 280n, 292
calculus (*see also* compensation of errors thesis; differential calculus; fluxions): apparently inconsistent principles in, 151, 183, 188, 193–95, 219–20, 244–45, 248–49, 253–54, 272–73, 276–79; Berkeley's alternative to, 160–62, 174–75, 212–15; Berkeley's case against, 3, 153, 157–61, 183–99; compensation of errors in, 162, 183, 199–215; confuses signs with things signified, 187–88, 220–21, 225; defined, 123; fundamental theorem of, 142–43, 149–50, 209; history of, 123–51; obscure object of, 184–89, 228, 242–43; principles and demonstrations of, 189–99, 227
calculus differentialis. *See* differential calculus
Cantor, Geoffrey, 180n, 182
Carnot, Lazare, 215
Cartesians, 20, 38
Cavalieri, Buonaventura, 62, 123, 133–36
Cheyne, George, 171–72
Cicovacki, Predrag, 17n
Clavius, Christoph, 12–13
Cleary, John J., 9n
Collins, John, 196
Colson, John, 259, 269–73, 280
compensation of errors thesis, 199–215, 238–40, 300; and Berkeley's alternative to the calculus, 212–15; and the elimination of infinitesimals, 204–6; shortcomings of, 208, 210–12, 227–28
compression, 125–26
Cooper, Thompson, 269n
Crapulli, Giovanni, 92
Crombie, A. C., 12n

Dancy, Jonathan, 2n
Dechales, Claude, 46
demonstration: abstractionist account of, 5–6, 19–20; Berkeley's account of, 36–38, 70; geometrical, 61–62, 113–14; in algebra, 114–15; in arithmetic, 112–13; in theology, 180–81
Descartes, René, 92
differential calculus, 138–43, 160, 186–88, 195–97; and compensating errors, 200–205; fundamental concepts of, 138–40; illustration of, 141–42
Digby, Kenelm, 29n
Dijksterhuis, E. J., 125n, 128
Doney, Willis, 21n
Dublin Philosophical Society, 53, 162, 165
Duhem, Pierre, 67n

Eames, John, 269n
Edwards, C. H., 123
Emerson, William, 231n
Epicurus, 63n
esse-percipi thesis, 2, 47, 54, 71–72, 95–96

Index

Euclid, 12–13, 49–50, 58, 62, 89, 101, 123, 134; and exhaustion method, 126–27; and theory of ratios 124–25; definition of angle, 156; definition of line, 5; on angle of contact, 156–57

Euclidean geometry: and infinite divisibility, 48–53; Berkeley's rejection of, 59–62, 75, 81–83

evanescent magnitudes, 149, 217–18, 237–38, 244, 246–47, 250, 275; compared with infinitesimals, 151, 195, 227, 230, 270–71

exhaustion method, 124–29, 265; basis of, 125; Berkeley's endorsement of, 185, 214, 229; limitations of, 128

extension, 44–45, 47, 54, 71–72, 80–83, 158–59, 214; and geometric magnitude, 48; and minimum sensible, 57; as object of geometry, 48, 61, 69–71, 79–80, 112–13, 214–15, 222

fallacia suppositionis, 194–95, 237–38, 261

Fauvel, John, 123n

Flage, Daniel, 57

fluxions, 143–51; as velocities, 144–45, 185, 216–17, 220–21, 226, 235, 246–47, 260, 274, 283; Berkeley's rejection of, 186; calculus of, illustrated, 148–50; contrasted to infinitesimals, 146–47, 154–55, 183, 191, 195–96, 227; defined, 144–45, 185; higher-order, 185, 235

Fogelin, Robert, 53n

Fontialis, Jacobus, 63n

formalism, 88, 106–14, 299; and large numbers, 108–9; Berkeley's denial of, in the *Analyst*, 219–23; defined, 106, 107; game, 107–8, 115; history of, 106–7, 118–20; problems with, 120–22

freethinking, 178–80

Frege, Gottlob, 103

Gandt, François de, 133n, 134n

Garth, Samuel, 179

geometry (*see also* calculus), 44–87; analytic, 90, 92–93; Berkeley's change in view on, 45, 69–70, 85–87, 196; Berkeley's critique of, 59–62, 75, 81–83; contrasted to algebra, 116, 222; contrasted to arithmetic, 89–93, 113–14, 159, 222, 297; infinite divisibility in, 48–53

Gibson, George A., 292

Gillispie, Charles C., 215n

Giorello, Giulio, 2n, 248n

Giusti, Enrico, 133n

Grabiner, Judith, 215n

Graeser, Andreas, 9n

Grattan-Guinness, I., 204n

Gray, Jeremy, 123n

Gray, Robert, 68n

Grayling, A. C., 2n

Guicciardini, Niccolò, 154n, 231n, 232n, 279n, 282n, 293

Guldin, Paul, 134n

Hall, A. R., 189n

Halley, Edmund, 179

Hanna, John, 258

Harris, John, 158

Hartley, David, 285

Hayes, Charles, 153–57, 159n, 192, 195n

Heath, T. L., 49n

Heine, Eduard, 106

Hilbert, David, 106

Hobbes, Thomas, 29, 39–41, 60n, 119–20, 299

Hodgson, James, 231n

ideas, abstract. *See* abstraction

immaterialism, 299

incommensurables, 50–51, 59–60, 74–75; and infinite divisibility, 51, 55; Berkeley's rejection of, 60–61

increments, evanescent. *See* evanescent magnitudes

indivisibles (*see also* Cavalieri; infinitesimals; method of indivisibles; Wallis): and minima, 62–65, 161–62; defined, 133; Newton on, 161–62

indivisibles, method of. *See* method of indivisibles

infinite divisibility, 7, 48–57, 154–55; and infinitesimals, 154–55, 169; Berkeley's apparent acceptance of, 53, 70, 164–66; Berkeley's case against, 53–57, 70–72; defined,

infinite divisibility (*continued*)
48–49; "paradoxical" nature of, 56–57, 72
infinitesimals, 7, 129–32, 154–55, 169, 256; and infinite divisibility, 52–53, 154; attitudes toward, 132; Berkeley's rejection of, 157–59, 164, 166–68, 186–87, 229; characterized, 130–31, 154; classical rejection of, 125; disputes over, 168–71, 173–74; higher-order, 131–32, 155, 169, 170
infinity, 164, 270–71, 280, 285; actual vs. potential, 51–52
instrumentalism, 74–78, 175–76, 223–26, 298; Berkeley's denial of, in the *Analyst*, 213–14, 223–26, 244; in Berkeley's account of geometry, 75–76; varieties of, 76–77
integration. *See* quadrature
intellect, pure, 32–33, 40, 186

Jesseph, Douglas M., 76n, 132n
Johnston, G. A., 2n, 160n, 268–69
Jones, J. F., 9n
Jurin, James, 199n, 231–50, 280, 286; and *New Theory of Vision*, 234; and *Siris*, 249–50; dispute with Robins, 249, 267–68; on abstraction, 240–43; on compensation of errors, 238–40; on conceivability of fluxions, 234–35, 246–47; on interpretation of Newton, 236–38; on moments, 236–37

Keill, John, 48, 52, 55–56, 66–67, 153, 155
Keisler, Gerald, 131n
Keough, Andrew, 154n
Kitcher, Philip, 198, 293
Klein, Jacob, 90, 92

Lasswitz, Kurd, 67n
Lear, Johnathan, 9n
Leclerc, Jean, 32–33
Leibniz, Gottfried W., 53, 84, 123, 163, 197; and the calculus 138–43; dispute with Nieuwentijt, 168–71; on Berkeley, 6–7; on infinitesimals, 170, 172
L'Hôpital, G. F. A. 139–41, 159, 171, 176, 186–87, 288n
limits, 198, 205, 251–52, 265–68, 286

Locke, John: on abstraction, 17, 19–20, 25–28, 42, 241–42, 245–46; on infinite divisibility, 52; on number and unity, 102–3; on infinity, 164, 280
Luce, A. A., 38, 41, 46, 51n, 160, 161, 163
Lucretius, 63n

Maclaurin, Colin, 226, 259, 279–85, 291, 293; adequacy of response to Berkeley, 284–85; and exhaustion, 284–85; on instantaneous velocity, 282–83
magnitude: and theory of proportion, 124–25; contrasted to multitude, 14–15, 48–49, 89; kinematic conception of, 143–44, 147, 260, 274–75, 286; nature of, 48; tangible vs. visible, 79–80
Mahoney, Michael S., 119n
Malebranche, Nicholas, 38–39, 85
Martin, Benjamin, 231n
Masi, Michael, 119n
mathematics (*see also* algebra; arithmetic; calculus; geometry): applied, contrasted to pure 13; history of, Berkeley's place in, 292–95, 299–300; object of, 13, 38, 89–93; philosophy of, basic issues in, 4–6, 297–98; pure, Berkeley's hostility toward, 117–18
McCracken, Charles, 39
method of indivisibles, 46, 132–38, 270; and classical geometry, 133–34, 137–38; and method of exhaustion, 135–36; reactions to, 134–36;
minimum sensible, 74–75, 79, 161, 175, 228–29; Berkeley's geometry of, 57–63; contrasted to Euclidean point, 58–59; whether coherent, 67–69; whether extended, 57
Moked, Gabriel, 2n, 57n, 62n 160n
Molyneux, Samuel, 108
moments (*see also* evanescent magnitudes; fluxions), 145, 160–61, 185–86, 190–91, 236–38, 250
Mueller, Ian, 9n, 49n
mystery, 178–82, 217; essential to religion, 180

Nagel, Ernest, 189n
Neri, Luigi, 48n, 55n

Index

Newton, Isaac, 14, 84, 123, 232; compared with Wallis, 196–97; on angle of contact, 153; on calculus of fluxions, 143–51, 185–86; on evanescent magnitudes, 245; on number, 101; on prime and ultimate ratios, 145–46, 217–19, 264; proof of "product rule," 190, 226–27; proof or rule for fluxion of any power, 193–94; rejection of infinitesimals, 144, 161–62, 172–73, 195–96
Newton-Leibniz priority dispute, 189, 195
Nicole, Pierre, 17–18
Nieuwentijt, Bernard, 163, 168–71, 192
nominalism, 94–95, 118–20
Norris, John, 18–19, 31n, 39
number (*see also* arithmetic; magnitude): as "collection of units," 98, 100–102, 104; as "creature of the mind," 91–92, 95–99; nonabstract nature of, 99–106

Olson, Mark A., 11n

Paman, Roger, 259, 285–93; definition of fluxions, 289–91; on first and last state of variables, 287–89; on maximinus and minimajus, 286, 288–91
Pardies, Ignatius Gaston, 158n
Pemberton, Henry, 199n, 268n
Percival, Sir John, 39
Pereira, Benedictus, 12
Philalethes Cantabrigiensis. *See* Jurin, James
Pitcher, George, 2n, 34n
platonism, 5, 9–10
Popkin, Richard, 51n
primary qualities, 96–99
prime ratios (*see also* evanescent magnitudes; moments; ultimate ratios), 145–47, 198; defined, 145
Proclus, 12, 92
proportion, 124
Pycior, Helena, 90n, 93n

quadrature: by exhaustion, 125–28; defined, 123; in differential calculus, 142–43; in fluxional calculus, 145, 149–50; in method of indivisibles, 136–37, 166–67; of the cubic parabola, 136–37; of the hyperbola, 166–68; of the parabola, 128

Raphson, Joseph 17, 89, 189n
ratios, 124; prime, *see* prime ratios; ultimate, *see* ultimate ratios
Raynor, David, 57
reductio ad absurdum, 125, 126, 262
representative generalization, 33–38, 69–70; and geometric demonstration, 74–75, 86–87, 174; and geometric truth, 77–78; and infinite divisibility, 73–74; and the *New Theory of Vision*, 80–81, 83
Resnik, Michael D. 107n, 119
responses to Berkeley, 198–99, 231–92; differences among, 259; inadequacy of, 199, 237, 247, 249, 258, 273, 277–79; on the continent, 231–32
rigor, 183–85; abstractionist criteria of, 5–6, 19, 61–62; and infinitesimals, 162–63, 168, 171, 187, 189, 229–30; Berkeley's conception of, 184–85, 196, 215–16, 243, 297–99; desirability of, 292–94; logical criteria of, 184; metaphysical criteria of, 184
Robins, Benjamin, 199n, 229n, 235n, 259–69, 280, 284, 286, 293; and exhaustion proofs, 260–64; and limits, 265–68; dispute with Jurin, 267–68; on prime and ultimate ratios, 264–66
Robinson, Abraham, 131n

Sageng, Erik L., 266n, 268n, 280n
Scholasticism, 3, 10–13, 39
sciences, hierarchy of, 298
Sherry, David, 189n, 192n, 198, 293
signs: and abstraction, 300; and representative generalization, 33–36; arithmetic as a science of, 105–13; as object of mathematics, 297; as the immediate objects of demonstration, 224–25; confused with things signified, 73, 187–88, 220–21, 225; geometric diagrams as, 37–38, 72–75, 79, 86–87, 174, 229; not to be used without ideas, 158–59, 164–66
Simpson, Thomas, 231n
Smiglecki, Martin, 16
Smith, G. C., 277n
Smith, James, 259, 277–79

Smith, Vincent E., 11n
Stammler, Gerhardt, 284
Steele, Sir Richard, 178n
Stock, Joseph, 179
subtangent. *See* tangents
surds. *See* incommensurables

Tacquet, André 46, 92–93, 101
tangents, 123, 139, 144; and compensating errors, 200–208; in differential calculus, 139, 141, 200–201; in the method of fluxions, 147–48
Taylor, Brook, 189n
theology, 178–83; contrasted to science, 180; revealed, 178, 180–81
Thomae, Johannes, 107
Thomason, S. K., 156
Turnbull, Herbert W., 280n
Tweedie, Charles, 280n

ultimate ratios, 145–47, 198, 208, 216, 217, 230, 237–38, 251–54, 271–72, 275–76; compared with infinitesimals, 146; defined, 145; Newton's defense of the doctrine of, 147
Urmson, J. O., 176

velocity, 281; instantaneous, 185, 216–17, 226, 230, 255–58, 282–83, 285
Vermeulen, Ben, 192n
Viète, François, 92

Wallace, William A., 12n
Wallis, John, 46, 60, 123; on abstraction, 14–17, 28–29; on method of indivisibles, 135–38, 166–68; on number, 101; on priority of arithmetic over geometry, 90–99; faulty inductions of, 138, 197; on the trinity 181–82
Walmsley, Peter, 183n
Walton, John, 232–33, 250–59; on evanescent magnitudes, 250–52; on limits, 251–52; on proof of the "product rule," 252–54
Warnock, G. J., 176, 222n
Weinberg, Julius, 21n
Whiteside, D. T., 128–29
Wilson, James, 280n
Winkler, Kenneth, 2n, 21n, 28n, 29n, 31n
Winnett, Arthur R., 41
Wisdom, John O., 210–11, 292
Wright, J. M. F., 199

Zeno 51n, 56, 63n

12